SOUTHERN
WONDER

Published in cooperation with The Nature Conservancy

The University of Alabama Press • Tuscaloosa

SOUTHERN

ALABAMA'S SURPRISING BIODIVERSITY

WONDER

R. SCOT DUNCAN

FOREWORD BY EDWARD O. WILSON

Typeface: Garamond Premiere Pro

Cover image: The vast salt marsh of Grand Bay Savanna, Mobile County; courtesy of Hunter Nichols/
Hunter Nichols Productions
Cover design: Michele Myatt Quinn

Library of Congress Cataloging-in-Publication Data
Duncan, R. Scot, 1969–
Southern wonder: Alabama's surprising biodiversity/ R. Scot Duncan.
    pages   cm
"Published in cooperation with The Nature Conservancy."
Includes bibliographical references and index.
ISBN 978-0-8173-1802-4 (trade cloth: alk. paper)—ISBN 978-0-8173-5750-4 (pbk.: alk. paper)—
ISBN 978-0-8173-8681-8 (ebook)
1. Biodiversity—Alabama. 2. Ecology—Alabama. I. Title.
QH76.5.A2D86 2013
333.95'1609761–dc23   2012050437

Research for this work was supported by the World Wildlife Fund.

Protecting nature. Preserving life.
Published in cooperation with The Nature Conservancy.

For all who labor to protect
and restore Alabama's biodiversity
so future generations can enjoy
the state's extraordinary natural heritage

The objective is to teach the student to see the land,
to understand what he sees, and enjoy what he understands.

—ALDO LEOPOLD
"The Role of Wildlife in a Liberal Education"
1942

CONTENTS

ILLUSTRATIONS

FOREWORD

EDWARD O. WILSON

In this one book, Scot Duncan has brought to a new level the understanding and aesthetic appreciation of Alabama's living heritage. This is no ordinary descriptive work. Alabama is near the top among US states for biological diversity, a result, Duncan makes clear, of its good fortune to have both a complex geology and a benevolent climate. Given that its fauna and flora remain some of the least explored by scientists, Alabama may one day be recognized as the most biologically diverse state in America.

Duncan gives us an exciting introduction to some of the more interesting inhabitants of Alabama. Alabama Beach Mice, West Indian Manatees, Black Bear, Red-cockaded Woodpeckers, Red Crossbills, Alabama Sturgeons, Paddlefish, Pygmy Sculpins, Bull Sharks, Alabama Cavefish, Two-toed Amphiumas, Cahaba Lilies, White-topped Pitcher Plants, Red Hills Azaleas, and Smallhead Blazing Star are among those pulled out of a very long roster by Duncan for special mention.

Many new species continue to be discovered. Duncan cites a glade along the Cahaba River which, over a few months' study, yielded eight new plant species and dozens of others that are rare and likely to be endangered. One of the latter, the Dwarf Horse-nettle, had not been seen for over 150 years.

How many species of plants and animals in all are native to Alabama? I will venture the following guess based on studies in other, better studied parts of the United States: if all plant and animal species, the latter including insects and other invertebrates, were known, the number would exceed 100,000.

Part of the reason for such a high conjecture is the extraordinary diversity of habitats in Alabama. A major strength of Duncan's review of the fauna and flora is the detailed treatment he gives of the principal ecosystems, each of which harbors its own distinctive assemblage of species. He further shows—and as decisively as has been done for any comparable region anywhere in the world—why it is necessary to know a lot about the soils, the water, the topography, and the geological history of Alabama, and some of the basic principles of its ecology, in order to understand fully the remarkable fauna and flora the state harbors.

This is a well-written book of Americana, made even more important by being the part hitherto least studied.

ACKNOWLEDGMENTS

I have many to thank for contributions of encouragement, knowledge, and resources. Most importantly I thank my wife, Ginger, for her unflagging support and patience. I will be forever grateful to my parents, Bob and Lucy Duncan, for inoculating me with the profound love of nature from which I drew inspiration for this work. My dear friends Ann and Dan Forster contributed generously to the Nature Conservancy in support of this project.

The Alabama Chapter of the Nature Conservancy (TNC) aided me from day one. I am especially indebted to Chris Oberholster for his leadership in building financial support for the publication. Birmingham-Southern College facilitated this project through a one-semester sabbatical and a travel grant. Travel funding was also provided by the World Wildlife Federation through its Southeastern Rivers and Streams Support Fund.

I am very grateful to Ed Brands for creating and donating all the maps. I offer special thanks to Bill Finch for several days of touring TNC preserves with me and thoughtfully answering my many questions. I'm also grateful to the many others who took me on field expeditions, including Dwight Cooley (US Fish and Wildlife Service), Jeff Gardner (US Forest Service), Mark Hainds (Longleaf Alliance), Francine and Bruce Hutchinson, Jack Johnston, Jim Lacefield, and Steve Miller (US Fish and Wildlife Service). I also want to thank the many I interviewed, including Forrest Bailey, Mark Bailey, Wayne Barger, Arvind Bhuta, Dave Borland, Ruth Carmichael, Doug Fears, Brittney Hughes, Paul Johnson, Bernie Kuhajda, James Lamb, Chris Oberholster, Al Schotz, Eric Soehren, Dan Spaulding, Sierra and Jimmy Stiles, and Keith Tassin. I also appreciate the two anonymous reviewers who provided insightful comments on the draft.

Alabama's nature photographers—both professional and amateur—are crucial to increasing appreciation of the state's natural beauty. Many of them donated images to the project, and though not all could be published, I am indebted to each contributor. I offer sincere thanks to a new group that shows great promise, the Conservation Photographers of Alabama.

Finally, I wish to thank the staff at the University of Alabama Press for their work on this project. I'll be forever indebted to Beth Motherwell (environmental editor) whose encouragement, guidance, and zeal were vital to the book's success. I also wish to thank Crissie. E. Johnson (managing editor) and Michele Quinn (book designer) for the careful attention and creativity they dedicated to the project. Special thanks also to Kathy Cummins (copy editor) for the extraordinary thoroughness with which she reviewed the manuscript. I also thank my brother, Will Duncan, for encouraging me to take on this project.

SOUTHERN
WONDER

1

Number One in the East

Honestly, I can't remember when I first heard the statistic, but I clearly remember my reaction—I didn't believe it. How could it be possible that Alabama, *Alabama*, could lead most of the nation, and *all* of the eastern United States in the number of species within its borders? I concluded that I had misheard the statistic or this was a result from some convoluted ranking scheme.

My skepticism was fueled by my overconfidence that I knew a thing or two about Alabama and southeastern biology. I grew up in the Florida panhandle raised by two renowned birders and naturalists; our family spent many spring and fall weekends looking for migrating birds on Dauphin Island and the Fort Morgan Peninsula. I knew Alabama had impressive bird diversity, but so do all the southern states. The statistic didn't mesh with my preconceptions of Alabama, and in my arrogance I dismissed it.

My initial ignorance about Alabama's biodiversity may be forgivable to a small degree. In 2002 I had just moved to Birmingham from Gainesville, Florida, and I was in my first year of teaching at Birmingham-Southern College. With a demanding teaching load, a young child, and my wife pursuing her career, I had little time to explore my new stomping grounds. After I settled into a routine and planned my lectures, I finally began meeting other biologists and conservationists. About this time Bruce Stein published his report "States of the Union: Ranking America's Biodiversity," which presented biodiversity statistics for all US states. Soon, Alabama's statistics began to come up in conversation with colleagues, though often with a degree of incredulity and reservation. This was, after all, a community of people accustomed to discouraging biodiversity data. After hearing the rumor several times, I looked it up myself. I easily found the report on the

Internet, and sure enough, there it was on page 12, Table A: "Alabama: rank 5, species = 4533."

Such a compilation of numbers is not easy to produce. Species inventories from states can be incomplete and out of date, and states may use different species classification systems. Yet, once reliable biodiversity statistics for a region are available, they begin to demonstrate their power. Researchers can find relationships in the data that inspire new ideas about patterns of species distribution, while conservationists can use the data to identify priority regions or taxa for protection.

Stein's report identified Alabama as a biological superstar on the North American continent. The only states surpassing Alabama's species diversity are California, Texas, Arizona, and New Mexico, all large western states with tremendous ecological range. Among the 26 states east of the Mississippi River, Alabama is at the very top. Furthermore, for several major freshwater taxa including fishes, snails, mussels, turtles, and crayfishes, Alabama leads the nation in biodiversity.[1]

The endangered Vermilion Darter, endemic to Turkey Creek, Jefferson County. Courtesy of Bernard R. Kuhajda/ Tennessee Aquarium Conservation Institute.

Stein also reported Alabama ranks highly for its endemic species, those restricted to a relatively small area, in this case just Alabama. With 144 endemics, Alabama ranks seventh in the United States. California leads with nearly 1,300 endemics, followed by Hawaii with 1,101. Other than fourth-ranked Florida (269 endemics), endemic totals among Alabama's neighboring states don't even come close:

Georgia (63), Tennessee (49), and Mississippi (23). A few Alabama endemics have achieved local or national fame and their names may be familiar, including Red Hills Salamander (the official state amphibian), Alabama Beach Mouse, and Alabama Sturgeon. Other endemics have unusual names rarely printed or spoken outside of professional circles: Armored Marstonia, Dark Pigtoe, Deceptive Marbleseed, and Pygmy Sculpin. Some are confined to a distinct geologic land form. Many are restricted to a single stream or river. Alabama's endemics, perhaps more than any other assembly of species, represent what is most unique and distinctive about the natural character of the state.

Sadly, 10 years after Stein's report, few of Alabama's citizens are aware that Alabama harbors one of the richest collections of biodiversity outside the tropics. This is regrettable, because Alabama's citizens and visitors enjoy the rewards of the state's biodiversity every day, and the abundance and diversity of life in its forests and streams constitute one of Alabama's greatest riches.

Why So Many?

Upon learning of Alabama's biodiversity rankings, I began wondering why the state has so many species. I wasn't alone—others began asking this, too. This question falls within the realm of biogeography, the study of geographic distributions of species. Several questions came to mind and I began investigating.

Is state size a factor? After all, the four states ahead of Alabama in Stein's US rankings are large western states among the top six states for land area. However, size alone isn't a sufficient explanation. Alabama is midsized, ranking 30th in land area, and the Alaskan goliath, our largest state, finishes nearly last in species diversity.

What about latitude, a measure of distance from the equator? The top seven states are aligned along the southern perimeter of the United States, and southern states have longer growing seasons and milder winters. Species diversity does generally correlate with latitude, with ecosystems at lower latitudes usually having more species than those at higher latitudes. It is true that if Alabama had Minnesota's latitude, it would have far fewer species than it does. Nevertheless, latitude alone cannot explain Alabama's biodiversity rank. Mississippi, which shares similar size and latitude, is 17th in the United States for species biodiversity. Florida is partly subtropical but ranks 7th for biodiversity, and Hawaii, our only fully tropical state, ranks dead last.

What about mountains and exposures of rock? Mountainous terrain, wherever found, supports ecosystems unlike those in the lowlands, and rock exposures create unique habitats. The Southern Appalachians begin at the heart of Alabama and mountainous topographies dominate 33 percent of the state's area. However, Tennessee, North Carolina, Virginia, and several western states are more topographically and geologically diverse than Alabama and have less biodiversity.

Could it be coastline? Alabama's position along the Gulf of Mexico and adjacent plains certainly increases its diversity of ecosystems and species. Then again, if coastal ecosystems were the overriding factor, Florida and Louisiana should rank higher for species diversity, and New Mexico and Arizona should rank lower.

With time, I realized no single feature was responsible for Alabama's biodiversity. Instead, three overarching natural factors have shaped the state's biodiversity. First, Alabama's warm, wet climate has nurtured productive ecosystems throughout the state. Second, and more influentially, geological processes guided the state to its current latitude, created a topographically diverse terrain, and exposed a great range of rock and soil types at the surface. Alabama's combination of a mild climate and diverse terrain supports 64 distinct terrestrial and wetland ecological systems. The more ecological diversity there is in a geographic area, the higher the species count. This is a portable truth about biodiversity you can apply anywhere at any scale.

A third factor, biological evolution, is a product of the other two. Radical changes in Alabama's climate and landscape have continuously forced its flora and fauna to adjust through evolutionary adaptations. Evolution is a complex process explored later in this book, but suffice it to say that these adjustments over millions of years led many populations to become distinct species living nowhere else but in Alabama.

There is a fourth reason that Alabama has so many species, but it fits awkwardly with the others. For reasons beyond the scope of this book, Alabama's boundaries were drawn such that within the state are several distinct geographic regions, each with a unique suite of ecosystems. These include a corner of an ancient mountain range, an exten-

Above: Talladega Mountains at sunset, Talladega and Clay counties. Courtesy of Frank Chitwood.

Left: Brackish coastal lagoon at Fort Morgan State Historic Site, Baldwin County. Photo by R. Scot Duncan.

sive swath of coastal plain, 50 miles of coastline, a piece of the largest temperate watershed in the world, and one of the most biodiverse watersheds in the temperate world.

WHAT'S AHEAD?

As I studied the scope and origins of Alabama's biodiversity, I encountered many amazing natural history stories. This book is an attempt to share these stories to help readers appreciate the range of biodiversity in the state. Some stories can excite the 10-year-old child in all of us. There will be strange and, sometimes, dangerous creatures lurking in the state's large rivers; sharks of the Gulf of Mexico and Alabama's bays; and tales about the first peoples of Alabama who just a few thousand years ago hunted bear-sized ground sloths and elephant-sized mammoths.

These stories will solve some of the state's natural mysteries: Why are shark teeth and other marine fossils found hundreds of miles from the coast? How could a blind fish species live underground in northern Alabama? What happened to the state's American Mastodons and Sabre-toothed Cats? Why are carnivorous plants abundant and diverse in southern Alabama?

And finally, these stories are for explorers who hunger to escape into wildernesses including the second-largest river delta in North America, the rugged Appalachians' forest-cloaked peaks, the labyrinth of box canyons in the Sipsey Wilderness, or the reefs at the edge of the Continental Shelf.

This opening chapter contains an overview of why Alabama has so many species. How ecosystems work is reviewed in chapter 2 by studying a critically endangered mouse and its unusual ecosystem. Chapter 3 explains Alabama's mild climate and why Alabama should be a desert, but isn't. In chapter 4, the geologic processes shaping the state's terrain are examined, from sedimentary rock formation to plate tectonics. Chapter 5 is an epic journey through millions of years to understand how the state's ecology, climate, and geology have changed, and how this history shapes today's biodiversity. Chapter 6 provides a review of the mechanics of evolution and why Alabama is at the epicenter of species evolution for several taxa.

The second part of the book is an application of the concepts from the first part via an exploration of the major ecological regions, or ecoregions, within the state. Alabama is a global hot spot for aquatic biodiversity, so chapter 7 begins with the state's rivers. Explorations of Alabama's terrestrial ecoregions begin in chapter 8 on the Southern Coastal Plain, the ever-shifting margin of the continent, a living mosaic of barrier islands, marshes, tidal swamps, flatwoods, oyster reefs, and seagrass meadows. Chapter 9 is an inland venture across the Southeastern Plains, including its fire-hungry pine woodlands, a poorly understood hill country, and black soil prairies that have nearly disappeared. Chapter 10 is a journey into the Southern Appalachian Mountains to explore the Ridge and Valley ecoregion, a geologically tortured terrain where sedimentary rocks hundreds of millions of years old support wildly diverse ecosystems.

Chapter 11 is a wandering through the Southwestern Appalachians ecoregion, an elevated plateau whose margins bear narrow canyons, labyrinthine valleys, steep escarpments, and underground caverns whose species and ecosystems seem otherworldly. The lonely Piedmont ecoregion is explored in chapter 12. Though still scarred from an abusive agricultural history, the Piedmont's old soils and ancient rocks harbor many surprises, including the lonesome monadnocks, Alabama's most prominent mountains. Chapter 13 is a tour of the Interior Plateau ecoregion, where the Tennessee River has ground down the region's bedrock, deposited fertile soils, and exposed another warren of underground caverns. The geographic expeditions finish in chapter 14 with a journey into Alabama's blue wilderness, the Gulf of Mexico. The northern Gulf supports far more species diversity than most appreciate. Areas of upwelling attract more than two dozen whale species, surface waters harbor menacing predators and the largest clonal organism on earth, and relic coral reefs sustain colorful sponge gardens and dense schools of fishes.

I conclude in chapter 15 with a look to the future of Alabama's biodiversity, including threats to the state's biodiversity, what's at stake for Alabama's citizens, and what's being done to protect the state's natural heritage. I'll explain why this may be a century of tremendous biological discovery in Alabama.

Does It Matter?

There is a small community of biologists, conservationists, and naturalists who are well aware of Alabama's exceptional natural heritage. We share a profound fascination with the seemingly infinite variety of species on earth and the ways they interact. With its rich biodiversity, Alabama is overflowing with ecosystems to explore and species to encounter.

However, for many of us there is a deeper, darker cause for our obsession. We who study the natural world also understand how the survival of our own species depends on the health of our ecosystems. These ecosystems function best when all of their component species are present and flourishing. To put it simply, people need biodiversity to survive. This is the most important lesson ecologists have provided to mankind. This assertion is backed by many millions of hours of empirical research, and tens of thousands of studies across the planet. What's more, archaeologists and anthropologists have amassed a large and sobering collection of stories of past civilizations that collapsed when they undermined the integrity of their ecosystems.

Despite our wondrous technological advances and wealth of knowledge, we are creatures dependent on our ecosystems for survival. This can be hard to fathom, especially for those of us who spend most of our life within cities, buildings, and cars. However, trace the production path of the book in your hand, the food on your plate, or the water in your cup, and soon you find a forest, field, or river where native species and natural ecological processes are essential for its production. The conversion of natural resources to food, fiber, and fuel tightly binds our fate to the land and ocean and is one of the most fundamental relationships on which our civilizations, and our individual lives, are built.

Examples of this dependence are everywhere in Alabama. Drinking water is taken from rivers and streams. Forests provide lumber, paper, and jobs in rural communities. Commercial fishers haul fish and shellfish from the rivers, bays, and Gulf. Farmers work soils deposited by seasonal river flooding. The benefits of Alabama's rich natural heritage extend beyond resource extraction and commodity production. Alabamians have a fervent love of outdoor recreation, and hunting,

Striped Mullet, the author's favorite catch. Courtesy of Lucy Rutland Duncan.

fishing, camping, hiking, and wildlife observation clearly depend on healthy ecosystems.

Seen or unseen, Alabama's thousands of native species play important roles in the ecosystems sustaining the state's economy and culture. Impaired ecosystems offer us little, and when they collapse there is loss of livelihood, property, and, sometimes, life. Consider flooding in overdeveloped watersheds or the trauma in coastal communities when fisheries collapse. When we lose biodiversity, we lose opportunity and ecological security. Ultimately, biodiversity protection is people protection.

2

A Tale of Two Mice

DESERT MICE

It's nearly midmorning at the edge of a continent. I stand on the crest of a primary dune overlooking the Gulf of Mexico near the end of the Pine Beach Trail in Bon Secour National Wildlife Refuge. Under clear bright skies I scan this marine wilderness with hopes of spotting dolphins, a school of mullet, or something more uncommon, perhaps a sea turtle or shark. Overhead, four Brown Pelicans spiral lazily, rising on a thermal to cooler heights from which they will cruise the shoreline looking for baitfish. Farther offshore, a flock of terns busily circles over a patch of disturbed water. Over and over, each hovers briefly, dives through the surface, then emerges with a struggling minnow in its beak. The terns are not hunting alone. The surface glints with silver as predatory fishes, probably Spanish Mackerel, attack the baitfish from below.

The sun is already strong, but the first gentle promise of a sea breeze is arriving from the Gulf to alleviate the heat. Around me, the tall stalks of Sea Oats sway in response. Among these grasses are various smaller plants, but mostly the dune's brilliant white sands are bare. Between some plants are winding trails of tiny footprints—the puttering of a Mourning Dove, the frenetic skitterings of a Ghost Crab, and the prints of a small but famous rodent, the Alabama Beach Mouse. These charismatic little mice have large eyes, tan or pale-gray fur above, and white feet and bellies. The mouse's native range is a slender 35-mile (56 km) stretch of sand dunes, from the western tip of the Fort Morgan Peninsula to the western shore of Perdido Pass and all of nearby Ono Island. As Alabama's only endemic mammal, the Alabama Beach Mouse is symbolic of the state's most unique biodiversity.

Unfortunately, there is double symbolism here, because the mouse has lost much of its habitat to urbanization and is one of the most en-

Above: Primary dune line and stormy surf of the barrier system. Photo by R. Scot Duncan.

Left: Tracks of Alabama Beach Mouse (left) and Ghost Crab (right), Bon Secour National Wildlife Refuge. Photo by R. Scot Duncan.

dangered species in Alabama. Many coastal residents know the mouse from the endless arguments and legal battles between those defending it from extinction and those seeking to develop highly coveted coastal properties. Listed officially as endangered under the US Endangered Species Act, the timid little rodent has found itself in the crosshairs of many heated arguments. Some people are furious that economic development can be hindered by a mouse. Others are outraged that coastal development has driven the mouse to the brink of extinction. Such views are not easily reconciled. The controversies surrounding the Alabama Beach Mouse represent the challenges facing much of Alabama's biodiversity.

Variation in the state's climate and geology has produced Alabama's great range of ecosystems. Before tackling these relationships, I'll review some basic ecology by explaining how the beach mouse can survive in one of the harshest ecosystems in Alabama. With sparse vegetation, hills of shifting sands, and intense summer heat, the dunes of the Gulf Coast are more similar to western deserts than to any southeastern ecosystem. That said, the regular lashing by cyclones propels this desertlike ecosystem into a league of its own. How can a mouse weighing less than four nickels endure such extremes? Ecology has the answer.

Notorious Inefficiency

From the largest whale to the tiniest beach mouse, the survival of all creatures is tightly bound to their ecosystem. Features like size, growth, reproduction, behavior, and diet are all linked to the resources in ecosystems and the impediments to obtaining them. An ecosystem is everything in an environment. This includes all living organisms plus nonliving components including bedrock, soil, atmosphere, water, and energy. Energy is particularly important. Differences among ecosystems are largely due to variation in the flow of energy through them. Generally, ecosystems with high energy flow, like wet tropical forests, have complex structure and high species diversity. Those with low energy flow, like dunes and deserts, have a simplistic structure and fewer species. The rules governing energy flow explain why.

Primary productivity is the ruler by which ecologists measure energy input in most terrestrial ecosystems. This input comes from plants. During photosynthesis, green plants gather carbon, hydrogen, and oxygen atoms from the environment, and with the use of energy captured in sunlight, assemble them into sugars. Each sugar molecule is a bundle of stored energy used for growth, reproduction, and tissue maintenance. Primary productivity is a measure of energy input into an ecosystem via photosynthesis over time. The qualifier "primary" denotes how plants provide the initial source of energy for food chains. Thus, an ecosystem's primary productivity influences its animal inhabitants. Ecosystems with many productive plants support more animals and animal diversity than those with less productivity.

Gazing over the dunes, I see every indication this is a stressful environment where primary productivity is low. Were this ecosystem typical of Alabama, the dunes would be crowded with trees, shrubs, and other plants and vistas would be measured in feet. Instead, vegetation is short and sparse, much of the ground is bare, and I can see hundreds of yards in any direction.

The rules governing energy flow through ecosystems determine why animal diversity is limited where there is low primary productivity. Wherever there are plants, there are organisms cleverly extracting the energy those plants have captured. Herbivorous animals, or primary consumers, are the first tier of animals in the food chain to consume other species. Unique adaptations enable each species to feed on plant tissues, especially roots, leaves, or seeds. In the food chain model, secondary consumers eat primary consumers, tertiary consumers eat secondary consumers, and so on. Through feeding, energy transfers from one organism to the next. The Alabama Beach Mouse is a primary consumer specializing on seeds. Occasionally, beach mice eat grasshoppers and become temporary secondary consumers. They even become tertiary consumers when they eat spiders or other predatory invertebrates. Such versatility in feeding roles among animals is the norm and confounds any desire to place species into tidy categories. Consequently, to portray the many feeding relationships connecting species in ecosystems, ecologists tie individual food chains together into a network of relationships called a food web. The food web is a road map for understanding energy's flow through ecosystems.

Energy transfer among creatures is not as simple as it might seem. The process of animals eating other organisms to obtain energy is notoriously inefficient. Consider a fox eating a beach mouse. Much of the mouse's energy is stored in indigestible tissues (e.g., bones, fur, and teeth), and the energy absorbed through digestion of muscle and soft organs will be mainly spent keeping the fox alive (e.g., helping it catch the next mouse). Though exact amounts vary widely, on average only a meager 10 percent of energy ingested by an animal can be converted into growth and reproduction. And regardless of how it is used, once energy is used by an organism it is forever lost to the environment as heat, a form of unorganized energy. Because of these limitations on energy transfer between organisms, there are always fewer organisms at each higher level on a food chain, whether quantified by number or by biomass. These relationships clarify much about dune ecosystems. With such limited primary productivity and inefficient flow through the ecosystem, there is relatively little energy to support the dune animal community.

Any animal that walks, hops, crawls, or slithers across the dunes leaves a trail. Curious about the other animals living here, I turn from the primary dunes and begin wandering into the secondary dunes. I'm careful not to trample the fragile dune plants as I scan for tracks. Northern Raccoon prints are among the most abundant, and after a while I also find the prints of Striped Skunk and Gray Fox. Most of these animals were in transit between the outer beach and the ecosystems behind the dunes to the north. There simply isn't much food for them in the dune ecosystem. Even beach mice tracks are relatively scarce. I do find a Great Horned Owl's feather, but it and other predators patrolling the dunes spend significant amounts of time hunting in adjacent, more productive ecosystems.

For the most part, beach mice subsist on the seeds of Sea Oats, Gulf Bluestem, and the few other grasses comprising a minor component of the dune plant community (in ecology, a community is an assemblage of species in any defined area). Scrub oaks—primarily Sand Live Oak and Chapman Oak—produce acorns eaten by mice when grass seeds are scarce. To find these trees, I head deeper into the dune system where few people venture. Dune slopes collapse easily underfoot, so

Left: The endangered Alabama Beach Mouse. Courtesy of the US Fish and Wildlife Service.

Below: Scrub dunes at Bon Secour National Wildlife Refuge, Baldwin County. Photo by R. Scot Duncan.

I wind through the low swales between them. Soon I am among the tertiary dunes (or scrub dunes), the largest of which are over 12 feet (3.7 m) tall. These older dunes have broad windward slopes facing the Gulf, and their steep leeward slopes shelter the scrub oaks. The oaks and a few other hardy trees, like Southern Magnolia and Slash Pine, are gnarled and stunted. Their dense canopies form a seamless continuation of the dune's profile. Some of the oaks may be hundreds of years old despite their stature.

Below a nearby oak are the sun-bleached fragments of last fall's acorn crop. At least some husks were left here by foraging mice. Beach mice will gather acorns in autumn and store them in their burrows for later months when fresh seeds are rare. Like oak species everywhere, these scrub oaks produce acorns in a gamble that at least some will be lost or abandoned by hoarding animals in a location suitable for a young seedling to establish.

The rules of energy flow only partially explain how the beach mouse survives in the harsh dune habitat. The beach mouse thrives in the dunes because of its tiny size. While small size might seem to make it vulnerable, smaller animals have lower total energy requirements relative to larger animals. This helps them survive in an ecosystem where food is scarce. Were the beach mouse larger, each mouse would need a larger territory to secure enough food. With dune habitat so limited (even before human occupation), the resulting population would be too small to survive.

DUNE DISSECTION

What limits primary productivity in ecosystems and why are dunes so sparsely vegetated? The answers are all around me as I meander among the secondary and tertiary dunes: the sun has shed its morning softness and risen high and hot; the sand is dry and shifts beneath every step; the Gulf's power is evident in the sea breeze now blowing steadily across the dune field and beyond.

Recall from chapter 1 that two of the fundamental natural forces shaping biodiversity patterns are climate and geology. These forces shape ecosystems by affecting the four key resources most ecological communities require—sunlight, heat, water, and soil. Climate and

geology also govern the frequency and intensity of ecosystem disturbances. This fifth factor, disturbance, alters the availability of the four
key resources. These five factors affect all of Alabama's ecosystems, including the dunes and the beach mouse.

Sunlight. Sunlight availability varies among Alabama's ecosystems.
Being north of the tropics, the sun in Alabama is never directly overhead and is always in the southern sky. The sun reaches higher midday positions in southern Alabama, resulting in this region's receiving
more sunlight than the north. Consequently, ecosystems in southern
Alabama tend to have higher primary productivity and longer growing
seasons than ecosystems farther north.

However, sunlight availability does not explain why dune ecosystems are sparsely vegetated. After all, as Alabama's most southern terrestrial ecosystem, the dunes receive more light than any other ecosystem. Furthermore, there are few indications that light is in short supply.
In most ecosystems, competition for light among plants is fierce, with
crowded plants jockeying for positions to maximize their harvest of
sun. Instead, most dune plants are widely spaced and intercept more
light than they can use. Could heat explain the dune's sparse vegetation?

Heat. The heat received by an ecosystem has a major influence on
the distribution, behavior, and growth of organisms. Life is sustained
by a complex suite of biochemical reactions occurring within a relatively narrow temperature range. Too much heat will damage proteins
and other molecules. With too little heat, biochemical reactions slow
and essential life processes fail. Freezing temperatures pose particular
problems since ice crystals cause significant tissue damage unless protective mechanisms are in place. Ecosystems with excessively high or
low amounts of heat usually support few organisms and species.

Most heat in ecosystems comes directly from the sun and factors
governing heat availability parallel those controlling sunlight. Southeastern ecosystems enjoy a comfortable range of temperatures throughout the year relative to regions farther north or south. Within Alabama,
average annual temperatures and their extremes are higher in the south
than in the north. Winter low temperatures are rarely a threat to beach
mice and other dune organisms, but daytime summer temperatures

can rise to levels jeopardizing survival. Consequently, dune organisms have adaptations to survive high temperatures. Some plants have narrow leaves that do not absorb much heat and easily radiate heat once absorbed. Others have on their leaves and stems a dense layer of hairlike fibers reflecting much of the incoming sunlight. Less obvious adaptations involve biochemical modifications allowing photosynthesis to proceed despite high temperatures. Animals also have adaptations to cope with high temperatures, many of which are behavioral. Beach mice are active only at night and near dawn and dusk, and they retreat into cool burrows during the day. The extreme summer heat is mitigated by the sea breeze arising nearly every summer day and the white dune sands that reflect sunlight back into the atmosphere.

So, while summer heat is intense, it is not sufficiently stressful to cause low primary productivity in the dunes. After all, nearby forests receive nearly the same amount of heat from sunlight and support far more species and biomass. The explanation for why the dunes are sparsely vegetated must be elsewhere.

Water. Under the stare of the midday sun I am sweating liberally. I have swallowed my last ration of water—like other organisms in this environment, I am using water to stay cool. The difference is that dune residents are stuck with what is here, and there is not much. Other than a few low swales that temporarily hold rainwater after a large storm, the coastal dunes are a dry ecosystem. The scarcity of water is an important factor keeping primary productivity low.

Water is essential for hundreds of biochemical processes and is used by many organisms to avoid overheating. Its most direct tie to primary productivity is its use in photosynthesis. As a step in capturing carbon dioxide from the air, plants uptake water from the soil, transport it to their foliage, and allow its evaporation from the leaf surface. This process, known as transpiration, draws carbon dioxide into the leaf, but also cools the leaf since evaporating water carries away heat.

In dry ecosystems, organisms have various adaptations for coping with water's scarcity. For dune plants, the strategies used to reduce overheating also reduce the need for water. Some small plants, especially succulents and cacti, will stockpile rainwater in modified tissues, while larger plants like scrub oaks grow deep taproots to reach ground-

Florida Rosemary's narrow leaves help it conserve water in the desertlike dunes. Photo by R. Scot Duncan.

water. Beach mice, like most rodents, are highly efficient with water derived from their diet, and avoid daytime activity to stay cool and reduce the need for water.

Why is water so rare in dune ecosystems when Alabama lies at the heart of one of the wettest regions of North America? The coastal counties receive more rain than all other counties in the state. Furthermore, the maritime forests and pine flatwoods adjacent to the dunes receive the same rainfall but have much higher primary productivity. Clearly, climate isn't forcing water limitation and low primary productivity in the dunes. As it turns out, the culprit is the dune sand itself.

Soils. Though most people don't think much about what's underfoot, a quick look into the soil reveals much about an ecosystem. Soils often govern primary productivity and species composition in terrestrial ecosystems. Soils store the water and many essential nutrients plants need for growth and survival, and give plants a place to anchor. So, why do dune soils limit the growth of vegetation?

A typical soil has an organic layer comprised mainly of decomposing plant materials, a topsoil layer of organic matter leached from above and of mineral particles, and a subsoil layer composed mainly of mineral particles. These mineral particles are sands, silts, and clays produced from the weathering of bedrock or other parent materials that lie below, though at low elevations, mineral soil may have been deposited by wind or water. Each parent material weathers into particles with distinctive chemical and physical aspects that influence ecosystems.

Alabama's dunes lack topsoil, save only for a thin layer beneath the scrub dune trees. Thus, dunes provide exceedingly low levels of many plant nutrients. Dune soils are chiefly composed of quartz, a stable mineral of little value to plants. When I scoop up a handful of dune sand, the grains slip easily through my fingers, illustrating their poor adherence to one another. This mobility allows sand to be easily transported by wind and sculpted into dunes. It is also why dunes need plant roots to reduce erosion. The sand grains are large and blocky compared to the fine sediments of most other ecosystems. The wide spaces between grains allow rain to quickly sink into the groundwater below and beyond the reach of small plants.

The combination of scarce nutrients, low adherence, and poor water retention is why soil factors are the greatest barrier to dune primary productivity. Nevertheless, the unique plants and animals inhabiting such extreme ecosystems contribute significantly to Alabama's biodiversity.

Disturbance. Frederick (1979), Elena (1985), Opal (1995), Ivan (2004), and Katrina (2005) are names and dates reviving unpleasant memories for most residents of the northern Gulf Coast. These are the hurricanes that inflicted the greatest damage on Alabama's Gulf Coast in the recent decades. Their combined economic damage approaches $150 billion, but the loss of lives and dreams defies measurement.

Hurricanes also wreak chaos on the dunes, and the damage and destruction of these increasingly rare ecosystems can be heart-wrenching. The primary dunes favored by beach mice suffer the greatest devastation. Abnormally high tides from the storm surge send wind-frenzied waves to pound away at dunes. The strongest storms can overwhelm any primary dune and quickly wash away what took decades to form. Where the primary dunes are breached, torrents of water can tumble inward and scour the secondary and sometimes tertiary dunes. In the worst wash-out areas, only a flat expanse of sand remains. During such storms, many beach mice of the primary dunes drown in their burrows or when attempting escape. Survivors endure months of famine since most seeds are blown or washed away and salt water infiltrating the groundwater kills many seed-producing plants.

Hurricanes are a type of ecological disturbance, defined as an event substantially altering the availability of the four key ecological resources (sunlight, heat, water, and soil). Disturbances can be large or small, fast or slow, intense or subtle, and caused by organisms or nonliving phenomena. Even in the midst of the most extreme disturbances, processes are begun that lead to the repair of a damaged ecosystem. Recovery may take months, years, or centuries, but eventually life will reclaim vacant space in any ecosystem.

Repair is made possible by organisms able to quickly establish themselves where disturbance creates a vacancy. Among plants, these species are known as pioneers or colonizers. The colonization of disturbed areas is known as ecological succession, a name referring to the sequence of species occupying a disturbed area at different times after the disturbance. While ecologists may use terms like *repair* and *reclaim*, ecosystems are not entities with self-healing properties.

Succession in dune ecosystems battered by a hurricane begins almost immediately. Ironically, a storm's damaging winds and waves can leave behind seeds and root fragments that initiate succession. Over the next months to years, more wind-blown seeds arrive from adjacent areas where the vegetation survived. As young plants grow in washed-out areas, they unintentionally begin restoring dunes by trapping wind-

Young dunes forming near storm-damaged primary dune. Photo by R. Scot Duncan.

blown sands. Plant roots and stems expand to match the growing dune, thereby trapping even more sand. With time, nascent dunes merge to form larger dunes that grow and accrue more of the vegetation lost during the storm.

Animals often play important roles in succession. Within the dune ecosystem, beach mice dispersing from surviving populations will colonize young dunes once enough shelter and seed are available. Mice bring seeds to their burrows, and some seeds will go uneaten, germinate, and contribute to the developing plant community.

Typically, pioneers exploit the disturbed area until more competitive species colonize and push them aside. Rates of succession vary, but it is usually faster in ecosystems with higher primary productivity and where disturbances are milder. In past decades many ecologists believed succession led to a climax community with stable resource availability and a fixed complement of species. Now we know that ecosystems never achieve full stability. Resources are always in flux, as is the roster of species present, their abundances, and their interactions. Succession doesn't lead to a climax community; it simply results in the restoration of complexity. This is certainly true of succession among the dunes of the northern Gulf Coast. For example, Sea Oats and other pioneer plants often retain their place in the dunes until the next disturbance.

The combined influence of sunlight, heat, water, soil, and disturbance shapes the dunes and all other terrestrial ecosystems. You can think of them as factors in a rough and dynamic equation that predicts ecosystem structure. The availability of the four resources are input variables on one side of the equation, while on the other side is an estimate of primary productivity. Disturbance is a wild-card variable, periodically altering resource inputs and, thus, the predicted outputs. This is an oversimplification but is a helpful conceptual tool for learning to study ecosystems.

DISAPPEARING MICE

Some dunes along the Alabama coast are crowned with grasses and oaks but are no longer inhabited by beach mice. The Alabama Beach Mouse population has been declining and extinction is a real possibility. Alabama has already lost the Perdido Key Beach Mouse, a species now surviving only on Florida's portion of Perdido Key. The chronicle of how both species declined in Alabama is a case study for understanding the factors governing population growth, the last ecological principle to review. It is also a cautionary tale about saving endangered species, for the survival of both mice largely depends on how quickly their populations recover after setbacks.

The Alabama and Perdido Key beach mice are subspecies of the Oldfield Mouse, a widespread southeastern species.[1] Typical populations are brown, but the several subspecies of beach mice along the Gulf and Atlantic coasts are lighter to match the dune environment. The Perdido Key Beach Mouse resides entirely on Perdido Key, a 15-mile-long barrier island shared by Alabama and Florida and separated from the Alabama Beach Mouse's historic range by only several hundred feet of water. Though genetically distinct, these beach mice are similar in morphology (size, shape, and color), breeding biology, and habitat requirements. The major forces threatening their survival are also similar—the expansion of human development and habitat damage caused by hurricanes.

The Alabama and Perdido Key beach mice occupied all of their native range prior to commercial and residential development of beach habitats. Occasional hurricanes would have wiped out segments of each population, but offspring of survivors in adjacent areas colonized rebuilt dunes. Humans transformed much of their original habitat into highways, condominiums, and parking lots—inhospitable habitats for beach mice. This resulted in isolated subpopulations that are largely confined to protected public lands.

Population fragmentation forces a species one big step closer to extinction. Isolated populations are vulnerable to being wiped out by a single catastrophic event like a hurricane. Fragmented populations with few individuals are also prone to inbreeding (the mating of related individuals). Inbreeding allows harmful genes that had previously been rare to spread quickly and cause declines in survival. Small populations can remain genetically healthy through periodic immigration of new individuals, but this is impossible for isolated beach mice populations. Such problems plague tens of thousands of species throughout the world.

Alabama's beach mouse subspecies have been fragmented for quite a while. By the early 1980s, following decades of development and extensive dune erosion caused by Hurricane Frederick, populations of both subspecies were suffering. Extinction was imminent for the Perdido Key Beach Mouse. Development wiped it out from the entire island except for a single Alabama population at Florida Point on the island's westernmost tip. In 1985, Hurricane Elena strafed the Alabama coast before its center made landfall in Mississippi. Although the population survived, fewer than 30 individuals remained by 1986. When the Perdido Key Beach Mouse was placed on the US Endangered Species List that year, biologists considered it the most endangered mammal in North America. Fortunately, the mouse had something going for it that many endangered species do not—it has a high r-value.

Population growth rates depend on two things: a species' potential maximum rate of reproduction, a value ecologists refer to as r, and resource abundance in the ecosystem. A suite of attributes collectively known as the life history determine the r-value for species and include reproduction frequency, number of offspring per reproductive bout, and the amount of resources invested by the mother in each offspring. Some species, like weedy pioneer plants, have high r-values. These *r*-selected species, as ecologists refer to them, do not invest in longevity. Instead, they have as many offspring as they can, as quickly as they can. While most offspring will not survive, the high number sent forth increases the chance some offspring will find a freshly disturbed area to colonize. It is a successful strategy, but there are costs. Making little or no investment in long-term survival, individuals do not survive long. While their populations grow rapidly soon after a disturbance, their numbers decline steeply when more competitive species colonize and resources become scarce.

You and I are members of a *K*-selected species. Such species have a low *r* and invest much of their surplus energy into becoming successful competitors with numerous defenses and long lives. *K*-selected species delay reproduction until they are ensured a sufficient supply of resources over the long term. Compared to pioneers, *K*-selected species make fewer, larger offspring, each of which has a higher chance of survival than an offspring of an *r*-selected species. The major cost to species with *K*-selected life histories is their slow population growth after a decline. This is why so many endangered species are *K*-selected species. However, once established, the population sizes of *K*-selected species are more stable than those of *r*-selected species. Over the long term, the population size of a *K*-selected species will average near the carrying capacity, the highest number of individuals the ecosystem can sustain for that species. Ecologists denote this value as *K*, the namesake variable of *K*-selected species. While many species have distinctively *r*- or *K*-selected life history strategies, some have life histories falling somewhere between these extremes.

Beach mice are more *r*-selected than most other endangered species in the United States. They reach sexual maturity within eight weeks of birth, produce four pups per litter, and give birth as frequently as once a month during peak breeding season. Historically, this capacity for quick population growth enabled them to recover quickly from hurricanes. Now, it is an asset for the mouse and those working to ensure its survival.

In 1986, with the Perdido Key Beach Mouse facing extinction and isolated at Florida Point, federal officials decided to establish a second population. Because the mice are *r*-selected creatures, the Florida Point population recovered quickly and provided enough mice over the next two years for the introduction of 15 pairs to a portion of Perdido Key in Florida, Johnson Beach in Gulf Islands National Seashore. The new population grew so rapidly that by the late 1980s it occupied nearly all suitable habitats in the 6.8 miles (11 km) of Johnson Beach. The future was looking brighter for the Perdido Key Beach Mouse.

In the fall of 1995, after a much-needed lull in hurricane activity, Hurricane Opal made landfall near Pensacola, Florida. At Florida Point the storm wiped out most of the remaining foredunes and extensively damaged the secondary dunes. Most of the beach mouse population was killed outright, and the few survivors soon died. Alabama had lost its last population of the Perdido Key Beach Mouse. More bad news was expected at Johnson Beach, which was much closer to where Opal's center made landfall. Instead, biologists found the population had survived. To the delight and relief of biologists and beach mouse enthusiasts, the decision to establish a second population had saved the subspecies from extinction.

Biologists then made an important discovery when studying why the Johnson Beach population survived and the Florida Point population failed. Unlike Florida Point, Johnson Beach had an area of tertiary scrub dunes that were well protected from the storm surge. These dunes were a safe haven for mice during the storm and provided food and shelter until the primary dune vegetation recovered. There had been no such refuge at Florida Point. Several years later, the US Fish and Wildlife Service added scrub dunes to the list of critical habitats for both subspecies.

Because the Perdido Key Beach Mouse was once again surviving in a single population, in 2000 biologists established a new population in Florida's Perdido Key State Park 2 miles west of Johnson Beach. Unfortunately, hurricanes Ivan (2004) and Katrina (2005) dealt punishing blows to the dunes of both populations, even damaging the scrub dunes. At the time of this writing, populations at Perdido Key State Park are extremely low, and those at Johnson Beach number just a few hundred. For the moment, the near-term fate of this species is far from certain and depends on how quickly ecological succession can produce more dunes.

Rampant development and the barrage of hurricanes in recent decades have taken a heavy toll on the Alabama Beach Mouse. By the early 1980s, residential development and Domestic Cats wiped out the mice from Ono Island, leaving only three isolated populations elsewhere. Since then, hurricanes eliminated the population in Gulf State Park in Gulf Shores several times. Currently, the park has no beach mice, and due to its isolation a population could only be reestablished with a deliberate reintroduction. A second population on the western end of the Fort Morgan Peninsula has survived despite substantial losses of dune habitat from recent hurricanes. I have been hiking today within the heart of the third population. Bon Secour National Wildlife Refuge's Perdue Unit has the largest surviving population of the Alabama Beach Mouse thanks to an abundance of tertiary dune habitats, but even this population suffered greatly from erosion and flooding from hurricanes Ivan and Katrina. These storms destroyed more than 90 percent of the primary and secondary dunes and damaged substantial portions of the tertiary dunes.

This destruction is evident as I hike out of the scrub dunes. It is early afternoon and temperatures have peaked, so I have decided it's time for a swim. Near the beach I enter a field of young dunes. They are populated with dune grasses whose stems are partially buried from the blowing sand they have trapped. The surviving primary dunes are substantially taller and starkly illustrate how much the young dunes must still grow.

Alabama has lost the Perdido Key Beach Mouse, possibly forever since development has eliminated most suitable habitat in Alabama. The Alabama Beach Mouse, however, has a fighting chance since large blocks of suitable habitat are protected. Both subspecies are doing their part to survive. Being resilient r-selected organisms, they quickly respond when conditions improve. With one surviving and one extirpated, the subspecies represent two possible outcomes for Alabama's many rare and endangered species. Stories similar to theirs are unfolding across the state, and all teach the same lesson—either Alabama will guide economic growth in ways ensuring its vulnerable species survive or the state will lose much of its unique biodiversity.

3
Anomalous Alabama

No doubt about it, Alabama is soggy. Winter and early spring weather is ruled by cold fronts and their soaking rains, summer brings thick humidity and drenching thunderstorms, and tropical storm systems come calling in late summer and early fall. This profusion of water is a blessing and a curse. It supports high farmland yields and abundant wetlands, but the humidity is burdensome and the frequent storms endanger life and property. Rain, dampness, and storms are such a basic part of Alabama life that most don't believe me when I assert Alabama should be a desert.

Nonetheless, it's true: Alabama should be desert. Its coastal plain should be dotted with cactus and sagebrush and its uplands a mix of chaparral and scrubby woodlands. And yet, Alabama and the Southeast enjoy a climate of mild temperatures and high precipitation. Climate, the long-term weather pattern of a region, is a chief influence on biodiversity patterns. By rationing sunlight, heat, and water, and by causing various ecological disturbances, climate has a firm grip on any ecosystem's primary productivity. Globally, climate produces broad ecological patterns like the treeless tundra encircling the Arctic and the equatorial tropical rain forest. At most locations sharing Alabama's latitude (a measure of the distance from the equator), sunlight is abundant, water is scarce, and arid deserts, grasslands, and scrublands prevail. Instead, Alabama is at the heart of one of the wettest regions on the continent. Why is Alabama an anomaly?

The explanation is complex and begins with the distribution of heat across the planet. In this regard, Alabama is a "Goldilocks" state—not too hot and not too cold. To the south is the unwavering heat of the tropics, and to the north are regions with short growing seasons and long, cold winters. Alabama lies between these extremes in the latitudes known as the warm temperate zone. By influencing the availability of heat and light, latitude has a tremendous bearing on a location's climate and biology. Light is captured by plants and stored as organic energy, and heat is needed to hurry along the biochemical processes on which life depends. Thus, the latitudes near the equator have the highest rates of primary productivity and biodiversity while the polar regions have the lowest. To understand what causes the relationships between latitude, primary productivity, and biodiversity, a view of our planet and its sun from space is helpful.

Not all places on earth enjoy the same share of the sun's energy. In the tropics, sunlight strikes the earth at nearly a vertical angle and delivers more energy than anywhere else on the planet. Due to its curvature, the earth's surface near the poles slants away from the sun, and incoming sunlight arrives at an indirect angle. Additionally, this incoming sunlight passes through more atmosphere than does sunlight arriving at the equator. Because the atmosphere scatters and absorbs some sunlight, incoming light near the poles has lost much of its energy. Thus, the tropics are hot, the poles are cold, and the temperate zone is mild. This also explains why Alabama's ecosystems have high primary productivity relative to northern states.

The imbalance of sunlight across the planet creates global air circulation patterns that at Alabama's latitude usually produce deserts. Each day, the sun heats the lands and waters of the tropics. This heat warms the lower atmosphere, and columns of heated air rise upward. Within this air is water vapor originating from photosynthesis and evaporation from soils and surface waters. As this warm air rises it also cools.[1] When cooled sufficiently, water vapor condenses into liquid droplets and forms clouds.[2] As heat intensifies during the day, the process accelerates, producing ever-larger clouds. Eventually, fine water droplets collide to form heavier droplets that drench the tropics daily. Generally, rainstorms created by rising columns of warm moist air are called convective showers.

Once rising tropical air has dumped its heat and water vapor, it ascends no more. Crowded from below by more rising air and unable to climb higher against the tug of gravity, these masses of high-elevation air flee sideways, away from the equatorial zone. Having cooled, these air masses contract in volume, become dense and heavy, and begin sinking at the margins of the tropics and latitudes just beyond.[3] The falling air warms under the increasing weight of the atmosphere above.[4] The air makes landfall at latitudes 30° to 35° north and south of the equator, regions known as the horse latitudes. Some of this air travels back toward the equator as surface winds that absorb heat and join new rising tropical air columns. These surface winds will complete a cycle within the great tropical atmospheric circulation loop known as a Hadley cell. There is a Hadley cell on each side of the equator around the circumference of the earth.[5] Winds from the horse latitudes heading away from the equator join similar atmospheric circulation cells at higher latitudes. Including the Hadley, there are three such cells in each hemisphere.

Finally, here's why Alabama should be a desert. The warm dry air descending in the horse latitudes produces a belt of arid ecosystems in both hemispheres. In the northern hemisphere, these include the deserts of the southwestern United States, northern Mexico, northern Africa, and southern Asia. Alabama sits squarely within this zone. Instead of being arid, it is at the center of one of the rainiest regions in North America.

The Gulf of Mexico has come to the rescue. Alabama and her neighboring states enjoy the maritime effect, where an ocean influences the climate of adjacent land. The Southeast's maritime effect begins hundreds of miles away as the Caribbean Sea absorbs vast amounts of heat from the tropics. The Loop Current brings the Caribbean's warm waters to the Gulf. Because water is thermodynamically stubborn—it is slow to warm and slow to cool—the Loop Current's waters retain their tropical heat as they reach the central and northern Gulf. Throughout the year, massive volumes of water evaporate from the Gulf's surface, drift over the continent, and cause much of the rainfall in places as distant as Oklahoma, Arkansas, Tennessee, and Kentucky. In total, the Gulf contributes half the summer rainfall in the eastern United States. Whether you love or hate the Southeast's muggy weather and frequent rains, it is the Gulf of Mexico that is to blame.

Our Slouching Planet

Our planet slouches, and this bad posture affects Alabama's ecological character via the changing seasons. Earth spins once each day about an axis tilted at a 66.5° angle with its orbital path around the sun. The axis aims at the same point in space all year, so for half of its yearlong orbit, the northern hemisphere aims away from the sun and has shorter days with less intense sunlight than the southern hemisphere. This is the northern winter, the season when primary productivity grinds to a near halt. The situation reverses during the other half of the year when the northern hemisphere tilts toward the sun. However, there is much more to seasonal change in Alabama than variation in day length and sunlight intensity. A generalized description of Alabama's climate through the seasons will illustrate.

The southeastern summer climate is stable. Each languid day is similar to the next. Temperatures are hot, humidity is high, and the only regular variation is an occasional thunderstorm. This stability is brought by an air mass known as the Bermuda High. Cool, dry air

Summer thunderstorms keep Alabama's ecosystems lush during the summer. Courtesy of Alan Cressler.

from the upper atmosphere falls into the horse latitudes mainly over a region of the Atlantic Ocean stretching from Bermuda, situated about 650 miles (1,046 km) east-southeast off the North Carolina coast, to the Azores, located 950 miles (1,529 km) west of Portugal. These descending masses of dense air create a zone of high atmospheric pressure. During late spring as the northern hemisphere warms, the Bermuda High's center migrates westward from the Azores to Bermuda. Once in place, it shields the Southeast from fronts and tropical storm systems until autumn. Another influence of the Bermuda High is its wind pattern. Due to the Coriolis effect, a complicated set of forces caused by earth's rotation about its axis, the surface winds of high pressure systems in the northern hemisphere flow clockwise around their centers. The western margins of the Bermuda High push water vapor and heat from the Gulf deep into the continent's interior. This is why high humidity and thunderstorms are common during Alabama's summers.

Extreme southern Alabama enjoys land and sea breezes. They arise because water retains heat more efficiently than land. Once the Gulf warms for the summer, it stays roughly the same temperature all day and night. Land surface temperatures are more variable. During a typical summer night, the temperature of the land and the atmosphere above it will drop below the temperature of the Gulf. Meanwhile, the nighttime Gulf's hot surface waters heat the lower atmosphere, causing surface air to rise. This differential pulls cool, heavy air from the continent onto the Gulf. The resulting land breeze begins late at night and continues into the early hours of the day.

The sea breeze is a land breeze in reverse. During the day, land temperatures become far warmer than the Gulf's waters. Sea breezes begin in midmorning when hot air rising over the land pulls in cooler moisture-laden air from the Gulf. The arriving air warms, rises, and supplies moisture to growing cumulus clouds, some of which will become convective rainstorms. These storms are so common that Mobile and Baldwin counties and adjacent areas comprise the wettest region of Alabama. More than 66 inches (168 cm) of rain falls annually and supports a degree of biodiversity and primary productivity found nowhere else in Alabama. These storms support a suite of unique wetland ecosystems and make southern Alabama one of the most lightning-rich regions of North America. And where lightning is common, so are

wildfires. Historically, many ecosystems near the coast burned every other year and sustained open pine woodlands carpeted with fire-adapted plants.

From mid-September to late November is Alabama's autumn, a time of gentle transition between summer and winter weather patterns. The Bermuda High shifts eastward, reducing the supply of moisture coming onto the continent. The days shorten, humidity drops, the sun's position in the sky shifts southward, and the Southeast cools. This cooling slows the cycling of water between the ocean, atmosphere, and land, and thunderstorms become infrequent, then cease altogether. Alabama usually experiences an autumn drought during the weeks of late September, October, and early November.

However, without the Bermuda High as a shield, tropical cyclones (a general term for tropical depressions, tropical storms, and hurricanes) can interrupt autumn's easy progression into winter. These storms can strike the northern Gulf Coast as early as late spring, but most arrive

Lightning-sparked wildfires keep many natural ecosystems healthy. Courtesy of George W. Ponder III/Beaux Point Photography.

Hurricane Ivan approaching the Alabama coast. Courtesy of NASA/Goddard Space Flight Center Scientific Visualization Studio.

from July to October. Tropical cyclones are extreme versions of low pressure systems, large areas of warm air rising into the upper atmosphere and producing convective showers. Tropical lows begin as small weather systems off the African coast near the equator. They travel west and then increasingly northwestward, all the while acquiring moisture and heat from the tropical waters of the Atlantic and Caribbean. Warm, humid air rising from the Gulf bolsters any storms passing over it. Low pressure systems can grow into tropical depressions, then tropical storms, and then hurricanes, with each transition marked by increases in storm size, organization, and sustained wind speed. Hurricanes develop a central cloud-free "eye," strong winds rotating counterclockwise around this opening, and arcing bands of storm clouds spiraling out from the center for hundreds of miles. When a tropical cyclone makes landfall, or even grazes the Southeast, it has tremendous impact on coastal and inland ecosystems. Strong waves and high tides batter the immediate coast, inland winds damage or down trees, and heavy rains cause erosion and flooding.

By late autumn the Southeast becomes an atmospheric battleground between the north and the south. Northern arctic air masses commonly called cold fronts begin invading regularly from the northwest. Cold fronts are high pressure systems of cold, dense air forming in the arctic. Ahead of fronts, a low pressure system develops with counterclockwise winds that pull in warm, humid Gulf air. Technically, the front is the contact zone between these air masses. As the high pressure system advances, its wedgelike leading margin of dense air pushes upward the warm moist air of the low pressure system. This triggers intense convective storms with lightning, downpours, and strong winds.

Cold fronts arrive all year, but during the warmer months they are weak and easily deflected or destabilized by the Bermuda High. By mid-autumn the Bermuda High has retreated and the fronts bring a day or two of crisp dry air and relief from the summer's muggy weather. By mid-December and the full onset of winter, cold fronts reach Alabama with more force. Fronts roll across the region roughly once a week from early winter to early spring. The rains produced are a major source of the region's yearly precipitation. After each front, cold, high-pressure air blankets the Southeast. Once the system shifts eastward, the Gulf restocks the region with heat and humidity until the next front arrives. Without the Gulf, winters in Alabama would be cold and dry, and primary productivity would be lower.

Winter usually gives way to spring during the last days of February. By mid-March and the spring equinox, the Southeast is warming rapidly. The Bermuda High creeps toward its summer position and begins slowing and deflecting arriving cold fronts. Before the Bermuda High can fully exert its dominance, spectacular clashes between similarly matched northern and southern air masses along advancing fronts spawn damaging winds and tornados.[6] The biological transitions are no less significant. Plants break winter dormancy and begin growth and flowering. Animals breed or migrate northward to their summer habitats. Rivers swollen with winter rains supply nutrients that stimulate marine productivity. With each passing week the influence of the winter weather pattern subsides and by early June summertime's stability has returned.

That's a quick portrait of the seasonal climate for Alabama and adjacent states. However, earth's climate is complex, and there can be substantial aberrations to the Southeast's typical seasonal pattern. A

A winter freshet builds on the Coosa River. Courtesy of Hunter Nichols/Hunter Nichols Productions.

shift in the summer position of the Bermuda High can be particularly significant. When it is farther east than normal, it cannot shield the Southeast from weak cold fronts and summers are wetter than usual. More importantly, the Southeast is more vulnerable to tropical cyclones. There is drought when the Bermuda High shifts farther west than normal and hovers over the Southeast. Fronts or other disturbances producing rains cannot penetrate the region, and the cool, dry air descending in the high's core prevents summer convective storms from forming.

Another important aberration is El Niño. This climate irregularity is caused by abnormally high water temperatures and a reversal of ocean currents in the equatorial Pacific Ocean. El Niño events occur every three to seven years and last from several months to a year. What triggers an El Niño is still not fully understood, but it clearly instigates worldwide climate deviations. For the Southeast, El Niño brings cooler temperatures and unusually high amounts of rainfall. If El Niño is present in late summer and autumn, it suppresses tropical cyclone activity in the Atlantic and Gulf.[7] Sometimes the equatorial Pacific becomes unusually cold and a La Niña weather pattern establishes, often following an El Niño event. In the Southeast, La Niña brings drought and an increase in tropical cyclone activity in fall.

Collectively, these deviations in the Southeast's seasonal pattern are ecological disturbances that alter resource availability. Summer droughts and associated high summer temperatures cause population declines in sensitive species and increase the chance for unusually intense wildfires. Periods of high rainfall can overwhelm streams and rivers and cause prolonged flooding. However, southeastern species have faced climate irregularities for tens of millions of years and most populations recover with the return of normalcy. Some species actually benefit from these extremes. For example, hardy plants surviving droughts face fewer competitors in the next growing season.

LOCAL CLIMATE VARIATION

Local variation in Alabama's climate also shapes the state's biodiversity patterns. Recall from chapter 2 that the heat delivered to an ecosystem affects the distribution, behavior, and growth of organisms. Alabama spans nearly five degrees of latitude. Thus, the temperature gradient from north to south in Alabama is one of the most influential of the local climate variations. Average annual temperatures in the southern part of Alabama exceed 68° F (20° C), while in the far north they are just below 60° F (15.6° C). The growing season, defined as the interval between the last frost of one winter and the first frost of the next, ranges from 270 days in the south to 190 days in the north, a difference of over two months. These differences enable southern Alabama to support greater primary productivity than areas farther north.

Though an index developed for farming, the length of the growing season is a good predictor of potential primary productivity. It is also one of the few carefully studied aspects of climate variation in Alabama. Growing season length does not decline smoothly from south to north. Some areas have growing seasons 10–30 days shorter than areas a dozen miles to the east or west. A major cause of this is the valley effect, where during winter nights cold surface air drifts down into large river valleys, especially the Coosa, Tombigbee, and Chattahoochee river valleys. Other areas, including the Black Warrior River Valley and the western portion of Alabama's Tennessee River Valley, have longer growing seasons for reasons not fully understood.

Local temperature variation can be exceedingly important for many organisms. Lands adjacent to wetlands and reservoirs can be warmed in winter as stored summer heat radiates from these waters. A canyon floor kept in shadow for part of the day is cooler than upland areas receiving more sunlight. South-facing slopes are warmer and dryer than north-facing slopes. Tall mountains have cooler temperatures than surrounding lowlands. These fine-scale differences matter to temperature-sensitive plants and animals.

Though all of Alabama receives more precipitation than most of North America, local variation in rainfall diversifies the state's ecology. Annual rainfall is highest near the coast and declines by 10 inches (25 cm) or more from the coast to the state's center. Drier conditions stretch across Alabama's midsection, averaging 54–56 inches (137–142 cm) of rain annually. Northern Alabama's rainfall patterns are highly variable, but generally range between 56 and 60 inches (142–152 cm) per year. These differences affect primary productivity in ecosystems and the diversity and abundance of wetlands.

Alabama's taller mountains produce localized areas of high precipitation as a result of orographic lifting. This occurs when warm,

Clouds forming at high elevations on Choccolocco Mountain, Calhoun County. Courtesy of B. William Garland.

moisture-laden air blown across the land is forced up mountain slopes or similar features and cools. Alabama lacks enough tall mountains to create broad areas of higher rainfall, but orographic lifting causes above-average precipitation on parts of Lookout and Sand mountains in the northeast, several mountains in northern Jackson County, and the tall mountains of Clay County.

THE CLIMATE ADVANTAGE

Alabama's warm, wet climate boosts primary productivity and its climate variation diversifies the state's ecosystems. But how important is climate in shaping species diversity patterns across a broader geographic scale? One approach to this question is to compare how climate shapes biodiversity patterns among US states.

It is no coincidence that the top seven states for species diversity are on the southern US border. The intensity of sunlight they receive gives them an advantage over states farther north. Latitudinal span is also important—tall states capture more climate diversity than squat states. California and Texas rank at the very top for species diversity and both have more latitudinal range than all states except Alaska. Climate variability resulting from topographic variation is another factor. The top four states for biodiversity have impressive mountain ranges producing high, cold ecosystems and strong rain shadows.[8] Overall precipitation is an important factor, but in complex ways. The top four states for species diversity are all predominantly arid, though California and Texas have wet regions. For these four states, montane climate variation makes up for having less precipitation than southeastern states. Conversely, Alabama and surrounding states lack extreme internal climate variation, but enjoy high precipitation.

Though climate has a tremendous influence on biodiversity patterns, it alone cannot explain why Alabama leads most of the nation for species diversity. Louisiana and Mississippi have climates similar to Alabama's, but they lag far behind Alabama for species diversity. The difference is that Alabama's geology is far more diverse than that of these and many other US states. In the next chapter, we'll explore the origins of the state's geologic diversity and how it contributes to the state's high rates of biodiversity.

4
Alabama Rocks

Birmingham Reef

To understand Alabama's biodiversity, it helps to know a little geology. This can be challenging if you happen to find salamanders far more interesting than chunks of rock. However, geologic processes are the ultimate causes of Alabama's high ranking for national biodiversity. Tectonic forces pushed Alabama into the warm-temperate zone; mountain building thrust dozens of ancient rock types to the surface; erosion carved a multiplicity of landforms; and weathering of rock and sediment deposition concocted soils varying in almost every possible way. With a superbly diverse geology compared to most states, Alabama is overrun with species tied to a particular rock formation, topographic feature, or soil. These include many fishes and salamanders, hordes of invertebrates and wildflowers, and a handful of feathered and furred creatures.

This chapter examines the geologic processes shaping Alabama's biodiversity. The overarching lesson is that any quest to fully understand the ecology of a place or a species is enriched with a study of the soils and/or bedrock beneath. To illustrate, I'll begin at one of my favorite spots in Alabama and examine how geologic processes shaped its biodiversity. The site presents several geologic puzzles whose solutions reveal how intimately geology and biodiversity are connected.

Hawk's View Overlook at Ruffner Mountain Nature Preserve in Birmingham is a spot most naturalists would enjoy. The steep trail to the overlook emerges from a shady broadleaf forest to a cliff and sweeping view of the city, nearby peaks and hills, and the distant terrain of the Cumberland Plateau. Just a few feet from the overlook is Salamander Cave. It is not much more than an exaggerated fissure, but within it are stalactites, roosting bats, and the Cave Salamander. The latter is a bright orange-red amphibian with fine black spots and an exceptionally long tail. While most salamander species inhabit wetlands or the shelter beneath leaf litter, logs, or rocks, the Cave Salamander's preferred habitat is near the mouth of caves in the limestone bedrock of the Southeast's uplands. The salamander's preferences are not unusual: dozens of Alabama's species—including fishes, crayfishes, shrimp, and pseudoscorpions—only inhabit limestone caves. Caves provide escape from surface predators, a near-constant climate, and just enough nutrients and energy to support a sparse ecosystem. The tight relationship between these species and their caves illustrates how geology has helped craft Alabama's rich biodiversity. Simply put, the presence of so many unique geologic features like caves has created an ample diversity of ecosystems and their dependent species. How did these unique ecosystems arise? The answer is revealed through a study of limestone's composition, origin, and weakness.

The state's geologic maps reveal that the limestone exposed at Ruffner has a name, the Chickamauga Limestone. Just as biologists name species, geologists name rocks. Adjacent rock layers sharing similar characteristics and origins comprise a named geologic "formation." Geologists usually name formations after a settlement or other geographic feature near where the rock was first discovered; Chickamauga is a town in northwestern Georgia.

The Chickamauga Limestone's description could seem far-fetched.

Cave Salamander, a karst-dependent species, Ruffner Mountain Nature Preserve, Birmingham. Photo by R. Scot Duncan.

Left chart — Phanerozoic

EONOTHEM / EON	ERATHEM / ERA	SYSTEM, SUBSYSTEM / PERIOD, SUBPERIOD	SERIES / EPOCH	Age estimates of boundaries in mega-annum (Ma) unless otherwise noted
Phanerozoic	Cenozoic (Ca)	Quaternary (Q)	Holocene	11,700 ±99 yr*
			Pleistocene	2.588*
		Tertiary (T) / Neogene (N)	Pliocene	5.332 ±0.005
			Miocene	23.03 ±0.05
		Tertiary (T) / Paleogene (P)	Oligocene	33.9 ±0.1
			Eocene	55.8 ±0.2
			Paleocene	65.5 ±0.3
	Mesozoic (Mz)	Cretaceous (K)	Upper / Late	99.6 ±0.9
			Lower / Early	145.5 ±4.0
		Jurassic (J)	Upper / Late	161.2 ±4.0
			Middle	175.6 ±2.0
			Lower / Early	199.6 ±0.6
		Triassic (Ŧ)	Upper / Late	228.7 ±2.0*
			Middle	245.0 ±1.5
			Lower / Early	251.0 ±0.4
	Paleozoic (Pz)	Permian (P)	Lopingian	260.4 ±0.7
			Guadalupian	270.6 ±0.7
			Cisuralian	299.0 ±0.8
		Carboniferous (C) / Pennsylvanian (P)	Upper / Late	307.2 ±1.0*
			Middle	311.7 ±1.1
			Lower / Early	318.1 ±1.3
		Carboniferous (C) / Mississippian (M)	Upper / Late	328.3 ±1.6*
			Middle	345.3 ±2.1
			Lower / Early	359.2 ±2.5
		Devonian (D)	Upper / Late	385.3 ±2.6
			Middle	397.5 ±2.7
			Lower / Early	416.0 ±2.8
		Silurian (S)	Pridoli	418.7 ±2.7
			Ludlow	422.9 ±2.5
			Wenlock	428.2 ±2.3
			Llandovery	443.7 ±1.5
		Ordovician (O)	Upper / Late	460.9 ±1.6
			Middle	471.8 ±1.6
			Lower / Early	488.3 ±1.7
		Cambrian (C)	Upper / Late	501.0 ±2.0
			Middle	513.0 ±2.0
			Lower / Early	542.0 ±1.0

Right chart — Proterozoic, Archean, Hadean

EONOTHEM / EON	ERATHEM / ERA	SYSTEM / PERIOD **	Age estimates of boundaries in mega-annum (Ma) unless otherwise noted
Proterozoic (P)	Neoproterozoic (Z)	Ediacaran	635*
		Cryogenian	850
		Tonian	1000
	Mesoproterozoic (Y)	Stenian	1200
		Ectasian	1400
		Calymmian	1600
	Paleoproterozoic (X)	Statherian	1800
		Orosirian	2050
		Rhyacian	2300
		Siderian	2500
Archean (A)	Neoarchean		2800
	Mesoarchean		3200
	Paleoarchean		3600
	Eoarchean		~4000
Hadean (pA)			~4600*

* Changes to the time scale since March 2007.

** The Ediacaran is the only formal system in the Proterozoic with GSSP. All other units are periods.

Geologic time divisions as approved by the US Geological Survey. Other than the Holocene, ages are in millions of years. Courtesy of US Geological Survey.

For starters, this rock is very, very old. It formed between 505 and 440 million years ago in a time known as the Ordovician Period.[1] Further, it formed in a saltwater ocean, though Birmingham is many miles from the coast. Even more perplexing is that the limestone formed in the southern hemisphere when most of Alabama was at the same latitude as New Zealand is today. How is all this possible? How could a half-billion-year-old rock laden with marine reef fossils and riddled with caves wind up high and dry in Birmingham, 217 miles (350 km) from the coast, 1,030 feet (314 m) above sea level, and some 5,000 miles (8,050 km) north of where it originated? The answers to these questions will help us understand the presence of the Cave Salamander on Ruffner Mountain and how Alabama came to lead most states in biodiversity.

Limestone and all other sedimentary rocks are the great storytellers of the earth's history. These rocks arose from particles, or sediments, deposited in aquatic environments, especially oceans but sometimes lakes, swamps, or larger rivers. As layers accumulate, older sediments become compacted under the weight of the water and younger sediments above them. Under this pressure and often with the help of minerals precipitating from the water, these particles become lithified, that is, cemented together to form solid rock. This only occurs when sediments lie undisturbed for exceedingly long periods.

Exposed layers of the 350-million-year-old Fort Payne Chert, Fort Payne, DeKalb County. Photo by R. Scot Duncan.

Because sedimentary rocks and sedimentary soils smother Alabama, geologists know much about Alabama's past. Geologists reconstruct this past by examining the structure and composition of these rocks, especially their particles, which range from pebbles to sand to minute silts and clays. By studying how particles of various sizes form and move in today's environments, geologists make inferences about the environments in which sedimentary rocks formed. For example, because sand is a heavy particle only moved by strong currents, geologists deduce that sandstones originated in turbulent environments. By examining a sandstone's chemistry, layering patterns, and geographic distribution, geologists reconstruct the environment in which sandstone formed, usually a large river or a coastal beach.

Sedimentary rocks often contain fossilized fragments of plants and animals. Paleontologists, scientists who study ancient life, can determine the types of ecosystems in which these creatures lived based on what they determine of a species' feeding habits, locomotion, and/or growth requirements. Geologists and paleontologists compare multiple rock layers and their fossils over large geographic areas to reconstruct the distributions of ancient ecosystems, land masses, rivers, and oceans.

The Chickamauga Limestone tells a colorful story. Limestone is essentially a giant marine fossil. Its primary chemical is calcium carbonate, a mineral commonly used by marine organisms to construct hard body parts. At Hawk's View Overlook and the abandoned quarries nearby, there are fossils of corals, snails, bryozoans, crinoids, brachiopods, cephalopods, and other reef creatures. In today's oceans, such reefs are found only in the tropics where enough sunlight can penetrate the water to support photosynthetic corals. Thus, geologists believe fossil-rich limestone layers are the remains of tropical reefs from shallow environments beyond the reach of sediment-laden river waters shed by landmasses. While the layers containing reef organisms are several inches thick, other layers of the Chickamauga are much thicker and have few visible fossils. This form of limestone originated from small planktonic organisms living near the surface far offshore over deep waters. As these creatures died, their microscopic shells drifted to the seafloor and accrued in thick layers that became limestone. But

Chickamauga Limestone with 440-million-year-old marine fossils, Ruffner Mountain Nature Preserve, Birmingham. Courtesy of Bob Farley/ F8PHOTO.

how did these and other fossilized chunks of Ordovician tropical sea-floor become stranded in central Alabama's uplands? To answer this, earth's tumultuous geologic nature must be confronted.

DRIFT

Most of us spend little time contemplating the ground beneath our feet. If anything, most assume it to be solid and immutable. In reality, the earth's surface is constantly moving and changing. Earthquakes and volcanic eruptions are spectacular, and often grim, reminders of this.

The earth's surface, or crust, is made up of over two dozen sections of cooled rock known as plates. North America is one of the seven largest of the plates. The plates float above the earth's mantle, a layer of semisolid, heated rock more than 1,700 miles (2,736 km) thick. If the mantle were stable, the plates would be nearly stationary. Instead, the mantle's unrest causes the plates to drift across the planet's surface. Currents of superheated mantle rising toward the earth's crust cause this instability. Most of this activity is beneath our large oceans where the earth's crust is thinnest. Where mantle material reaches the surface, it cools to form new seafloor. This occurs along the oceanic ridges stretching for thousands of miles beneath our major oceans.

The plates are set in motion by this process because the continuous upwelling of hot mantle from below forces the mantle near the surface to spread away from the zone of upwelling. This seafloor spreading pushes away plates adjacent to the oceanic ridge at rates averaging about 2 inches (5.1 cm) per year. Because near-surface mantle currents travel for thousands of miles away from the oceanic ridges, they carry along the plates floating upon them. The processes associated with crustal plate movement, known as plate tectonics, propelled the North American plate and the Chickamauga Limestone into the northern hemisphere. Over the past 400 million years, North America drifted from latitude 40° south to 34° north. The idea that Alabama has wandered so greatly may seem incomprehensible, but continents creep at rates averaging about 1 mile every 32,000 years. Thus, Alabama had plenty of time for this journey.

Since earth isn't getting bigger, the formation of new crust must be offset. This adjustment occurs where drifting plates collide and one plate is driven beneath another, a process known as subduction. The submerged crust melts and becomes fresh mantle material.

Subduction is messy. Earthquakes are triggered when crust shifts suddenly in response to pressure. Volcanoes arise when subducted material melts and rises to the surface; igneous rock is formed when it cools. Mountains form when some crust fails to be subducted and heaps up along the collision zone. Metamorphic rock forms when igneous or sedimentary rock trapped in the collision is transformed chemically and physically under extreme pressure and heat. These tumultuous processes are intimately tied to southeastern biodiversity. Plate collision and subduction formed the Southern Appalachian Mountains (explained in chapter 5). The diverse landscapes created help Alabama and other southeastern states sustain dozens of ecosystems and thousands of species that wouldn't be here otherwise.

LIFT

A remaining question is how marine fossils became stranded so high above sea level. Land rises and falls for various reasons, the most dramatic involving continental collision. For a long time after its formation, the Chickamauga Limestone rested peacefully beneath the ocean, disturbed only by the mounting weight of younger sedimentary rocks forming above it. When plate collision created the Southern Appalachian Mountains about 100 million years later, the Chickamauga Limestone and many other ancient rock formations were thrust to the surface, where they now rest, high above their original seafloor elevations. The Chickamauga bears signs of this collision. Wherever exposed, including the quarries on Ruffner Mountain, one can see its layers are no longer horizontal, but are strongly tilted upward.

Sedimentary rock and the ocean where it formed can separate for other reasons. Coastlines constantly shift as the amount of water in the oceans changes. During cold geologic periods, sea level drops as water accumulates in glaciers and polar ice sheets. During warmer periods, sea level rises. These are absolute changes in sea level.

Relative changes in sea level occur when land rises or falls independent of oceanic volume, such as with the uplift of the Chickamauga Limestone and much of Alabama during formation of the Southern Appalachian Mountains. Crust also rises when bearing less weight due to massive sediment erosion or glacial retreat. Crust can subside, or sink, when taking on new weight from glacial growth, mountain building, sediment deposition, or inundation by rising seas. Changes in absolute and relative sea level often occur simultaneously, such as when absolute sea level drops and exposed seashore and adjacent lands rise with the removal of the ocean's weight. Relative and absolute changes in sea level are why the Chickamauga Limestone includes layers derived from both offshore planktonic deposits and shallow-water reefs.

The Cave Salamander exists at Ruffner Mountain thanks to the Chickamauga Limestone—its formation, its long-distance drift from the southern hemisphere, and its stranding high above Birmingham. There is one more geologic question left unanswered: how did the salamander's cave form?

KARST WORLD
Caves are a topographic feature, a landscape element created by geologic processes. Familiar topographic features include hills, mountains, valleys, rivers, and plains, and more unusual features like plateaus, oxbow lakes, and dunes. Most topographic variation in a landscape is created when weaker materials (soil or bedrock) erode faster than adjacent materials. The resulting variation in slope, exposure, and elevation influences biodiversity patterns by mediating availability of sunlight,

Neversink Cave Preserve (Southeastern Cave Conservancy), Jackson County. Courtesy of Alan Cressler.

heat, water, and soil. Topography also influences the frequency and intensity of many ecosystem disturbances. For example, canyon forests are sheltered from strong winds.

Karst is a unique landscape of sinkholes, caves, and springs developing wherever limestone is at or near the surface. These features form because limestone's calcium carbonate is an alkaline mineral that dissolves when exposed to rainwater, which is naturally slightly acidic. Below ground, rainwater is further acidified by absorbing acids left in the soil from plant decomposition. As groundwater infiltrates the limestone bedrock, it dissolves calcium carbonate, turning small fissures into channels, and channels into streams. These streams formed the vast underground cave systems of northern Alabama. When the roof of a near-surface cavern collapses, the ground above slumps to form a sinkhole or pond. Elsewhere, underground streams break through the surface to form springs. These karst features are unique ecosystems used by hundreds of plant and animal species.[2] Many, like the Cave Salamander, cannot survive elsewhere.

The rocks of the Cave Salamander's ecosystem bind it to a dynamic geologic history and link it to ancient marine species that lived over 400 million years ago. Much of Alabama's high biodiversity ranking is similarly tied to the state's geology.

SWEET SPOT

The geologic map of Alabama is absolutely stunning. Across the state, over 125 colors sweep in bands, tight stripes, and complex eddies. Though resembling abstract art, this is a technical document through and through. Each color indicates the surface distribution of a different rock or mineral soil formation named for out-of-the-way locales like Hillabee, Halawaka, and Hatchetigbee. Those familiar with Alabama can become lost in the colorful swirls while contemplating how this complex geology has influenced their favorite places. The influence has probably been profound, for each geologic formation offers a different combination of environmental variables.

The diversity of formations displayed on Alabama's geologic map indicates the state's great soil diversity. As discussed in chapter 2, soils are exceptionally important for shaping ecosystems and are also products of geologic processes. A soil's mineral component originates from the weathering of bedrock or old sediment deposits, and a soil's landscape position results from erosion, transport, and deposition.

Soil chemistry is particularly important. Most southeastern soils are slightly acidic, largely due to the organic acids accumulating from plant decomposition. Soils derived from limestone are alkaline. Though

A geologic map of Alabama with rock/sediment types originating in the same time period combined. Courtesy of Ed Brands/University of Minnesota Morris.

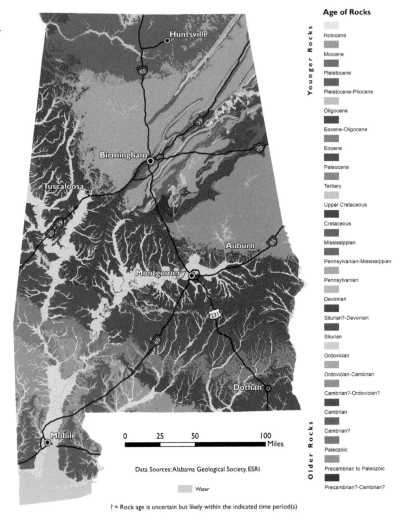

Age of Rocks

Younger Rocks
- Holocene
- Miocene
- Pleistocene
- Pleistocene-Pliocene
- Oligocene
- Eocene-Oligocene
- Eocene
- Paleocene
- Tertiary
- Upper Cretaceous
- Cretaceous
- Mississippian
- Pennsylvanian-Mississippian
- Pennsylvanian
- Devonian
- Silurian?-Devonian
- Silurian
- Ordovician
- Ordovician-Cambrian
- Cambrian?-Ordovician?
- Cambrian
- Cambrian?
- Paleozoic
- Precambrian to Paleozoic
- Precambrian?-Cambrian?

Older Rocks

0 25 50 100 Miles

Data Sources: Alabama Geological Society, ESRI

Water

? = Rock age is uncertain but likely within the indicated time period(s)

laden with calcium, karst soils are deficient in other nutrients and many plants cannot grow on limestone-derived soils. Despite this, alkaline soils boost regional plant diversity by supporting plants adapted to these soils. For example, the forest surrounding Hawk's View Overlook harbors Eastern Redbud, Shumard and Chinkapin oak, Eastern Red Cedar, and Slippery Elm; all are trees tightly associated with limestone-derived soils. Forests on nearby acidic soils are populated by their counterparts: Sparkleberry, Northern Red and Chestnut oak, Loblolly Pine, and Winged Elm.

In such ways, Alabama's diverse geology has created a multiplicity of topographic features, soil types, and resulting ecosystems. These relationships reveal one of the great lessons of biogeography—the more natural environmental variation there is within any geographic area, the more ecosystems and species are present. This is why Alabama hosts over five dozen types of terrestrial ecosystem. Another lesson for us is that geology trumps climate in its influence on shaping Alabama's biodiversity, for geologic processes pushed the North American continent to its warm-temperate zone position.

Eastern Redbud blooming in abandoned quarry, Ruffner Mountain Nature Preserve, Birmingham. Courtesy of Bob Farley/F8PHOTO.

How important is geology for promoting Alabama's biodiversity over that of other states? An easy approach to answering this question uses the physiographic system of land classification developed by geologists and geographers. In this scheme, adjacent geologic formations sharing a similar origin and topography are grouped into areas called physiographic districts. Districts are the smallest of four nested categories that include, from smallest to largest, districts, sections, provinces, and divisions. Physiographic provinces are handiest for comparing geologic diversity among states in relation to biodiversity, for they minimize subtle variations between rock and mineral soil formations and emphasize geologic history and topographic diversity.

Alabama has five physiographic provinces. Three of them, the Piedmont, Valley and Ridge, and Appalachian Plateaus, are landscapes created by the formation of the Appalachian Mountains. At the top of the state is the Interior Low Plateaus province.[3] These four provinces comprise about 40 percent of Alabama, leaving the remainder for the East Gulf Coastal Plain. Each province has its own unique geology and ecology. For example, the Valley and Ridge comprises elongated linear mountains and valleys, while the Coastal Plain is made up of sedimentary deposits that have never fully lithified.

When comparing states, a clear link emerges between the numbers of physiographic provinces and numbers of species. While most states (60 percent) have three or fewer physiographic provinces, seven of the top 10 states for biodiversity have four or more provinces. Even more convincing is a comparison between Alabama, Louisiana, Mississippi, and Georgia. All have coastline and are similar in size and climate. Yet, Louisiana and Mississippi each have only one physiographic province and are respectively ranked 17th and 18th for biodiversity. Sixth-ranked Georgia, like Alabama, owns a piece of the Appalachian Mountains and has five physiographic provinces.

There are limits to how well physiographic and biological diversity are correlated: Florida has one province yet is ranked seventh for species diversity, the top four states for biodiversity all have fewer provinces than Alabama, and New York has the highest number (eight) of provinces yet ranks 22nd for biodiversity. Climate causes these anomalies. Florida and the top four states for biodiversity are "tall" states enjoying a range of climates, including warm zones where primary productivity can be high. New York has a short growing season and less primary productivity than southern states. Thus, Alabama sits in a sweet spot. The combination of rich geologic diversity and a warm, wet climate provides the state with scores of productive ecosystems and a rich array of species.

The creation of rock, plate tectonics, and geologic forces—all have shaped Alabama's ecosystems and biogeography. But the complete

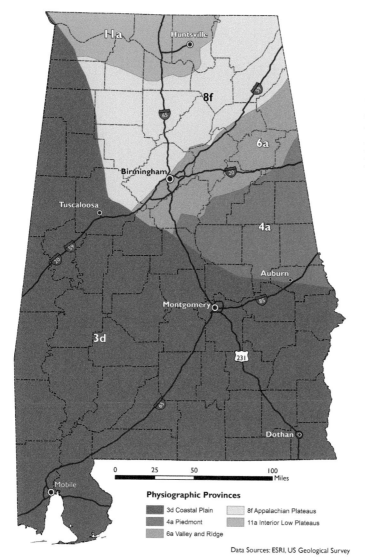

The physiographic provinces of Alabama. The Fall Line separates the Coastal Plain from the other regions. Numbers correspond to the system of classification adopted by geologists. Courtesy of Ed Brands/University of Minnesota Morris.

Physiographic Provinces

3d Coastal Plain
4a Piedmont
6a Valley and Ridge
8f Appalachian Plateaus
11a Interior Low Plateaus

Data Sources: ESRI, US Geological Survey

story of how modern Alabama's landscape and its biodiversity arose is even more wondrous. It is a turbulent and fascinating history involving mass extinctions, ice ages, and the handiwork of the first North Americans. Exploring this deep history is our next step.

5
Biodiversity through Deep Time

IN THE BEGINNING

Alabama's deep history holds the secrets to how the state's biodiversity arose. It's a colorful history involving the genesis of the Southern Appalachian Mountains, wild climate swings, asteroid impacts, and Spanish pigs. From 4.6 billion years of history, I've chosen to relate those events having the most influence on modern biodiversity in Alabama and the Southeast.

Every species in Alabama has ties to this long history. Consider one of the state's rarest fishes, the Pygmy Sculpin. Coldwater Spring, its only known home, emerges from ancient rocks in Alabama at the base of Coldwater Mountain just southwest of Anniston. The rocks comprising this mountain, including the Weisner Formation, Shady Dolomite, and Chilhowee Group, are among the state's oldest. They originated 542–488 million years ago (mya) in the Cambrian, the first period of the Paleozoic Era (542–251 mya). The connection between Alabama's deep history and today's biodiversity is clear: without these Cambrian rocks, there would be no Coldwater Mountain, no Coldwater Spring, and no Pygmy Sculpin.

Little that is tangible remains in Alabama from the 4 billion years prior to the Cambrian. Ancient rocks from elsewhere reveal that during these pre-Cambrian times, the young planet cooled, oceans and lands formed, continents grew rapidly, and single-celled organisms proliferated in the oceans. Much more is known about the Cambrian, since great stacks of sedimentary rock from that time still remain. They reveal the Cambrian was a stable time, and this calm fostered the evolution of a tremendous diversity of animal forms in the shallow oceans. The seafloor accumulation of their remains and sediments eroded from the young continents became the rocks now comprising parts of the Valley and Ridge physiographic province. Among these rocks is the sandstone

of the Weisner Formation, through which a fault allows groundwater
to emerge at Coldwater Spring. Most other springs in the province
arise from karst valleys of Cambrian-aged limestone and dolomite, and
a few support other rare creatures like the Watercress Darter.

APPALACHIA RISING—THE PREQUEL

The geologic and topographic diversity of the Southern Appalachian
Mountains supports dozens of ecosystems and hundreds of species
that would not be in Alabama otherwise. Thus, the origination of
these mountains is a pivotal story in the creation of Alabama's modern biodiversity. The Appalachians did not arise suddenly. Rather,
they formed over a period of about 170 million years during a series
of mountain-building events, or orogenies, that reconfigured eastern
North America.

The first of these orogenies occurred in the Ordovician, the period following the Cambrian. The Ordovician (488–444 mya) was an
unstable time for earth. Crustal plates collided, mountains rose, and
earthquakes rumbled across the planet. During this time Alabama was
mostly tropical seafloor on a tectonic plate in the southern hemisphere
known as Laurentia.[1] The Chickamauga Limestone featured in the

previous chapter formed beneath these shallow seas during this time. During the Middle to Late Ordovician, tectonic processes sent a chunk of oceanic crust crashing into Laurentia in areas adjacent to Alabama. The collision created the Taconic Mountains, which stretched from New England to northern Georgia and eastern Tennessee and were a foundation on which more mountains would be formed. An archipelago of volcanic islands created from this collision stretched into central Georgia and their volcanic activity periodically smothered Alabama's reefs with ash that appears in the Chickamauga Limestone as traces of chert.

By the mid-Silurian Period (444–416 mya), the first vertebrates were swimming in the oceans and vascular plants had colonized wetlands near coasts and rivers. Most of Alabama was still a shallow tropical sea into which rivers from nearby volcanic islands and the Taconic Mountains dumped sediments. Some of these sediments were iron rich and became the Red Mountain Formation, a band of purplish-red rock containing hematite (iron ore). Millions of years later, the extraction of hematite and its smelting to produce pig iron sparked the establishment and growth of Birmingham.

During the Ordovician and Silurian, Laurentia slowly drifted northwards on a collision course with the northern European continent known as Baltica. This slow-motion collision began in the Late Silurian, ended in the Devonian Period (416–359 mya), and produced a supercontinent known as Euramerica (also called Laurussia). The collision triggered the Acadian Orogeny and growth of the Northern Appalachian Mountains. The mountain building forced uplifting of Alabama's crust and the disappearance of its oceans during the first half of the Devonian. However, these disturbances were trivial compared to what was coming. A massive supercontinent named Gondwana was steaming northward toward Euramerica, and what is now the Southeast was dead center in its sights.

MISSISSIPPIAN EDEN
Before Gondwana collided with Euramerica, a magnificent reef ecosystem flourished over much of Alabama. The limestone created from this ecosystem now supports some of the world's most spectacular karst

terrain and cave biodiversity. The caverns and springs of northern Alabama are home to such rarities as the Alabama Cavefish, Spring Pygmy Sunfish, Gray Bats, and dozens of cave-dwelling invertebrates.

The reef originated during the first half of the Mississippian (359–318 mya), the first of two subperiods of the Carboniferous Period. During this time much of Alabama was beneath the shallow, tropical waters of the Rheic Ocean. The Rheic was an underwater Eden. Tropical heat and sunlight blended with nutrients shed from the nearby continents and sustained reefs teeming with corals and fishes. Nearby, dense beds of clamlike brachiopods or gardens of flowerlike crinoids sieved the water for plankton. The debris from these ecosystems yielded the thick limestones—most notably the Tuscumbia, Bangor, and Monteagle—that became the karst wonderland of modern northern Alabama.

Eden's days were numbered. Throughout the Late Mississippian, Gondwana was racing toward Euramerica. The first indications for Rheic reef creatures were earthquakes and ash raining down from volcanic eruptions beyond the southern horizon. As Gondwana advanced, sediment eroded from newly minted mountains built an extensive coastal plain with ecosystems much like those along today's coast. Sediments washed offshore by rivers smothered reefs not overrun by the advancing coastline. Though the magnificent Mississippian reefs were snuffed out, another complex ecological feature emerged—the Southern Appalachian Mountains.

APPALACHIA RISING
Gondwana was a supercontinental mash-up of Australian, South American, African, Indian, Arabian, and Antarctic continental plates. Its fusion with Euramerica created Pangaea, the largest supercontinent to ever exist. Though the Allegheny Orogeny began in the Mississippian Subperiod, mountain building peaked in the Pennsylvanian Subperiod (318–299 mya). The colliding supercontinents created three of Alabama's most important regions. The Valley and Ridge physiographic province was the site of intense sidelong pressure causing the crust to fold in long lines parallel to the core collision zone. Elsewhere in the province, the crust fractured and mountain-sized slabs of rock were

thrust up and onto adjacent areas. The Piedmont bears the deepest scars of orogeny. Tall mountains were created here from displaced seafloor, continental crust, and volcanic islands. Pressures were so intense that rocks trapped in the collision were melted and squeezed into metamorphic rock. The Appalachian Plateaus province arose from more gentle processes. As the Southern Appalachian Mountains grew, their weight caused adjacent crust on Euramerica to sag. New shallow seas received sediments washed from the growing mountains. Later, at the end of the Allegheny Orogeny in the Permian Period (299–251 mya), tectonic pressures in the greater region shifted and sedimentary rocks of this basin rose and became a patchwork of plateaus. By the end of the Allegheny Orogeny, the Southern Appalachians were a mountain range whose stature and complexity would rival those of any mountain system on the planet today.

COAL FORESTS

Portions of the Appalachian Plateaus province have thick coal deposits formed in the Pennsylvanian.[2] Between the shallow sea basin and the young Appalachians was a coastal plain built of sediments washed from the mountains. These low plains were dominated by swamp forests stretching for hundreds of miles and persisting for millions of years. Irradiated by the tropical sun and nourished by abundant rains and flooding rivers, these forests were bursting with ferns, club mosses, and horsetails (scouring rushes), many growing in tree-sized proportions. The biomass production in these swamps must have been unlike anything in existence today. We know this because the remains of these swamp forests survive as deep coal deposits.

Coal formation begins in oxygen-starved swamp waters. Without oxygen, decomposition by microbes is slow and unable to keep pace with the biomass falling from above. Additional biomass would accumulate during periodic inundation by seawater and subsequent mass die-offs of plants. Plant debris would then be covered by sand and other river deposits. Protected and compressed by these sediments, the trapped biomass became coal. The swamp forests persisted until the end of the Allegheny Orogeny, when Alabama became landlocked. In the center of the largest supercontinent of all time, rainfall declined and the swamps dried up.

A seed fern fossil from
Alabama's ancient coal
forests. Courtesy of
James A. Lacefield.

Coal's direct influence on Alabama's ecosystems is minimal because
sandstone and shale layers cover most deposits. However, extensive
coal mining in north-central Alabama has destroyed or degraded many
of the region's forests, creeks, and rivers.

APPALACHIA RISING—THE SEQUEL?
Recent geological research challenges the long-held view that the Allegheny Orogeny was the only period of mountain building for the
Southern Appalachians. Instead, the mountains may have been worn
down and uplifted since the Paleozoic, possibly several times. What's
more, some uplift may have occurred just a few million years ago.
This conclusion would have profound implications for understanding
southeastern biodiversity. Biogeographers and evolutionary biologists
would need to reconfigure hypotheses about patterns of evolution and
speciation, and many unexplained distribution patterns of modern
species might be resolved. However, the recent uplift hypothesis is new
and these considerations have only just begun.

The evidence for recent uplift is varied. Some erosion rates in the
Appalachians are so rapid that the mountains could not have survived
this long unless there was additional uplift or initially they were ab-
surdly tall. The routes of many southeastern rivers and creeks cut across

mountain ridges, suggesting their paths preceded recent uplift. Geologists have found Cretaceous Period (146–65 mya) marine sediments at high elevations in the Appalachians, suggesting a phase of uplift sometime after the Cretaceous. Massive amounts of marine sediments dating to the Miocene Epoch (23–5 mya) have accrued off the Atlantic and Gulf coasts, suggesting this was a time of substantial uplift and erosion.[3]

Providing a mechanism to trigger a post-Paleozoic uplift has been the greatest challenge to acceptance of the recent uplift hypothesis. Critics point out the Southeast has lacked active plate margins since the Paleozoic. One proposed mechanism involves upward adjustments of the southeastern crust as the northern Gulf seafloor sank under the growing weight of sediments deposited by the Mississippi River. Another proposal is that there was a warming phase in the mantle below the Southeast. At the surface, the heated crust expanded, became more buoyant, and readjusted upward. The recent uplift hypothesis needs more study, but some geologists are already convinced the Paleozoic-only model for the creation of the Southern Appalachians is insufficient and a new paradigm is needed.

The Gulf Is Born

The Gulf of Mexico has a huge impact on Alabama's biodiversity by flooding the Southeast with water vapor. The story of its influence begins with the end of Pangaea. The formation of Pangaea early in the Permian created a continent stretching from the planet's top to bottom. For about 50 million years, Pangaea was a solid supercontinent with Alabama landlocked at its center. Early in the Mesozoic Era (251–65 mya), Pangaea began to rift apart. Ever so slowly, the continents of today's world broke free and began migrating toward their current position.

During the Triassic (251–200 mya), the Mesozoic's first period, North and South America began separating and a young, narrow Gulf of Mexico was born. Throughout the Triassic and into the Jurassic Period (200–146 mya), the widening Gulf underwent phases of drying and refilling as its connection to the growing Atlantic Ocean was affected by changing sea levels and the distributions of land masses. Eventually, the Gulf broadened such that its shoreline reached Alabama and began influencing the state's biodiversity in lasting ways. This included the creation of the Fall Line.

FALL LINE

A prominent feature on many maps of Alabama is the Fall Line. Stretching from northwestern Alabama to the state's center and then eastward, the line is a demarcation of cultural, biogeographic, and geologic significance. Early European and American settlers depended on large rivers for transportation. At a certain point on each river, large boats met waterfalls and rock outcrops and further transportation into the interior necessitated an overland route. The geographic line linking these points was dubbed the Fall Line, and river intersections with it became hubs for settlement and commerce.

The distributions of many species, including plants, amphibians, reptiles, and hundreds of aquatic species, end or begin at the Fall Line. Above the line are uplands whose rocky, shallow rivers and streams harbor many of Alabama's endemic freshwater species. Below the line,

Fall Line on Chestnut Creek, Chilton County. Courtesy of Paul D. Johnson/ADCNR.

rivers are deeper, muddier, and generally populated by species with widespread distributions.

The Fall Line's origination dates to the Cretaceous, when the climate was so warm that glaciers and ice packs were rare and global sea levels were extremely high. Eastern and western North America were divided by a shallow sea that, together with the Gulf, gnawed on the Southern Appalachians and prepared lower Alabama to become an extensive coastal plain when sea levels dropped. The conventional view is that the highest extent of the Cretaceous oceans etched the Fall Line across the Southeast. However, as introduced above, recent findings suggest the high-water mark during the Cretaceous was further inland. Regardless, the Fall Line is an undisputed ancient shoreline and is an important ecological barrier for many of Alabama's species and ecosystems.

INCIDENT AT CHICXULUB

The Cretaceous was arguably the most marvelous ecological episode in earth's history. Lands supported herds of herbivorous reptiles and the dinosaurs that hunted them. Oceans teemed with fishes, huge sharks, and a diversity of large, predatory reptiles. An exceptionally warm climate made possible this magnitude of biological productivity.[4] Balmy conditions fostered the evolution of life forms whose ancestors dominate today's ecosystems. Flowering plants and their codependent relationships with pollinators and seed-dispersers emerged and diversified, pushing aside nonflowering plants as the most abundant plants on earth. Smaller forms of the Mesozoic's reptiles evolved into the first birds and mammals.

The Cretaceous ended abruptly 65.5 million years ago with the K-Pg extinction (*K* for the German spelling of Cretaceous, and *Pg* for the next period, the Paleogene). The agent of doom was an asteroid striking near Chicxulub in Mexico's Yucatan.[5] This was one of the most dramatic moments in earth's history. The asteroid was 6–8 miles (10–13 km) in diameter and hit at a speed of nearly 16 miles per second (26 km per sec). Had we been present nearby, we'd first be blinded by a quick flash of light filling the sky, then the sky would fill with superheated rock and darken. Debris would rain down for hours on surrounding regions. The heat blast sweeping outward from the impact site ignited firestorms that vaporized forests. Tsunamis of horrific size and force tumbled over nearby lands, including much of the Southeast. All this was the dramatic beginning to a saga of destruction.

For weeks, pulverized crust and volcanic ash rained down at distances many thousands of miles away. For the large reptiles, the most deadly debris was the smallest. Fine particles from the collision lingered in the upper atmosphere for months, possibly years, blocking sunlight and severely hindering photosynthesis by land plants and oceanic plankton. Death was everywhere. Plants died, animals starved, and ecosystems crumpled. By the end of the Cretaceous, 75 percent of terrestrial and marine species, including most large reptiles, had disappeared forever. The survivors, however, inherited a world of evolutionary opportunity.

TERTIARY CHILL

In the Tertiary Period (65–1.8 mya) that followed the Cretaceous, earth's ecosystems rebounded and steadily began to resemble those of today. Three Tertiary events have had lasting impact on Alabama's biodiversity. The first was a cooling climate. During the Tertiary's first 20–30 million years, the earth's climate returned to the hot, stable condition prevalent in the Cretaceous. Then the planet began to cool due, in part, to widespread volcanism in western North America and atmospheric ash reflecting light back into space. Meanwhile, North America was drifting from the tropics into the temperate zone. Colder climates generate less precipitation, and the cooling and drying forced Alabama's tropical forests to be replaced by temperate forests, then woodlands, then grasslands. Many modern species evolved from their ancestral stock during these transitions. In short, the cooling forced ecological and species diversity in the Southeast to become much closer to what they are today.

The second influential Tertiary event was a falling sea level caused by thermal contraction of the ocean and ice accretion on Greenland and Antarctica. There were several periods when the oceans temporarily retook lost territory, but ultimately the retreating shoreline in Alabama revealed marine sediments as diverse as sands, clays, gravels, chalks, and marls. Today, these sediments comprise the Coastal Plain and support thick forests, open woodlands, shady ravines, sunlit prairies, and a dizzying array of wetlands. Each of these ecosystems harbors unique flora and fauna, which is why the Coastal Plain is so important to Alabama's biodiversity.

The third important Tertiary event for the Southeast was the origination of the Loop Current in the Gulf of Mexico. Today, the Loop Current supplies heat and water vapor to the Southeast and is largely responsible for the region's lush climate. Early in the Tertiary there was no connection between North and South America and warm tropical waters of the Atlantic and Caribbean escaped into the Pacific. Near the Tertiary's end, volcanism created a land bridge between the continents as the Pacific plate subducted beneath the Caribbean plate. The barrier shunted the Caribbean current into the Gulf, where it became the Loop Current.[6] The current's influence on the Southeast was immediate. As the Tertiary climate cooled, the current's warmth and moisture helped protect the region from one of the most brutally cold episodes of earth's history, the Pleistocene ice ages.

ICE AGE LEGACIES

Earth's entrance into the Pleistocene Epoch (2.6 mya to 11,700 years ago) was rapid. During the Late Tertiary, polar ice accumulation increased and planetary cooling accelerated. Thick glaciers spread across the north, then advanced southward and decimated all ecosystems in their path. Throughout the epoch, glaciers advanced and retreated repeatedly, which explains why the epoch is nicknamed the "ice ages."

The greatest consequence of the ice ages on Alabama's modern biodiversity is that the glaciers failed to reach the Southeast. The ecosystems in Canada and most northern US states are relatively young, having developed in the few thousand years since the glaciers retreated. In contrast, southeastern ecosystems survived this calamity, and Alabama's biodiversity is vastly higher as a result.

Though southeastern ecosystems survived, the new climate modified them radically. Whenever the glaciers advanced, the Southeast experienced short cool summers, long cold winters, and relatively dry conditions. During these episodes, forest species retreated to sheltered positions in the river valleys and mountains, while the ranges of wood-land and prairie species better suited to drier conditions expanded. Changes along the coast were also extreme. Lower sea levels pushed the coastline of Alabama 60 miles (97 km) south of its current position. During interglacial periods the southeastern climate temporarily warmed, sea levels rose, and forests expanded. This ebb and flow continued throughout the Pleistocene's 2 million years.

As the ice ages reconfigured Alabama's ecosystems, immigrant species from the north were arriving. Fossil collections reveal that many plants and animals migrated to the Southeast seeking refuge from the deteriorating conditions to the north. Many fleeing species used the Appalachians as a migratory freeway. Trees that today are distributed far to the north invaded northern Alabama, including species of spruce and fir.[7] More than a few immigrants, including Eastern Hemlock and Black Birch, linger and augment Alabama's species list.

MEGAFAUNA

During the Tertiary, mammals diversified and assumed many of the ecological roles formerly occupied by the Mesozoic's large reptiles. The onset of climate change in the mid-Tertiary forced these species to adapt to increasingly cooler conditions and prepared them for the Pleistocene. By the time the ice ages began, North America swarmed with wildlife—BIG wildlife.

As revealed by fossil discoveries in the Southeast's springs, lakes, rivers, and caves, Alabama's Pleistocene fauna was spectacular. The region was populated by a mix of year-round residents and seasonal migrants driven south by the brutal northern winters. Alabama's prairies supported large herds of Southeastern Elk, American Bison, Long-horned Bison, and several species of horses and camels. There were family groups of the Columbian Mammoth on the prairies, a cousin to the elephant looming 13 feet (4.0 m) tall at the shoulder and wielding two huge spiral-curved tusks. Ten-foot-tall (3.0 m) American Mastodons, another cousin to the elephant, roamed the woodlands and forests in small groups. Mastodons used their trunks and tusks to snap large tree branches and strip them of their leaves. Another tree-eater was Jefferson's Ground Sloth, a species that could rear back on its hind limbs to heights of 10 feet (3.0 m) to browse tree canopies beyond the reach of the many smaller herbivores. The wetlands and woodlands were also home to several species of tapir and giant beaver. Grazing in Alabama's marshes and nearby grasslands were five species of glyptodonts, large mammals resembling a cross between an armadillo and a giant tortoise.

All these herbivores supported a formidable lineup of predators in the Southeast including Cougar, Jaguar, American Lions (which were 25 percent larger than African Lions), and Dire Wolves. The most formidable hunter was the Sabre-toothed Cat. They were nearly twice as large as the other big cats and sported 9-inch-long (23 cm) canines. Paleontologists believe they were ambush predators using their bulk to pin down prey and their canines for making killing bites to the neck.

Another enthralling aspect of the Pleistocene is how recently megafauna roamed the Southeast. We've mostly studied past events occurring millions, even billions, of years ago. It's difficult to fully fathom the

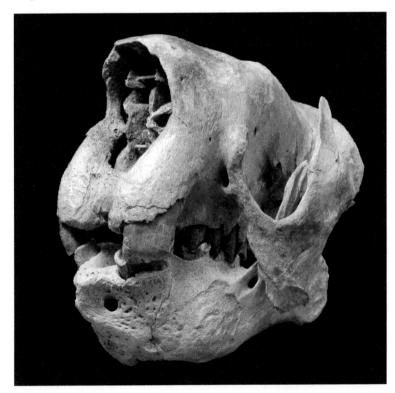

Jefferson's Ground Sloth, a large Pleistocene herbivore (skull height = 9.3 inches, 23.5 cm). This specimen from Colbert County is 125 thousand years old. Photo by R. Scot Duncan.

vast amounts of time that have since elapsed. In contrast, the Pleistocene ended merely 11,000 years ago, a time scale measurable in terms of human generations. There is a hidden implication here for Alabama's modern biodiversity. Most species alive today evolved before the ice ages and would have coexisted with Late Tertiary and Pleistocene megafauna. Thus, today's species evolved alongside a mammal community and in a climate quite different from those of today. Herbivory by the megafauna must have been intense and may explain some of the puzzling features of today's plants, like the dense clusters of long thorns on the trunks of Honeylocust trees and the spiny branches of locusts and hawthorns dangling high above today's tallest leaf-eaters.

QUATERNARY KILL

After enduring many episodes of glaciation and warming, dozens of megafaunal species began disappearing near the end of the Pleistocene. For decades, their disappearance was a mystery. Many speculated the stress of a warming climate at the end of the epoch caused populations to crash. After all, these species were well suited for the cold, not the mild temperatures that followed. Though some paleontologists still favor this hypothesis, skeptics argue that climate-driven extinctions should have claimed fewer species. They also point out that mild climates prevailed between the ice ages, yet little extinction is known from these intervals. Furthermore, if the disappearances were climate driven, extinction rates should have kept pace with the warming trend at the end of the Pleistocene that lasted several thousand years. Instead, most species disappeared during a brief 600-year period.

A new hypothesis for the Pleistocene extinctions in North America has emerged from the fields of paleontology and archaeology. The culprit was a bipedal mammalian predator armed with spears, refined hunting skills, and a measure of intelligence the continent had never before seen. The evidence is overwhelming that the North American megafaunal extinctions occurred during and just after the spread of Paleoindians across the continent.

Excavations of Pleistocene soils at dozens of sites show an abundance of megafaunal bones in the deeper layers, but these bones become uncommon once spear points appear in younger layers. This transition appears at different times across the continent and suggests the arrival of hunters and ensuing extinctions spread concurrently. Other evidence linking Paleoindians to the extinctions include spear points embedded in the skulls of fossils, scraping-tool marks on bones, and artistic carvings on bones and tusks.

The arrival of humans in the Late Pleistocene was one of the most influential events in the history of North American biodiversity. The archaeological record reveals that 13,000–14,000 years ago a nomadic hunting culture known as the Clovis spread across North America. The leading hypothesis suggests they entered Alaska from Siberia by crossing a land bridge exposed by low sea levels during one of the glaciations. As they spread across the continent, they left behind a trail of distinctive stone tools, especially large, well-crafted spear points necessary for hunting big animals. The absence of long-term encampments suggests to scientists the Clovis were nomadic and followed wildlife herds during their seasonal migrations.

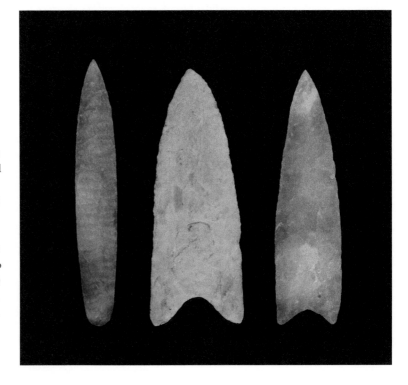

Paleoindian spear points similar to those collected in Alabama. Types illustrated: Agate Basin (left) and Clovis (center and right). Points were not photographed in relative scale; lengths from left to right are 6.3 inches (16.0 cm), 3.4 inches (8.8 cm), and 4.5 inches (11.5 cm). Courtesy of Jun Ebersole/McWane Science Center.

The Clovis were present in the Southeast during the Late Pleistocene and into the early Holocene Epoch (11,700 years ago–present day). Most Clovis artifacts in Alabama have been recovered in the Tennessee River Valley lowlands. No species was too big for Alabama's Clovis to hunt. Mammoth bones with scraping-tool marks alongside Clovis artifacts have been found at several sites. Oddly, researchers have recovered relatively few artifacts near the Gulf Coast, suggesting the Clovis used this region infrequently.

Scientists assumed for many decades the Clovis were the first to colonize North America. However, in the past quarter century, studies from fields as diverse as archaeology, anthropology, linguistics, geology, and molecular biology have produced a convincing stack of evidence that humans arrived in the New World long before the Clovis, perhaps as much as 43,000 years ago. While there is lively debate as to when, where, and how often prehistoric people colonized North America, there is now broad agreement the Clovis were not the first. Perhaps in the coming decades scientists can offer insight into the impacts these earliest people had on North American ecosystems and what role they played, if any, in the megafaunal extinctions.

Ultimately, a full understanding of how megafaunal extinctions shaped today's southeastern ecosystems will be elusive. Nevertheless, consider that dozens of herbivores disappeared, including many large species with big appetites. In Africa, the only place where a similar community has survived, there are wetlands, woodlands, and forests with open structures maintained by the munching and trampling of large animals. Perhaps the North American megafauna would be having a similar influence on the Southeast had they survived.

UNDER NEW MANAGEMENT

The Holocene brought a warming climate and the establishment of today's ecosystems. The thawing began in the Late Pleistocene, about 20,000 years ago. Glaciers gradually retreated, the oceans warmed, and sea levels rose to their present positions. In the Southeast, the distributions of plants and animals shifted as species sought optimal growing conditions. Many species migrated northward. Alabama's modern ecosystems reached their present composition and distribution about

6,000 years ago once warm air masses from the Gulf of Mexico were able to fully penetrate the continent. Without the human presence, the Southeast's ecology during the next few thousand years would have been stable.

What is known about pre-Columbian cultures in North America is from artifacts found during the excavation of sediments at former encampments or settlements. Artifacts are uncovered by scraping away thin layers of sediment, and their relative age is inferred from their vertical position in the soil. Actual ages are estimated by radiocarbon-dating bone, shell, or charcoal left in each layer. Usually, only durable materials resisting decay have persisted, principally pottery, bone, shell, and rock. Recovered artifacts include tools, refuse, ceremonial objects, and human remains. From them, archaeologists make inferences on everything from diet to spiritual beliefs. It has become clear that southeastern Native American culture changed with time and at each stage had a profound influence on Alabama's ecosystems.

The southeastern Paleoindian culture persisted for about a thousand years after the end of the last ice age. What followed in the Archaic Cultural Period (10,500–3,000 years ago) was the transition from a nomadic to a more sedentary way of life. Archaic people camped in valleys during the summer, where they hunted, fished, and gathered plants.[8] In fall, they relocated to upland camps to harvest nuts of oak, hickory, and chestnut. Recovered tools help archaeologists interpret how they used the landscape. These relics include fishhooks, sinkers for woven fishnets, and adzes for shaping wood. The Archaic people used atlatls to hurl spears farther and faster than is possible with an unaided arm. Reflecting the absence of the megafauna, their spear points were smaller than those of the Paleoindians. The preponderance of White-tailed Deer bones reveals this was a favorite prey.

Technological and social innovations mark the transition into the Woodland Cultural Period (3,000–1,200 years ago; 1000 B.C.–A.D. 800). Many seasonal villages became permanent as the use of natural resources became more efficient. On floodplains, slash-and-burn agriculture was used to create small garden plots for crops that likely included tobacco, maize, beans, sunflowers, squash, and small-seeded grains. Despite the innovation of agriculture, there was still a great reliance on hunting and gathering. The bow and arrow came into use and drastically increased hunting efficiency.

This sophisticated use of the landscape and the population growth it fueled put increasing pressure on ecosystems near villages. The Woodland people's use of fire for landscape management likely had a more widespread impact on biodiversity. Fire was used to drive game during hunts and stimulate the emergence of tender plant shoots to attract herds of deer, bison, and elk. Fires also opened the forest for travel and the detection of approaching enemies. The landscape-wide use of fire by the Woodland and subsequent cultures would have augmented the natural frequency of wildfires ignited by warm-season thunderstorms.[9] These natural fires facilitated the emergence of many unique ecosystems, and southeastern ecologists have energetic discussions over whether Native American use of fire facilitated the spread of these ecosystems or simply helped sustain them. We'll likely never know for sure.

Near the end of the Woodland Period, there was a mysterious decline in larger settlements and an increase in smaller villages. Archaeologists have recovered fewer ceremonial artifacts from this era, suggesting that cultural traditions also deteriorated. Scientists have proposed several reasons for these transitions, most involving a per capita decline in resource availability. Larger Woodland villages may have surpassed their carrying capacity, that is, the number of people the surrounding environment could support. Some scientists suggest this resulted from overhunting following the introduction of the bow and arrow. Others propose there was a decline in the primary productivity of the forests and garden plots due to a minor climate cooling trend occurring in the northern hemisphere during this time. Whatever the cause, the shift to smaller villages promoted the spread of Woodland settlements into many previously unsettled areas of the Southeast.

MOUNDS OF MAIZE

The Woodland and Archaic peoples' impact on the landscape was light compared with the Native American culture that followed. During the Mississippian Cultural Period (1,200–500 years ago; A.D. 800–1539) an entirely new pattern of living emerged, one characterized by large permanent settlements, extraordinary population growth, and rich cultural development. All of it was fueled by the cultivation of maize. Unlike the crops in limited use during the Woodland Period, maize production yielded much more food for the effort invested. Squashes, beans, and sunflower were also important mainstays, but maize was by far the principal crop.

Large villages and small cities arose in floodplains and terraces adjacent to rivers where the soil was fertile, moist, and easy to work. The initial surplus of food stimulated population growth and sociopolitical organization in the form of increasingly large city-states with distinct social classes. Mississippian people constructed huge flat-topped mounds of sediment that took years to build and were only possible with the surplus of labor and food. Archaeologists believe the mounds provided a location for ceremonies and housing for priests, shamans, and other leaders.

As the Mississippian cities grew, so did their impact on nearby ecosystems. Villagers felled large tracts of forest to provide agricultural land, fuel wood, and construction timber. Some accounts by early European explorers and estimates by archaeologists suggest that 6–16 square miles (16–41 km²) of deforestation surrounded larger cities. Meat was frequent in the diet, so the impact on fish, shellfish, and game species near population centers was intense. The burning of forests waned during the Mississippian, but a new form of forest management took shape. Vast tracts of forest were converted into managed nut orchards of hickories, oaks, butternuts, hazelnuts, pecans, walnuts, chestnuts, and other trees producing edible nuts.

Mississippian settlements were located all along Alabama's large Coastal Plain rivers. One site is preserved at Moundville Archaeological Park near Tuscaloosa on the Black Warrior River. Moundville is the second-largest Mississippian city ever discovered. The impressive ruins consist of 20 mounds, the largest towering 60 feet (18 m) above an expansive central plaza. The city would have supported 3,000 people, but today the mounds rest silently in a large open field, kept company by curious visitors who scale them and stand where chiefs and priests once presided.

The maize culture nearly extended to the coast. The second-largest known Mississippian city in Alabama is Bottle Creek, in the heart of the Mobile River Delta. It consists of 18 mounds, the largest rising 45 feet (13.7 m) above the floodplain. Bottle Creek may have been the seat of power for the Pensacola Culture, a series of chiefdoms centered on northern Gulf Coast embayments. Bottle Creek was unusual among Mississippian cities due to the influence of water. The city was on an island surrounded by several large rivers and numerous creeks. Travel by canoe was part of daily life and winter and spring flooding would have presented numerous dangers and logistical challenges. However, the benefits of a life at Bottle Creek are obvious to those who know the delta. Maize grew quickly in its fertile soils and long growing season, game was abundant, and the mingling of marine and river water supported a rich harvest of fish and shellfish.

Native American ruins at Moundville Archaeological Park, Tuscaloosa County. Note tall mound at lower right. Courtesy of Moundville Archaeological Park, University of Alabama.

SPANISH PIGS

Nurtured by rivers and fueled by maize, the Mississippian people of eastern North America flourished until about 500 years ago when their way of life unraveled. Mississippian people abandoned their cities and resettled along the coast and coastal plain rivers. What happened? What could topple a culture that had mastered the cultivation of maize, a crop that is a staple for cultures around the world? Archaeologists have long struggled to find a durable hypothesis. Finally, a well-reasoned explanation has emerged, a story involving Spanish conquistadors, a lost city of gold, and pigs.

Hernando DeSoto and his army were the first Europeans to explore the Southeast's interior. DeSoto was one of many Spanish explorers of the New World with an insatiable hunger to find gold. Unfortunately, DeSoto's brutality was as excessive as his greed. Landing in 1539 near present-day St. Petersburg, Florida, DeSoto led an entourage of 600 soldiers and a mass of servants, slaves, priests, horses, hogs, mules, and bloodhounds on a rampage through the Southeast to find El Dorado, the fabled city of gold. As detailed in chronicles kept by his officers, DeSoto and his men murdered, raped, slaughtered, sacked, and kidnapped their way through the Native American populations they met. El Dorado didn't exist, which is why DeSoto found that no amount of bullying and butchery could force the befuddled natives to reveal its whereabouts. Unfortunately, DeSoto survived three years before he succumbed to illness on the banks of the Mississippi River. If there is a bright side to the DeSoto expedition, it would be that Europeans were so intimidated by his epic failure that none attempted to explore the interior of the Southeast for nearly 150 years afterward.

DeSoto's debacle marked the beginning of the Protohistoric Period (A.D. 1539–1800) for southeastern Native Americans. During this time, culturally advanced civilizations with millions of inhabitants across the Americas disintegrated following their first contact with Europeans. Unexpectedly, it wasn't European swords, cannons, or military tactics that brought these cultures down—it was the diseases they unwittingly unleashed. The Europeans brought an array of illnesses the likes of which Native American immune systems had never before encountered. These diseases had originated from European livestock centuries earlier. Over the ages, the diseases adapted to European immune systems and vice versa (successful diseases keep their hosts alive long enough for them to spread the disease). Native Americans had few domesticated animals and their immune systems were completely unprepared. Viruses causing limited mortality in Europe decimated the Native Americans.

DeSoto's expedition brought the first wave of diseases to the Southeast. While there would have been some human-to-human transmission, much of the spread may have been done by DeSoto's pigs. DeSoto brought swine with him for provisions, but some of his herd escaped and established feral populations that still persist in the Southeast (and cause havoc on ecosystems). Swine are known carriers of many human and livestock diseases, and scientists believe some spread into Native American populations when they began hunting the feral hogs.

Thus, European diseases likely brought down the Mississippian culture. Even though some Native Americans survived, large-scale agriculture became nearly impossible without a large and healthy labor force. And for a culture with no written language, the deaths of artisans, elders, and leaders meant the loss of talent, wisdom, and centuries of knowledge. Archaeological evidence suggests survivors abandoned the Mississippian villages and reverted to a lifestyle similar to that of the Woodland Period. By the time Europeans ventured back into the Southeast, the Mississippian culture had vanished. Familiar with DeSoto's expedition, these later explorers were perplexed when they

Diseases from feral pigs helped decimate Native American populations; today the pigs severely disrupt ecosystems. Courtesy of Sara Bright.

found a landscape devoid of the large and numerous villages that had been chronicled.

The demise of the Mississippian culture dramatically affected southeastern ecosystems. Forests and rivers teeming with human enterprise, culture, and trade were emptied. Abandoned farms and villages filled with invading brush. Nut orchards became choked with shrubs and trees. Populations of shellfish, fish, and game species rebounded, especially deer, bison, elk, and turkey.

By the beginning of the 19th century, the human influence on the Southeast became distinctly European. As the new colonists settled in full force, the wilderness they conquered wasn't pristine and unsullied, as we have so often been led to believe. Instead, it was a landscape that had been actively managed for thousands of years.

6

Speciation Southern-Style

Every Species Has a Story

The last puzzle piece for understanding why Alabama has so many species is a study of biological evolution. Every species in Alabama has an epic evolutionary story involving survival through cataclysms such as mountain building, ice ages, and asteroid impacts. These tales reveal how populations adapt to an ever-changing environment and how adaptations can lead to the evolution of new species. Evolutionary biologists decipher these stories from the evidence around us, but appreciating these grand stories necessitates understanding the small processes creating genetic diversity.

Genes and Genetic Variation

All organisms on earth—from bacteria to us human beings—have one thing in common. We all carry a full set of instructions on how to grow, maintain ourselves, and reproduce. Biologists have named this owner's manual the genome, a name derived from its basic unit of information, the gene. Each gene is a subsection on an elongated molecule known as deoxyribonucleic acid, or DNA. Each gene either supplies the information needed for building a particular type of protein or regulates protein production by other genes. Proteins are used as building materials and as workhorses for conducting the elaborate chemical reactions sustaining life. Fish muscle fibers, rattlesnake venom, and fox pheromones—all are built of proteins.

Although any two species have some genes in common, each species has its own unique genome. This is why individuals within a species are similar in appearance and behavior. There is, however, variation *within* the genome of most species. For any one gene, there may be two or more versions that when used (expressed) yield slightly different results. This variation creates differences among individuals within a species.

We excel at spotting variation among humans, but save for a few sharp-eyed biologists, most of us are lousy at detecting such differences within other species. Consider the Alabama Beach Mouse from chapter 2. Several male beach mice coaxed to sit side-by-side would look very similar to you or me, but to a bachelorette beach mouse, these males might vary tremendously. One has pleasingly light fur to help with camouflage in the dunes; another has charmingly large forepaws to aid with digging new burrows; while a third is delightfully plump, indicating he's an effective forager. Genetic variation within the species is at the root of such differences, and this variation is the fodder for evolution.

Natural selection is the core evolutionary process through which some genes are more successful than others through time. Some individuals have genes making them better suited to the environment than others. If beach mice with lighter fur survive longer than good burrowers and foragers, then the better-camouflaged mice will be more likely to reproduce and send their genes to the next generation. If lighter fur continues to be advantageous, then the gene or genes coding for lighter fur will become more common with each generation, eventually spreading to all members of a population. Thus, the bachelorette mouse should choose carefully!

Genetic variation arises through changes to DNA known as mutations. Mutations can happen when DNA is exposed to radiation or certain chemicals, or when mistakes are made while new DNA is prepared for new cells. The odds are strongly against a mutation ever having a beneficial impact on a population for a string of reasons. Most importantly, mutations arise randomly. Random changes to the genome are likely to wreak havoc, which is why DNA mutations are the cause of all cancers. Mutations are so dangerous that cells are tremendously efficient at detecting and repairing or eliminating mutations, regardless of whether or not the change would be advantageous. Thus, efficient policing and DNA repair by the cell prevents most mutations from affecting organisms or their populations.

Another implication of the random nature of mutation is that organisms cannot create them out of necessity. Even when the environment changes and organisms become ill suited for survival, the likelihood of a mutation arising to benefit its carriers remains the same as ever. If organisms could conjure up beneficial mutations at will, life on earth would be completely different.

More barriers to the emergence of beneficial new genes involve heritability. For sexually reproducing species, the mutation must arise in a sperm or egg so it can be part of an offspring's genome and potentially expressed. Mutations elsewhere in the body, no matter their effect, are evolutionarily unimportant. An added hurdle to a beneficial mutation entering a population's genome is that a sex cell with a mutation is, like all sex cells, very unlikely to be one of those successful in reproduction.

Despite the long odds, new mutations sometimes arise in the DNA of a parental sex cell, evade DNA policing, and sneak into an offspring. It then must survive the scrutiny of natural selection before becoming part of the genome. Natural selection quickly weeds out most harmful mutations, since individuals expressing them have lower survival and fewer offspring. Exceptions include sneaky mutations expressing themselves after an organism has reproduced and passed the mutation to its offspring. Thus, only rarely does a mutation emerge as a new gene, provide an advantage to its inheritors, and spread through a population. It's no wonder that only an estimated 0.1 percent of new mutations are beneficial. Ironically, life's slight vulnerability to mutation is the engine producing new traits for populations and generating the great diversity of life on our planet.

Natural selection and mutations are not the only forces shaping genetic variation within populations. Genes are lost from populations through several random processes collectively known as genetic drift. All involve the loss of a gene, or genes, from a population simply because it was not passed on to the next generation. Perhaps its lone carrier died or failed to reproduce for reasons independent of natural selection. During what is known as a founder effect, a new population is established but the colonists don't carry all the genes of the source population. If the colony has no further contact with the source population, the new population is genetically distinct. During a genetic bottleneck, a population is thinned by environmental pressures that kill individuals regardless of their genetic composition, and genes are randomly lost because they were not carried by survivors. Small populations are highly vulnerable to this latter form of genetic drift, explaining why many endangered species have lost considerable genetic variation.

Natural selection and genetic drift impose a reduction in genetic diversity, but only the former benefits a population's survival. Why? Consider how genetic drift is playing with the fate of the Alabama Beach Mouse. When a cyclone wipes out a portion of its population, some genes are lost regardless of how helpful they might be. Bright, dim, fast, slow, brave, wary—an 8-foot storm surge doesn't discriminate. In contrast, natural selection eliminates disadvantageous genes in a population, thereby fortifying future populations.

SPECIATION

Over long periods of time, the factors affecting genetic variation drive speciation. Speciation involves two steps: reproductive isolation and genetic divergence. First, a population must become completely unable to interbreed and exchange genes with other populations of its species. Allopatric speciation, the most common mode of species formation, involves a geographic barrier preventing gene flow. Tall mountains can divide lowland species, while mountain peaks and valleys harbor isolated ecosystems where populations can become trapped. Shifts in large rivers can split populations of small terrestrial animals. Rising seas can divide species by turning high-elevation coastal lands into islands and by isolating river species in the upper parts of watersheds. Climate warming can force temperature-sensitive species into the isolated highlands of mountain ranges. Climate drying can drive forests and their denizens off the plains and into river valleys.

The second step in speciation is for an isolated population to accrue enough new genes to become genetically distinctive. New genes may be associated with adaptations to the local environment or may result from changes in mate choice.[1] Either way, complete reproductive isolation from other populations is necessary for speciation. Otherwise, any new genes would be carried by emigrants to other populations and the genetic distinctiveness of the semi-isolated population would be lost.

Given how slowly new beneficial mutations arise, thousands of generations may pass before the isolated population has sufficiently diverged to become its own species. Thus, the geographic barrier isolating the population must persist for a long time.

When does a genetically diverging population become a new species? This simple question has bedeviled biologists for centuries. Biologists initially classified species based mainly on appearance. These subjective assessments instigated debates over whether populations exhibiting subtle differences should be classified as two species or as variants of a single species. Sometimes taxonomists erroneously classified males and females of the same species as different species. As the disciplines of taxonomy, genetics, and evolutionary biology matured, many adopted the biological species concept (BSC) as a clearer guideline for distinguishing species.

The BSC states that a species is a group of actually or potentially interbreeding natural populations that are reproductively isolated from other such groups. This definition is complex, but the key issue is reproduction. If members from two populations can mate and produce fertile offspring, biologists applying the BSC consider them to be the same species. Barriers preventing interbreeding between two similar species include different courtship rituals, attraction pheromones, or even mismatched genitalia. Should individuals from different species mate, the BSC predicts no offspring would be produced or hybrid offspring would be sterile and ill equipped to survive.

The BSC has its shortcomings. It doesn't help distinguish asexually reproducing species, which include many plants, microbes, and invertebrates. Furthermore, many distinct species, especially plants, occasionally interbreed and produce fertile offspring—several southeastern oaks are notorious for this. What about populations geographically isolated from one another and not interbreeding? Are they "potentially interbreeding"? Captive breeding experiments won't help. Two captive animals might or might not mate regardless of whether they would mate in the wild.

The species problem, as it's known, ultimately arises because populations exhibit varying degrees of genetic divergence from one another, and the BSC does not address this variation. Even today, when biologists can measure genomic differences between two populations, there is no convention on when there is enough genetic divergence for two populations to be considered separate species.

Biologists have proposed many alternative species concepts. Among them, the phylogenetic species concept (PSC) most gracefully resolves the species problem. According to the PSC, a species includes all individuals sharing a genetically unique history. It does not matter whether they can interbreed with other species or whether they will one day merge with closely related populations through reestablished gene flow. Thus, the PSC embraces the full range of biodiversity produced by evolution. However, there are substantial practical impediments to its adoption. Because the PSC would classify two populations as separate species based on a single unique gene, its use would vastly increase the number of species to track with our nomenclature, texts, and laws. The debate among proponents of the BSC, PSC, and other species concepts will likely continue for as long as there are biologists. For now, the BSC remains the most widely accepted approach.

PEERING INTO THE PAST

If the events leading to the emergence of Alabama's species occurred in the distant past, how can evolutionary biologists reconstruct the history of speciation? Biologists have developed increasingly sophisticated methods to peer into deep time. Analyses of the fossil record were the beginning. Paleontologists have scrutinized fossils for well over a century and these records in the rock are still our primary means of studying extinct species. From fossils and their host rock, paleontologists can determine when, where, and how a species lived.

Unfortunately, the fossil record has major shortcomings for studying the Southeast's evolutionary past. Small species and those lacking hard body parts are disproportionately underrepresented. Terrestrial species are often poorly represented because fossils are found in sedimentary rock, most of which is marine in origin. The scarcity of rocks formed in the past 65 million years has hampered reconstruction of the recent past. That said, paleobotanists have recovered pollen from the Tertiary and Quaternary periods from lake sediments and revealed how southeastern plant communities responded to climate changes.

Paleozoologists have recovered thousands of bones and teeth of Pleistocene species from depositional traps including springs, sinkholes, and surface caves.

Biologists also study the past by extracting clues from living species. An early approach was to draw inferences about evolutionary steps based on the geographic distributions of related species. In theory, regions that were a source of multiple allopatric speciation events, areas known as speciation centers, can be identified by the presence of numerous closely related species. Biologists have used this approach to distinguish speciation centers for many of the Southeast's most species-rich taxa. This approach is weak when used on its own because closely related species may be found near one another regardless of where they originally evolved. More careful reconstructions of the evolutionary past also consider the Southeast's geologic and climate history. By assimilating geologic information on the past distribution of rivers, tectonic uplift, changing sea levels, and shifting climate, biologists can study potential speciation centers in the context of events that may have promoted speciation.

While these techniques give biologists a set of tools for reconstructing broad evolutionary patterns through geologic time and across geographic space, they cannot illuminate the fine-scale evolutionary relationships between species. With the advent of molecular biology and DNA sequencing in the late 20th century, biologists could test hypotheses about the evolution of species by directly studying their genes. Today, researchers measure genetic differences among species and draw inferences about evolutionary relationships. For instance, two species with few genetic differences probably diverged more recently than two species with many genetic differences.

Still, this approach has its weaknesses. Chiefly, measures of genetic differentiation say nothing directly about the exact timing of speciation, since genetic mutations may accrue at different rates in different species. This hinders placing patterns of evolutionary relationships in the context of deep time.

One solution is to examine the tiny molecular clocks found in every organism. Within cells are small structures known as mitochondria. Mitochondria have their own set of DNA, denoted as mtDNA, that is distinct from the genomic DNA. Because mothers pass along unaltered mtDNA to offspring via the egg, the mtDNA of you, your mother, your mother's mother, and so forth, is the same. However, mutations accumulate at a fixed rate in mtDNA over long periods of time. Once biologists determine this rate for a taxa, they can estimate the timing of divergence between any two species based on the number of unique mtDNA mutations in each species.

The combined use of all the above techniques now enables biologists to construct evolutionary models with a precision like never before. Despite widespread appreciation of such efforts, relatively few scientists work on reconstructing the Southeast's evolutionary history. Practical problems, like preventing extinction or managing dwindling natural resources, attract most researchers and research funding. Consequently, evolutionary biologists are still in the early stages of examining the Southeast's detailed evolutionary history. Still, decades of work by paleontologists, geologists, and evolutionary biologists have produced general explanations of the events crafting the evolution of the Southeast's unique and highly diverse flora and fauna. In lieu of reviewing all of the state's taxa, I'll review four evolutionary case studies. The processes they illustrate were likely mirrored in less-studied taxa with similar ecologies and geographic distributions.

A PLETHORA OF PLETHODONTIDS
When you think of quintessential Alabama fauna, salamanders might not—but should—come to mind. The United States has more species of salamander than any other country in the world, and that diversity is centered in the Southern Appalachian Mountains. Northeastern Alabama is home to many of Alabama's species, which number more than 40 species and subspecies. Alabama's salamanders include such oddities as the Tennessee Cave Salamander, which lives entirely below ground, and the eel-like Two-toed Amphiuma, which lives in Coastal Plain swamps and grows nearly 4 feet (1.2 m) long. Most salamanders, however, live in streams or moist environments including leaf litter, rotting logs, and rock outcrops. While largely unnoticed, salamanders are the most common vertebrate in many forests of the Southern Appalachians.

One family of salamanders, the plethodontids, has undergone speciation like no other. This is the most diverse salamander taxa in North America and includes over half the salamander species in Alabama. Many salamander taxa depend on wetlands to survive, but sometime in the Late Jurassic Period, in a region not yet known, the plethodontids adapted to drier, terrestrial life. Though they still depend on moist microhabitats, this evolutionary step enabled their spread across North America, Europe, and Asia over the next 100 million years.

Recent studies involving genetic diversity analysis, molecular clock methods, and the latest models of past climates and geologic history have refined our understanding of how the plethodontids diversified in the Southern Appalachians. Speciation of the taxa in the Southern Appalachians has spiked twice. The first instance was during a period of exceptionally warm and wet conditions in the Late Paleocene and Eocene epochs when plethodontids colonized new areas, especially mountain valleys. With an inability to survive in dry habitats, many populations became isolated at high elevations in these valleys later in

Plethodontid salamanders from Alabama. *Clockwise from top left:* Northern Red, Northern Slimy, Longtail, and Three-lined. Courtesy of Alan Cressler, Kevin Messenger/Alabama A&M University, Jimmy and Sierra Stiles, and Alan Cressler.

the Tertiary as the climate became cooler and drier and moist forests contracted. This isolation and ensuing genetic divergence led to the evolution of several important plethodontid genera.

A second phase of diversification dates to the Miocene. It is speculated that renewed uplift of the Southern Appalachians during this time again fragmented plethodontid populations and promoted more speciation. These periods of diversification are examples of adaptive radiation, when many new species evolve over a short period of time. They are also classic examples of how geographic isolating mechanisms and a changing climate often combine to encourage species evolution.

The Asian Connection

The deciduous forest plants of the Southern Appalachians are another famously diverse group. Though this assemblage includes thousands of species of widely varying ancestry, all share a peculiar evolutionary history. In the late 18th century, botanists noticed remarkable similarities between the deciduous forest plants of the eastern United States and eastern Asia. Despite the vast distances between them, the two regions share over 100 plant genera, including ginseng, mayapple, jack-in-the-pulpit, and trees like beech, catalpa, chestnut, gum, oak, persimmon, tulip-poplar, and witch hazel. The two floras are more similar to each other than either is to the neighboring floras of western North America and Europe. For over 250 years, the origin of this pattern has spurred discussion and speculation among biologists, including Carl Linnaeus and Charles Darwin.

Scientists now believe this pattern arose in the Early Tertiary with the establishment of a deciduous forest across northern regions of North America, Europe, and Asia. These plants migrated between the continents on land bridges arising periodically due to tectonic uplift and sea level change. The opportunities for intercontinental exchange ended with the cooling climate during the Late Tertiary and the disappearance of deciduous forests in the north of all the continents. With disconnection from their populations of origin and adaptation to their new homeland, many immigrant populations evolved into new species.

The story doesn't end there. The climate cooled further as the planet entered the Pleistocene's ice ages and deciduous forest species migrated

southward. In western North America, a long-term drying trend eliminated most traces of its deciduous forest. In Europe, plants migrated southward, but many went extinct when the Mediterranean Sea and east–west trending mountains like the Alps blocked their retreat. In eastern Asia and eastern North America, north–south trending mountains and a sufficiently moist climate provided avenues for southward escape. Many plants colonized the Southern Appalachians and Alabama, where they currently enjoy southern hospitality in the form of warm, humid conditions. Some taxa split into more new species as they colonized and adapted to the Southeast's diverse soils and topography.

DOIN' IT IN THE DARK

Deep below the surface is one of Alabama's most outstanding but least celebrated stories of speciation. Northern Alabama is one of the world's hot spots for caves and their fauna. The diversity of cave animals completely adapted to subterranean life exceeds that of all other locations in North America and ranks third among nearly two dozen cave-rich

Above left: Like many Appalachian plants, Mayapple's close relatives are in eastern Asia. Courtesy of Bob Farley/F8PHOTO.

Above right: A Phantom Cave Crayfish (top) and an Alabama Cave Shrimp (bottom), two species endemic to the state. Courtesy of Danté Fenolio.

regions in the northern temperate zone. Alabama's caves are home to one species of salamander, two fishes, and dozens of invertebrate species, including spiders, shrimps, and crayfishes, plus less well-known taxa like pseudoscorpions and flatworms.

Biologists believe species tiptoe their way toward cave living and trogloxenes represent the first evolutionary step. Trogloxenes use caves temporarily and rarely venture beyond where daylight reaches. Examples include birds nesting near the cave entrance; the Allegheny Woodrat, which lives in colonies near cave entrances; and dozens of bat species using caves for roosting, breeding, and hibernating. Very few trogloxenes will further adapt to life underground. Troglophiles are midstride in an evolutionary adjustment to cave ecosystems. They show partial adaptation to cave life, such as reduced vision and reduced pigmentation. Troglobites have inhabited the cave environment for so long, they have fully adapted to subterranean life. Most troglobites are ghostly white and have sophisticated nonvisual sensory systems—including exceptionally long legs and antennae—to navigate and find food. Their bizarre appearances arose through millions of years of small evolutionary steps, each one improving survival and reproduction. Troglobites are unusually common in the cavern systems of northeastern Alabama.

Once a new troglobite species has evolved, further speciation is facilitated by the underground topography. Recall from chapter 4 that karst terrain—the limestone or dolomite landscapes in which caves are most common—is riddled with underground fissures and streams. Though rarely extending more than a few miles in any direction, these openings are conduits for dispersal and colonization. As new passages open and old ones close due to changes in groundwater supply and tectonic adjustments, populations of troglobites can easily become isolated. With time, these isolated populations can diverge genetically from their parent population and become new species.

Northern Alabama is a hot spot for cave biodiversity because of its abundant karst topography and mild climate. The karst is a relic from the eons Alabama spent as reef-encrusted seafloor during the Paleozoic. The warm and wet southeastern climate contributed to cave formation by accelerating the dissolution of limestone. Because of this combination, Alabama has a higher density of cavern systems than most other regions in the world. The climate also supports cave ecosystems by supplying groundwater laden with energy and nutrients from the decomposition of vegetation at the surface. Drier and cooler regions of the world have less primary productivity at the surface and cannot sustain as much cave biodiversity.

Freshwater Fish Frenzy

Speciation among southeastern freshwater animals—particularly turtles, fishes, crayfishes, snails, and mussels—has bolstered the region's biodiversity more than any other ecologically affiliated group of organisms. The Southeast, with Alabama at its heart, is often *the* center of global diversity for these taxa. What pushed this frenzy of aquatic species evolution in the Southeast, and in Alabama in particular? Evolutionary biologists are still discovering answers to this question. The most advanced research is on the evolution of freshwater fishes. Given their shared dependence on freshwater ecosystems, the processes and events shaping the evolution of freshwater fishes are probably quite similar to those influencing the evolution of the Southeast's other diverse freshwater taxa. So what have evolutionary biologists revealed about fish evolution in the Southeast?

The absence of Pleistocene glaciation and comparatively mild conditions during the Late Tertiary and Pleistocene allowed many southeastern aquatic ecosystems and species to thrive. Additionally, the region's complex, mountainous topography fragments it into numerous large watersheds that provide opportunities for population isolation and subsequent genetic divergence. Because each watershed drains a unique combination of rock formations, its topography and water chemistry usually differ from those of adjacent watersheds. The Cahaba River, for example, is much clearer and carries a higher mineral content than does its western neighbor, the Black Warrior River. Such variation encourages species evolution when fish populations adapt to local conditions.

The above generalities only go so far. Basic geologic and climate histories of the region fail to explain many unusual fish distributions. The most puzzling involve cases in which populations of a species, or two closely related species, live in different watersheds separated by inhospitable habitat. Consider the Mountain Shiner. This little minnow inhabits the headwater streams of both the Tennessee River watershed and the Upper Mobile River Basin. Did the fish evolve in one headwaters region and migrate to the other? Not likely. In today's landscape, the two populations are separated by 2,000 miles (3,219 km) of inappropriate aquatic habitat, including over 100 miles of the Gulf of Mexico. Did the initial formation of the Southern Appalachians split the population? Again, not likely. The mountains first arose in the Paleozoic, several hundred million years before most modern fishes evolved. Furthermore, the region occupied by the shiner was a marine sea during the first phase of mountain building.

The best explanation for these distributions involves thievery among watersheds. On rare occasion, one watershed steals a portion of another, including its aquatic animal community. Several processes cause stream capture, or stream piracy, including when a headwater stream erodes through the barrier separating it from an adjacent watershed. The invading stream then captures the flow of a neighboring stream from the adjacent watershed. Tectonic forces also cause stream capture when uplifted lands send streams spilling into new areas. With fish populations of the pirated stream now isolated from their parent populations, they can evolve into distinct species. Resulting pairs of closely related species are known as sister species. Other divided populations like the Mountain Shiner remain as disjunct, but relatively unchanged, populations of the same species.

Fish biologists believe a series of stream piracy events during the Tertiary triggered the evolution of many southeastern species residing above the Fall Line. In Alabama, a series of piracies seems to have occurred between the Mobile River Basin and neighboring watersheds. The most dramatic proposed piracy in the Southeast is the capture of the theorized "Appalachian River" by the Tennessee River. Prior to the capture, the headwaters of the Appalachian River were what is now the Upper Tennessee River above Chattanooga, Tennessee. The remainder flowed through the Mobile River Basin via the Coosa, Cahaba, or Black Warrior River. In the Early Miocene, the Tennessee River captured the headwaters of the Appalachian River and forever kidnapped populations of species that evolved in the Mobile River Basin. Though scientists debate the existence of the Appalachian River, many biogeographers are fervent believers due to the strong pattern of sister species between the Upper Tennessee River and Upper Mobile River Basin.

A process other than stream capture must be called upon to explain the evolution of Coastal Plain species. Many fish biologists believe this process involved sea level changes in the Oligocene, Late Miocene, and Pliocene. During the Oligocene, the earth entered a phase of rapid cooling and sea levels dropped markedly. Freshwater fish species along the continental margin colonized adjacent watersheds as formerly isolated rivers became connected on the emerging plain. Sea level rise in the Early Miocene pushed freshwater species higher in their watersheds. For six million years during the Middle Miocene, sea level was 263–328 feet (80–100 m) higher than in the Oligocene. This isolated populations, and some diverged genetically and became new species. Lower sea levels in the Late Miocene allowed another round of fish exchange between neighboring watersheds. A smaller (164–236 feet, 50–80 m) and shorter (one million years) sea level incursion occurred during the Pliocene and further diversified southeastern fishes.

Sea level fluctuations also occurred during the Pleistocene's glaciation cycles and for many years biologists speculated this caused the diversification of southeastern Coastal Plain fishes. Critics pointed out these Gulf incursions were less dramatic (increase in sea level of 33–66 feet, 10–20 m) and shorter (30,000–100,000 years) than those of the Miocene and Pliocene. Two recent tests of these models used molecular clock methods to estimate the divergence times of related southeastern fishes. Results of a study of eight basses conformed to the pre-Pleistocene model of speciation, but a study of 10 logperches supported the Pleistocene model. While the pre-Pleistocene model still has the most evidence and backing, the latter findings illustrate that our understanding of fish speciation in the Southeast is still maturing.

A final factor contributing to the high fish diversity in the Southeast is the widespread presence of several taxa with exceptionally fast speciation rates. The minnows, fishes of the family Cyprinidae, account for about 30 percent of Alabama's fish species, and members of the darter family, Percidae, follow close behind. The state's fish diversity would be much diminished were these taxa not present in Alabama.

The Story Continues

What lies ahead for evolution in Alabama? In the near term and probably for much longer, species will adapt to the inescapable influence of humans on the atmosphere, waters, and land, or they will succumb to extinction. Over the longer time frame, surviving species will contend with natural climate swings, complete erosion of the Southern Appalachians (or another period of tectonic uplift), and the global wanderings of the North American continent. No one knows how much of this future we and our fellow species will witness. But assuredly, as long as there is life on earth, the evolutionary story will continue.

Though this completes the review of the processes shaping and maintaining Alabama's biodiversity, the story continues in the next chapters. What follows are explorations of the state's major ecological regions—from its tallest mountain peak to the depths of the Gulf of Mexico—and encounters with some of the state's most spectacular plants and animals.

7
The Aquatic State

One of the greatest biodiversity marvels of North America comes to an end at Mobile Point. This windswept and wave-battered promontory is the eastern mandible to the mouth of Mobile Bay. Here one can watch the final moments of a river's seaward journey. Though these waters have mingled with salt water since arriving in the river delta to the north, at Mobile Point their freshwater character is finally lost. The waters of the Mobile River Basin, a collection of watersheds draining nearly two-thirds of Alabama, have passed over and through countless fishes, plants, mussels, snails, crayfishes, insects, and many species few humans ever see.[1] Some are rare; many are common. Some may soon go extinct; others are being saved. Most have never been carefully studied, and a few have yet to be discovered. Considering the biodiversity these waters have nurtured, Mobile Point is a profoundly meaningful place on Alabama's natural history map.

The rivers of the Mobile River Basin play a major role in shaping the aquatic biodiversity of the Southeast, a region whose aquatic fauna is globally celebrated. The Southeast has more species of freshwater snails, mussels, crayfishes, freshwater turtles, and fishes than any other temperate zone region of the world. Alabama is at the epicenter of this species diversity and leads the United States or ties for first place for each of these taxa. More than any other ecological grouping of species, aquatic biodiversity pushes Alabama near the top among states ranked for total species.

The significance of the state's rivers transcends their biodiversity. Colorado has mountains, Arizona has desert canyons, Florida has beaches, and Alabama has rivers. Alabama's 77,000 miles (123,919 km) of waterways distinguish the state more than any other natural feature. Long before roads and railways crisscrossed the state, the rivers were the major thoroughfares and settlement zones for Native Americans, then Europeans, then Americans. Even today, Alabamians depend on rivers for commercial transportation and as a source of water for drinking, irrigation, and industry. River recreation is also economically and culturally important. Many spend long hours each year fishing in, boating on, hiking near, or simply sitting by and watching Alabama's rivers. The collective biological, economic, and cultural significance of its rivers and other freshwater ecosystems has inspired some, including renowned biologist and author Edward O. Wilson, to suggest that Alabama be known as "The Aquatic State."

This chapter is a submersion in Alabama's rivers. We'll drift down a river, from its mountainous headwaters to its terminus, learning how river ecosystems support a diversity of species, then we'll meet the most bizarre and intimidating creatures in the state's rivers. Finally, we'll tour Alabama's major rivers and examine their most distinctive ecological features.

Fast and Shady

Stream habitats can differ widely. Some are deep channels where large fishes lurk, while others are swift shallows where shiny minnows rush about. Ecologists use a model known as the river continuum concept to explain this variation. The concept divides rivers into three zones, the headwaters, mid-reaches, and lower reaches. In each zone, topography, water throughput, streambed composition, and inputs of energy create a different set of stream habitats.

Each river has a thousand humble beginnings. These are the points in the hills or mountains where surface water from rainfall and groundwater seepage combine to form narrow rivulets. Most are intermittent, flowing only during or after a rain. Where enough water comes together, a permanent stream is born. As a young stream winds downhill, it gathers size and momentum as it receives more groundwater and joins with other streams. In steep terrains, headwater streams zip past boulders and spill over falls. Their swift currents sweep away fine sediments, so the water is clear and the substrate, or stream bottom, is often rocky.

As headwater streams grow, so do their number of aquatic animals.

Headwaters of Turkey Creek, Jefferson County. Photo by R. Scot Duncan.

These narrow streams are shaded by forest canopy, so there are few stream plants to provide energy for a food web. Ironically, the trees intercepting the sunlight provide the energy for headwater stream ecosystems. Microbes, mostly bacteria and fungi, colonize leaves and branches falling into the stream and salvage remnant nutrients and energy. They transform this detritus into a tasty mush devoured by aquatic insects. Members of the shredder guild (a "guild" is a group of organisms performing a similar ecosystem function, regardless of their taxonomic affiliations) consume the detritus for energy and nutrients.

Shredders generate a lot of fine organic particles as waste that drifts downstream. Another guild, the collectors, gathers these particles and harvests from them what energy and nutrients remain. Caddisflies are often the most abundant collectors in headwater streams. A larva will attach to a submerged rock or log and build a cocoon of silk and stream debris. It then spins a silk net at the cocoon entrance to capture fine organic particles. When mature, the larva pupates and emerges as a winged adult that lives only a few days to reproduce. Like the shredders, the caddisfly and other collectors in the headwaters will generate waste supporting collector species far downstream.

A third guild in the headwaters comprises the predators, the animals hunting the collectors and shredders. Because organisms transfer energy inefficiently through the food chain, it takes many prey to support one predator. Thus, the predator guild makes up only about 15 percent of the headwaters' fauna, measured by number or weight. Shredders make up about one-third and collectors about half of the headwaters community.

Headwater streams may appear to be simple ecosystems of rock and flowing water, but closer examination reveals different habitats, each supporting different creatures. Ecologists specify four stream habitats: riffles, runs, pools, and glides. These habitats vary in depth, slope, current speed, stream width, substrate, and oxygen availability.[2] Pools are slow, deep sections often at the outside of a stream bend. The slow current lets fine sediments and leaves accumulate and shredders are common. With little turbulence, dissolved oxygen levels are low. Many stream creatures are well suited for pools, including turtles, bass, sunfish, and other larger animals needing the space pools provide.

Riffles are shallow, steep habitats where the stream tumbles over rocks. The swift current prevents accumulation of fine sediments and leaves, so riffle fauna are mostly collectors or small predators awaiting drifting prey. Riffle dwellers depend on high oxygen levels created by the turbulence. Being shallow, riffles harbor small animals, including tiny fishes. Invertebrates cling to the underside of rocks to avoid predators and the strong current.

Glide habitats are transitions between pools and riffles, and run habitats are transitions between riffles and pools. Both are intermediate between pools and riffles for stream characteristics. Many species prefer these intermediate conditions, and mixed-species minnow schools are common.

WIDE AND SUNNY

The river's ecology shifts as it emerges from the headwaters and enters the mid-reaches. Less confined by bedrock and steep slopes, the river begins meandering across the landscape. The river deepens and widens to accommodate its larger size. When it meets resistant bedrock, it widens and creates shallow, rocky habitats known as shoals. Shoals are like the riffles of smaller streams, but larger and wider. Pools, runs, and glides are also present.

Mid-reaches of Little River, Little River Canyon National Preserve, Cherokee County. Courtesy of Alan Cressler.

Throughout the mid-reaches, sunlight strikes the water for many hours each day and raises water temperatures. Supplies of nutrients from the headwaters plus abundant sunlight and heat encourage aquatic plant growth. Drifting microscopic plants (phytoplankton), algae, and a few vascular plants thrive in the mid-reaches and boost the energy available for food webs. The most stunning plant of the southeastern mid-reaches is the Shoals Spider-lily, known in Alabama as the Cahaba Lily. Late each spring in the shoals of the Cahaba and other southeastern rivers, lilies sprout and flaunt large white blossoms. In places like the Cahaba River National Wildlife Refuge near West Blocton, large shoals and their chattering streamlets become spectacular fields of greenery and showy flowers.

The new conditions of the mid-reaches prompt big changes in the aquatic fauna. Due to the mid-reaches' size, there is relatively little plant debris and shredders largely drop out of the aquatic community. Grazers, animals that eat living plants, flourish in the mid-reaches and average a third of the fauna. Snails are the most numerous grazers, and in some Alabama streams there can be hundreds in a square yard of shoal. The collectors still comprise half of the community, feasting on the fine organic matter streaming from the headwaters. Predators hold steady at about 15 percent of the animal community. Collectors and predators in the mid-reaches are usually different species from those in the headwaters. This turnover ramps up total diversity of a watershed. All the above statistics say nothing about the overwhelming abundance of life in the mid-reaches. Several hours of collecting on a shoal could yield enough mollusks, fishes, insects, and crayfishes to keep a small army of taxonomists busy for months.

Deep and Muddy

The archetypical image of a southeastern river is a wide expanse of muddy water lazily drifting to the coast. This imagery matches a river's lower reaches. In Alabama, the transition between the mid-reaches and the lower reaches occurs at or near the Fall Line, that biogeographically and historically important point where rivers break free of the rocky uplands and spill into the lowlands. No longer confined by rock and hill, large rivers meander wherever they please.

Left: The endangered Anthony's Riversnail of the Tennessee River watershed. Courtesy of Thomas Tarpley/ ADCNR.

Below: Alabama River at Haines Island Park, Monroe County. Photo by R. Scot Duncan.

Big rivers are naturally muddy. They carry a high load of fine clays and silts stripped from their eroding banks and gathered from upstream watersheds. Habitat diversity is a function of depth, current speed, and substrate type. Sand and gravel habitats at the inside of a bend provide shallow zones that are important spawning habitats for many fishes. The outside of a bend where the river cuts into the adjacent terrain is deep and the substrate is often compacted mud or clay. Sometimes gravel accumulates here and provides mussel habitat. Rooted plants in the channel are absent, but the bottom collects waterlogged tree trunks. It takes years for them to decay in the low-oxygen environment, and in the meantime they provide important fish habitat.

Big rivers lack enough detritus and surfaces for algal growth to support shredders and grazers, respectively. Instead, they receive from upstream a generous supply of fine organic particles and nutrients that supports large populations of collectors. Collectors comprise about 85 percent of the fauna. Many live in burrows to evade the current and predators, the latter still comprising 15 percent of the community.

The delta at the bottom of the lower reaches—where the river meets the ocean—adds much habitat and species diversity to the river system. Here, the volume of water and sediment a river carries is at its peak. As the river approaches sea level, the maritime waters offer resistance and the river slows, even reversing course on incoming oceanic tides. The slowing river drops much of its sediment, thereby contributing to the formation of new land. Over the decades, land growth forces the river to adjust course. Bends and twists are common, and the river often splinters into multiple channels. As new conduits are created, old ones are abandoned and become sloughs, oxbow lakes, and swamps. Seasonal flooding is essential to the delta's survival. The marshes, swamps, and floodplain forests created by rivers slowly sink as their soft sediments are compacted. Winter and spring freshets rescue the delta with sediment-laden waters. Eventually, the river reaches the bay, where it begins to relinquish control to the tides. The brackish waters born from the mingling of river and marine water will eventually spill into the ocean on an outgoing tide.

DAM THOSE RIVERS
From its headwaters to its delta, a river system supports a tremendous

range of wetland ecosystems. Each and every one of these ecosystems depends on a continuous supply of water, sediment, energy, and nutrients from lands upstream. With Alabama being so rainy, you might think its river ecosystems are well nourished and thriving. They are not.

Last century we built dozens of dams on Alabama's large rivers. Many generate virtually pollution-free electricity and aid the passage of commercial vessels by raising water levels. Dams can reduce flooding by controlling water levels during periods of high river volume. The reservoirs created by dams provide a reliable source of water and encourage recreation and development in rural areas with few other means of economic growth.

Unfortunately, dams prevent the natural flow of river resources and impair a river's ability to sustain its ecosystems. Floodplain forests and barrier islands become sediment starved. Discharge patterns don't match the seasonal rhythms that once triggered breeding or migration in aquatic species. Dams fragment populations and block migratory fishes from reaching spawning and feeding habitats. Shallow, rocky, flowing habitats upstream become deep, silt-filled, and stagnant reservoirs. The changes caused by damming are profoundly injurious to biodiversity. Nearly half of all US extinctions since the arrival of Europeans are of species living in the Mobile River Basin, and most of those extinctions were caused by damming.

Despite this difficult history, there is still a fabulous diversity of animals and plants living in Alabama's rivers. We'll next meet a few of Alabama's most interesting river creatures, ones chosen to represent each of the three river zones.

Minners

Wade into any headwater stream and you'll soon spot a school of small fishes zipping through the current. Most folks would call them minnows, perhaps "minners," and they'd be right. These fishes are members of the family Cyprinidae, the minnows. Cyprinids are the most diverse fish family in both North America and Alabama, comprise nearly a third of the state's fish diversity, and include shiners, chubs, minnows, and stonerollers. Cyprinids can be found in all portions of the river, but in the shallow habitats of the upper reaches they are the most conspicuous fishes present.

A male Rainbow Shiner, Big Canoe Creek, St. Clair County. Courtesy of Emily Horton Reilly.

Minnows enjoy the greatest perk of stream life—they let the currents bring food to them. However, maintaining one's position in the current requires a lot of energy and oxygen consumption. Thus, most minnows prefer habitats where a swift current brings enough dissolved oxygen and food to fuel their active lifestyles. What's more, they have evolved a narrow, streamlined body form that reduces drag in the current.

Most minnows are gray or silver, often with a highlight of black as a side streak or tail spot. But during the spring and summer, older males will sometimes erupt into bright colors that among the shiners inspired names like Bluestripe, Cherryfin, Emerald, Golden, Ironcolor, Rainbow, Scarlet, and Warpaint. Breeding males will also grow strange bumps, or tubercles, on their head, body, or fins. These colors and ornamentations are the products of sexual selection, wherein female preference for the most extravagant males shaped evolution.

Darters approach stream life differently. Members of the family Percidae are the second most diverse fish family in both North America and Alabama. Lacking swim bladders, they rest on the bottom in eddies, then dart out to capture small drifting invertebrates. Their short movements don't attract attention and they are supremely well camouflaged when viewed from above. However, in many species the males sport vivid breeding colors on their faces, fins, and flanks that have earned them names like Blueface, Goldline, Greenbreast, Halloween, Harlequin, Lollypop, Rainbow, and Vermilion. Like shiners, many darters are at home in headwater streams and mid-reach shoals. The clear waters of these habitats help them spot prey and select the most colorful mates.

Though their habits differ, darters and minnows share a high sensitivity to stream degradation. Ten of Alabama's species are on the US Endangered Species List and dozens more are classified as vulnerable to extinction. Sediment pollution is a major threat to many darters and shiners, especially those needing clean gravel or other crevice-rich habitats in which to deposit eggs. Fine sediments eroded from upstream developments have smothered prime breeding habitats for many species.

FLEXIN' MUSSELS

The most bizarre life histories that can be told of southeastern streams involve ostensibly unglamorous creatures, the freshwater mussels. Mussels bury themselves in the stream bottom and spend their days peacefully filtering the passing waters for plankton, drifting algae, bacteria, and fine organic particles. Even a small mussel can filter more than 12 gallons (45 l) of water each day, and the collective feeding of a thriving mussel community can keep river waters clear. It's a good life, and ages of 30–70 years for mussels are not uncommon.

The diversity of mussels in Alabama is impressive. The state's 182 species represent 22 percent of the world's species and about 60 percent of US species. Their outer shells may be rounded, oblong, squarish, or triangular; textures may be smooth, rippled, or knobbed. Some are thick-shelled giants once collected for use as hoe blades by Native Americans. Mussel biologists have enjoyed bestowing on them whimsical names such as the Southern Fatmucket, Purple Wartyback, Cumberland Monkeyface, Inflated Heelsplitter, and Fuzzy Pigtoe.

To protect valves from abrasion and dissolution in acidic stream water, mussels coat their valves with a sheet of olive, brown, or black protein. Otherwise, shells are composed of nacre, a smooth, shiny material of which calcium carbonate is a major component. Nacre can be blue, purple, pink, orange, or white, and its color and durability have attracted humans for millennia. Native Americans fashioned mussels into tools, ornaments, and jewelry, while industrialists harvested southeastern mussels by the millions to make pearl buttons or, in recent decades, the "seed" used for growing cultured pearls.

While all this is noteworthy, mussel reproductive behavior is nothing short of astonishing. Many aquatic animals with limited or no mobility, like mussels, reproduce by spawning. Successful reproduction depends on males and females responding to environmental cues and spawning at once. Sperm and egg set adrift quickly find each other, and the resulting larvae grow and drift for days or weeks until the current or tides bring them to the right habitat and they drop to the bottom and start a more sedentary life.

River mussels, however, can't follow this plan exactly. The current would whisk larvae into the ocean or unsuitable river zones, and upstream mussel populations would be gradually forced downstream since young mussels would never settle upstream of their parents. Clearly, mussels have evolved another reproductive strategy, and here is where the story gets weird.

Mussel valves from the Cahaba River, Bibb County. Photo by R. Scot Duncan.

Male mussels do release sperm into the current, but females take in sperm for internal fertilization. The young larvae that result are parasites, known as glochidia, ready to attach to the gills or fins of fishes. Many species simply release glochidia into the current in hopes they will encounter a fish host. Most mussel species use several host species, but a few species are quite selective. Successfully attached larvae become encased in a cyst formed by the host's immune response. Inside, the glochidium will transform into a juvenile mussel and then drop off the host. Fish transport some of the hitchhikers upstream from the parent mussels' location, thereby unintentionally helping the population maintain its stream position.

Even more bizarrely, some mussels evolved elaborate tricks to help their larvae infect hosts. Some species bundle glochidia into packets, called conglutinates, resembling food items for fishes, including worms, leeches, insects, fish eggs, and small fishes. When a naïve fish bites, the conglutinate explodes, releasing glochidia into the fish's mouth and gill chamber. Some mussels even dangle the bait on a transparent mucus tube up to 8 feet (2.4 m) long. The lure appears to swim as it sways in the current. Others display a fleshy flap around their valve opening resembling a small fish, worm, larval insect, or crayfish. Mussels wiggle lures to attract hungry fish and spray glochidia in their faces when the dupes get close.

Many mussel species are highly vulnerable to stream degradation, especially damming and sediment pollution. Of the 300 or so species in the United States and Canada, a quarter are federally listed (on the US Endangered Species List), half are of high conservation concern, and a sobering 6 percent are extinct. Alabama is ground zero for this crisis, hosting 48 of the federally listed species—more than any other state. It's not all dire news, however. There are mounting efforts to safeguard streams with mussel populations, repair degraded streams, dismantle derelict dams, and breed rare mussels for release in the wild.

RIVER MONSTERS, *REALLY*?
Alabama's big rivers host many large and intimidating creatures. The generous delivery of nutrients and energy to the lower reaches fuels a

productive food web whose top predators can be huge. Defined by size alone, they are truly "monsters." Yet which among them, if any, are a threat to humans?

The Blue Catfish is a species eagerly sought by commercial and recreational fishers. Catfish carry poison-tipped spines on their dorsal and pectoral fins, lack scales, and sport short, fleshy whiskers around their mouth. Blue Catfish range throughout the central United States, hunting in deep pools and strong currents for smaller animals and nutritious debris. They definitely grow to monstrous sizes. Huge specimens were documented in the 19th century, the largest weighing 150 pounds (68 kg). Much larger specimens, one said to weigh more than twice this amount, have been reported but were never confirmed. Colossal lunkers are still out there. In 2012 an angler landed a 120-pound (54 kg) Blue Catfish, the largest ever confirmed in Alabama. Drawing from the catholic diet and large size of this fish, river folklore is rife with tales of giant catfish attacking humans. Stories tell of narrow escapes, drownings, and even children being swallowed. True or not, most catfish these days are caught and eaten long before they attain sizes where a human would be a tempting meal.

A more obscure large fish of Alabama's big rivers is the Paddlefish. This awkward fish superficially resembles a shark and has a broad oar-like snout nearly one-third its length. The largest Paddlefish captured in the Southeast was 142 pounds (65 kg). While impressive, Paddlefish pose no threat to any creature larger than the period at the end of this sentence. They eat tiny zooplankton, collecting enough to support their large size by cruising slow-moving waters with their oversized mouths held wide open. The paddle's function is unknown, but may be used to locate food patches.

A much larger odd fish from Alabama's big rivers is the Gulf Sturgeon. With rows of armored bony plates along their length, sturgeon look primitive. Indeed, paleontologists have found fossil sturgeon similar to modern species in Cretaceous sediments. The Gulf Sturgeon, a subspecies of the Atlantic Sturgeon, is the largest of four sturgeons historically inhabiting Alabama. It attains lengths up to 10 feet (3 m), and the largest documented in Alabama weighed 850 pounds (386 kg). With a diet of invertebrates dug from the mud and sand, sturgeon may seem to pose little threat to man. However, they have a penchant for leaping out of the water and have unintentionally injured many boaters.

Gulf Sturgeon mainly dwell in bays and in the Gulf's shallows. Each spring some will migrate up rivers to spawn over gravel, cobble, or rock outcrops. Before spawning, eggs comprise up to 25 percent of a female's weight. This unusually large reproductive investment forces females to recuperate for several years after spawning. Ironically, it has also been the downfall of many sturgeon species. Sturgeon eggs are an exceptionally coveted and expensive form of caviar. Globally, fisheries supporting the human fixation on caviar have decimated many species, including the Gulf Sturgeon. Though federally protected and no longer fished, Gulf Sturgeon struggle to rebound because dams block access to many historic spawning sites.

If the Gulf Sturgeon isn't a river monster, perhaps another sturgeon qualifies as a river ghost. The Alabama Sturgeon is a strictly freshwater fish less than 3 feet (0.9 m) long. Once, commercial fishers landed thousands of them annually. Today, it is one of the rarest animals on the planet. Overharvesting, dams, sediment pollution, river dredging, and gravel mining have pushed it to the very edge of existence, and possibly over. Despite periodic surveys, biologists have seen only a hand-

The highly endangered Alabama Sturgeon. Courtesy of Paul D. Johnson/ADCNR.

ful since the mid-1980s. There is little to be done besides hoping the Alabama Sturgeon has survived and can begin a recovery. For now, it is a ghost on Alabama's species list—unable to be found, but not yet declared extinct.

The Alligator Gar is another primitive fish of the Lower Mobile River Basin. Gars have a torpedo-shaped body up to 10 feet (3 m) long, allegedly longer, and a tapered snout brimming with needle-sharp teeth. They slowly cruise swampy backwaters, but with lightning speed they can seize and gulp down live ducks and fishes one-third their size. There are a few accounts of a gar biting a hand or foot, probably because it mistook the unfortunate appendage for a fish. Still, Alligator Gars are not a serious threat. In fact, it's the other way around. Overfishing for sport and habitat alteration have caused population declines throughout its range.

Large sharks patrol some Alabama rivers. Bull Sharks are common in coastal estuaries and they occasionally venture up rivers for unknown reasons. In a famous account, one was captured on the Mississippi River in central Illinois. The Bull Shark is responsible for most southeastern shark attacks, and it may deserve status as a river monster. That said, its freshwater wanderings are believed to be rare, so the chance of being attacked while swimming in a river is slim (see chapter 14).

Of the large beasts lurking in Alabama's rivers, the one most deserving of our anxiety is the American Alligator. The alligator's hunting skills and physique have been honed by tens of millions of years of cutthroat evolution. One end of this reptile is fitted with sharp conical teeth for grasping, puncturing, and crushing, and the other is armed with a fiercely muscular tail studded with sharp scales for crushing and lacerating. In between, it is armored with bony plates, thick scales, and short spines. All in all, the American Alligator is a perfect ambush predator. From seemingly tranquil waters, an awaiting alligator can leap from the water, seize with precision, and drag a hapless creature into the depths to drown, all within the time it took you to read this sentence.

Though sometimes inhabiting the main river, alligators prefer the backwaters, where ambushing prey is easier in the slow current. Young alligators hunt crayfish, insects, and fish, while adults take larger prey

including snakes, turtles, and mammals. At the top of the food chain, alligators keep prey populations in check. Once they surpass about 3 feet in length, the only animal an alligator needs to fear is a larger alligator. Cannibalism can account for over half of all alligator mortality.

Alligator populations plummeted early last century due to harvesting for skins and meat. Alabama was the first of many states to protect alligators, and the Mobile River Delta was an important refuge during this time. State and federal protection eventually gave alligators the chance they needed to rebound, and the species no longer faces extinction.

Should we be wary of alligators? Certainly. Fear them? Perhaps. Since 1948, attacks on people in the Southeast averaged seven per year with a steady increase over time. Most occurred in Florida, where people and alligators are numerous and live in close proximity. Attacks on people approaching or attempting to handle alligators make up the largest single category of incidents, but the overall majority of attacks are on people doing normal activities (swimming, wading, fishing, etc.) in or near the water. Very few alligators in Alabama reach a size such that they would attack an adult human without provocation, though

American Alligator, a perfect ambush predator. Courtesy of Alan Cressler.

children are more vulnerable. And most people who spend years hunting, fishing, or paddling through southeastern wetlands do so without incident through vigilance and staying out of the water when alligators might be near.

So, do Alabama's rivers harbor monsters? By some definitions—those emphasizing large size or the threat of attack—yes. But most big river critters are gentle giants and the few occasionally attacking humans are not malevolent brutes. They are creatures simply doing what millions of years of evolution have prepared them for—survival. Attacks happen when humans venture into their world and get in their way. Considering how we've often treated these species and their ecosystems, are we the *real* river monsters?

The Mobile River Basin

For our whirlwind tour of Alabama's major rivers, I've adopted the framework aquatic biologists often use to group Alabama's rivers into four groups: three large basins discharging into the Gulf and the small lower Coastal Plain rivers. The table provided summarizes geographical and biodiversity data about these watersheds for quick reference.

The Mobile River Basin (MRB) is the crown jewel of southeastern aquatic biodiversity. A hydrologic powerhouse, the basin discharges three times more water into the Gulf than the Mississippi River Basin when scaled for land area. Though the latter is vastly larger, it drains many dry regions, whereas the MRB is centered in one of the continent's rainiest regions. The abundant rain and the diverse geology and topography have produced a basin unmatched by any temperate aquatic ecosystem in the world. Many of its species are endemic, including 40 fishes, 34 mussels, and 110 aquatic snails. For comparison, consider its 220 native fishes. The Colorado River Basin, the largest basin in the southwestern United States, drains 12 percent of the United States, is 5.7 times larger, and yet historically supported only 36 native fishes.

Mississippi and Alabama share the Tombigbee River, the westernmost of the MRB's major rivers. It is entirely within the Coastal Plain physiographic province, and consequently the river bears a high sediment load. Beginning in northeastern Mississippi, the Tombig-

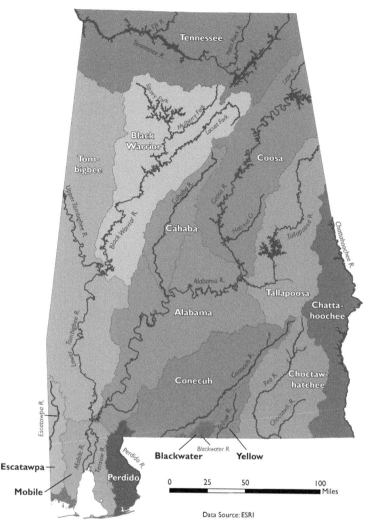

Alabama's major rivers and their watersheds. Large reservoirs created by damming are also illustrated. Courtesy Ed Brands/University of Minnesota Morris.

bee flows south-southeast until joining the Alabama River to form the Mobile River. The Tombigbee's largest tributary is the Black Warrior River, and their confluence marks the division between the Upper and Lower Tombigbee River watersheds. Until the 1970s, we spared the Upper Tombigbee River the damming that devastated big river habitats throughout Alabama. Then construction began on the Tennessee-Tombigbee Waterway (or "Tenn-Tom Waterway"), a project undertaken to facilitate commercial shipping between the

Alabama Watersheds

Basin and Watershed	Area in Square Miles (km²)	Percent Area of Alabama Drained	Number of Dams
Mobile River Basin	32,207 (83,416)	61	21
Tombigbee	7,694 (19,927)	15	4
Black Warrior	6,274 (16,250)	12	4
Cahaba	1,818 (4,708)	4	0
Coosa	5,400 (13,986)	10	6
Tallapoosa	4,022 (10,416)	8	4
Alabama	6,023 (15,599)	11	3
Mobile	>300 (>777)	1–2	0
Mississippi River Basin			
Tennessee	6,800 (17,612)	13	3
Apalachicola River Basin			
Chattahoochee	3,090 (8,003)	6	5
Gulf Coast Drainages	9,240 (23,931)	18	2
Escatawpa	767 (1,987)	1	0
Perdido	840 (2,176)	2	0
Conecuh	3,848 (9,966)	7	2
Blackwater	148 (383)	<1	0
Yellow	507 (1,313)	1	0
Choctawhatchee	3,130 (8,107)	6	0

Basin and Watershed	Number of Select Aquatic Taxa			
	Fishes	Mussels	Snails	Crayfishes
Mobile River Basin	220	73	114	51
Tombigbee	133	52	17	27
Black Warrior	127	51	24	22
Cahaba	125	50	36	14
Coosa	113	54	86	17
Tallapoosa	119	36	17	15
Alabama	142	51	23	23
Mobile	136	14	14	19
Mississippi River Basin				
Tennessee	165	93	68	36
Apalachicola River Basin				
Chattahoochee	86	30	17	12
Gulf Coast Drainages	124	32	20	31
Escatawpa	58	0	14	14
Perdido	82	0	13	8
Conecuh	82	30	14	17
Blackwater	21	0	12	17
Yellow	49	15	13	17
Choctawhatchee	73	21	16	12

Data for basins with only one watershed are combined on one line. Area data are only for Alabama's portion of the watershed. Dam totals include only those in Alabama on the watershed's primary river. Species counts are based on the most recent compilations published, and these will become out of date as new discoveries are made.

Tennessee River and the MRB. Six dams and a lot of dredging later, much of the Upper Tombigbee is now a deep canal. Miles of habitat were destroyed, and the canal now allows the unnatural exchange of species between the Mobile and Mississippi river basins (the Tennessee belongs to the latter). Three fishes have colonized the Tennessee River from the Tombigbee River. Of much greater concern is the threat of Asian Carp and Zebra Mussels invading the MRB. These exotic invasive species are wreaking biological havoc on Mississippi River Basin fisheries, industries, and ecosystems, and the Tenn-Tom Waterway is an open-door invitation.

The Black Warrior River is the Tombigbee's kid sister to the east. Its headwaters arise from the Cumberland Plateau physiographic province and drain the Bankhead National Forest. Its headwaters are known for the spectacular cliffs created where they have sliced through more than a hundred feet of the plateau's sandstone cap. This cap reduces rainwater infiltration and headwater streams rise quickly during a rain and subside soon thereafter. The Black Warrior crosses the Fall Line near Tuscaloosa, then meanders through the Coastal Plain until joining the Tombigbee near Demopolis. As with other rivers crossing the Fall Line, endemism and total species diversity of the river is high. Nevertheless, the Black Warrior has taken some hard knocks. Channel modifications and extensive damming began in the late 19th century, in part to service the export of coal from the upper watershed. One headwater dam created Lewis Smith Lake, one of Alabama's largest reservoirs. These modifications helped force the extirpation of over two dozen fishes from the Locust and Mulberry Forks, two of the river's three largest watersheds. Many of the Black Warrior's other species are federally listed.

The next river to the east is the Cahaba, a stream globally famous as an aquatic biodiversity hot spot. Originating in the Valley and Ridge physiographic province, the Cahaba crosses the Fall Line at Centerville, then joins the Alabama River near Selma. The river supports a tremendous number of aquatic species for its small size. For comparison, the neighboring Coosa watershed, whose original biodiversity is similarly diverse, is over 5.5 times larger in area. Much of the Cahaba's biodiversity survived because this river was too small to justify damming. Instead, the Cahaba faces a different threat. The Birmingham metropolitan area sprawls across large portions of its headwaters and injects the river with damaging amounts of sediment and nutrients. Water itself becomes a pollutant when stormwater systems shunt rainwater from developed areas to the river and its tributaries. These floods erode habitats, injure wildlife, and trigger riverbank collapse. Today, nearly 20 of the Cahaba's aquatic animals are federally listed, and at least 15 of its mussels are extinct or extirpated.

The Coosa River to the east was the setting for the largest multispecies extinction event known from US history. This story begins with the river's geology. Its watershed collects great volumes of water by draining portions of the Blue Ridge, Valley and Ridge, Piedmont, and Coastal Plain physiographic provinces in Tennessee, Georgia, and Alabama. The Coosa's volume peaks in Alabama while passing through mountains, steep hills, and deep valleys before ending at the head of the Alabama River. The river's mid-reaches were an aquatic Eden, harboring several hundred species, including dozens of endemics. As historic photographs heartbreakingly attest, the river spilled over stunning falls and sweeping shoals. One set of falls, the Devil's Staircase, was so powerful its roar was heard a mile away. Unfortunately, large mountainous rivers were tempting targets for hydropower dams during the past century. Damming of the Coosa River by Alabama Power Company began in 1906 and continued through the 1960s. The impoundments drowned falls and shoals and wiped out nearly 40 endemic mollusk species. There has been progress in protecting the Coosa's surviving species. Alabama Power Company and governmental agencies have arranged for the restoration of shallow river habitats below two of the Coosa's dams. These sections now support recovering populations of rare and endangered species.

The Tallapoosa River is often overshadowed by the dazzling biodiversity of the Cahaba and the Coosa. The upper Tallapoosa drains portions of the Piedmont in Georgia and Alabama, while the lower third drains the Coastal Plain. Above the Fall Line the river had many impressive shoals and falls, sometimes cutting through 300 feet (91 m) of metamorphic rock. Though its biodiversity doesn't match that of the Coosa and Cahaba, the Tallapoosa is more diverse than most similarly sized US rivers. Dams constructed for hydropower generation impound most of the Tallapoosa, including one constructed in the 1980s. Martin Dam created a huge reservoir with over 700 miles (1,127 km) of shoreline and is one of the most popular reservoirs for vacation and recreation in Alabama.

The Alabama is a river between rivers. It forms where the Tallapoosa and Coosa rivers converge near Wetumpka and ends when it joins the Tombigbee River 30 miles (48 km) north of Mobile.[3] The Alabama resides entirely in the Coastal Plain, where soft sediments allow it to meander wildly. To illustrate, the direct distance between the Alabama's beginning and end is 136 miles (219 km), but its actual path is 304 miles (489 km). Abundant sand and gravel shoals were important spawning habitats for sturgeon and other fishes, but damming for navigation in the 1960s transformed the Alabama into reservoirs and deep river channels.

At the bottom of the MRB is the Mobile River, a waterway destined to fail hydrologically. Created where the Alabama and Tombigbee rivers converge, the Mobile River accepts all the river water and sediments of the above watersheds. The overburdened river splinters into a throng of smaller rivers and streams to distribute the load. On their 30-mile (48 km) trip to the coast, these streams create the Mobile River Delta, a realm of floodplain forests, swamps, marshes, and backwaters where few humans venture. It is also where marine and river water mingle, and it's not uncommon to find saltwater fishes, like flounder, needlefish, and mullet, swimming past freshwater fishes like sunfish, bass, and gar. In sum, the Mobile River's failure to carry its load maintains one of the Southeast's finest ecological wonderlands.

MISSISSIPPI RIVER BASIN

The Tennessee River drains a small fraction of Alabama, but its watershed adds disproportionately to the state's species roster. The river is Alabama's only connection to the largest river basin on the continent, the Mississippi. Like the MRB, the Mississippi River Basin has its own endemic species. Many are Coastal Plain creatures of the basin's lower reaches, and their distributions extend into Alabama.[4] The Tennessee River also has an extensive stream network draining the western slopes of the Southern Appalachians, where many headwater species evolved. These species make the eastern section of the Tennessee River, including northeastern Alabama, one of the most diverse regions for aquatic biodiversity in the United States. Thus, the Tennessee supplies several hundred species not found elsewhere in Alabama, including 73 fishes, 74 mussels, and over 20 crayfishes.

The Tennessee River's biodiversity story has a tragic chapter. The entire 200-mile path (322 km) of the South Bend, the section looping through Alabama, is a series of deepwater pools created by the Tennessee Valley Authority for hydropower and navigation. Previously, the river was broad, with numerous shoals and islands. One continuous stretch of shallow rocky habitat known as Greater Muscle Shoals stretched over 50 miles (81 km) in length. The shoals were so expansive and shallow that the river could be crossed on foot. Greater Muscle Shoals teemed with aquatic species and it would be among the preeminent natural wonders of the Southeast had it survived. The damming triggered an economic boom for the region, but nine fishes and over a dozen mussels in Alabama are now extinct or extirpated. Other species found refuge in the smaller rivers feeding the Tennessee, but impoundment waters trap, isolate, and further endanger many of them.

APALACHICOLA RIVER BASIN

A war is being fought over the Apalachicola River Basin along Alabama's eastern border. For too many years, Alabama, Georgia, and Florida have battled over rights to the basin's waters. The list of stakeholders involved is dizzying and includes state and federal agencies, dam operators, municipalities (most notably Atlanta), developers, conservationists, agriculturalists, and oyster fishermen in Florida. In short, the demand for water has surpassed the basin's capacity to supply it. During skirmishes of recent years, the basin's biodiversity has been on the front lines. How did all this come about?

A little battlefield geography helps with understanding the biology and politics. The Apalachicola River Basin is mostly within Georgia, where its headwaters drain the Blue Ridge and Piedmont physiographic provinces. The basin's largest rivers are the Chattahoochee and Flint, with the latter entirely in Georgia. Atlanta sprawls across the top of the basin, and its rapid growth and water demand have caused most of the conflict. Though the basin drains only 5 percent of Alabama, the state withdraws water for its southeastern cities and farmers. At the three-way intersection of Alabama, Georgia, and Florida, the Chatta-hoochee leaves Alabama and flows southward to join the Flint River. Their confluence forms the Apalachicola River, which bisects the Flor-ida Panhandle before emptying into the Gulf. Florida's Chipola River is the last major tributary to the Apalachicola, and its headwater creeks originate in southeastern Alabama. Florida's most lucrative oyster fishery is in Apalachicola Bay. Oysters need the right blend of marine and river water, and excessive water withdrawal throughout the basin threatens the fishery.

Despite its six dams, the Chattahoochee River section within Al-abama supports a moderate diversity of river fauna. Residing here are ten fishes not found elsewhere in Alabama, eight of which are basin endemics. Four of Alabama's mussels are endemic to the basin, but two are extinct and two are struggling to survive.

Species conservation in the Chattahoochee is central to debates over water rights. The diminishing flow of water each year endangers many species, especially mussels and fishes. Though federal and state laws mandate the protection of listed endangered species, these laws compete with an entanglement of other laws governing the distribu-tion of water. Wildlife biologists pulled into the fray have been bur-dened with calculating the volume and timing of flowing water needed to sustain endangered species. The fate of many species and river eco-systems depends on the settlements emerging from these water wars.

GULF COAST DRAINAGES

Six small coastal rivers—the Escatawpa, Perdido, Conecuh, Blackwa-ter, Yellow, and Choctawhatchee—drain much of the state's southern third. Two words sum up their ecology: free and dark. "Free," because unlike most other Alabama rivers, they are encumbered with few dams. Their river valleys are too shallow and lack the topographic restriction points where large dams could be placed. To improve navigation on several rivers, work crews long ago removed gravel and sand shoals, overhanging trees, logs, and other "obstructions" providing import-

ant aquatic habitat. Compared to damming, these modifications were mild, and the free-flowing Gulf Coast rivers and their aquatic fauna are nearly intact.

Coastal rivers are "dark" because of their chemistry. They and their tributaries are known as blackwater streams, so named for their dark tea-colored waters that are more acidic and have fewer nutrients and suspended sediments than rivers originating farther inland. The sandy soils of the lower Coastal Plain lack nutrients and minerals that other sediments offer. Furthermore, unlike clay-based soils, sand doesn't react with compounds released through plant decomposition. Groundwater carries these compounds through the sandy soils and into streams, especially acidic compounds known as tannins that stain blackwater streams a dark golden-brown.

Alabama's coastal blackwater rivers are depauperate of freshwater species relative to more inland rivers. Scientists offer several potential

A blackwater stream of the lower Southeastern Plains ecoregion (Coldwater Creek, Santa Rosa County, Florida). Photo by R. Scot Duncan.

explanations. One involves stream acidity. Mollusks make their shells with calcium carbonate, a mineral gradually dissolved by acidic water. To survive, mollusks continually replace dissolved shell with new material. Shell dissolution in blackwater streams may have been too high for some species and may explain why three of Alabama's Gulf Coast rivers lack mussels entirely. Scientists also speculate that repeated flooding by marine water over the past few million years may have reduced freshwater biodiversity.[5] Despite having fewer species, Alabama's coastal blackwater rivers host 10 fishes, 11 mussels, and 6 crayfishes found here and nowhere else in the state. And by surviving the past two centuries more unscathed than any other rivers in the state, these dark and free rivers are ecological treasures.

On to Dry Land

Having completed our tour of the state's rivers, met many delightful and deadly species, and studied river ecology, it should be clear why rivers are at the heart of Alabama's spectacular biodiversity. Next, we'll explore the Alabama coastline, a chaotic region, where river, ocean, land, fire, and storm regularly collide in ways creating some of the most dynamic and diverse ecosystems Alabama has to offer.

8
Southern Coastal Plain

The Chaotic Coast

The youngest and most dynamic region of Alabama is the land fringing the Gulf of Mexico and its embayments. The state's 500 miles (805 km) of shoreline are continuously changing—land where one stands today could be gone in a year. Barrier islands are reworked subtly by each day's currents and waves and drastically remodeled by periodic cyclones (tropical depressions, tropical storms, or hurricanes). Mainland flatwood forests and grasslands are waterlogged for months at a time and were once charred by wildfire several times each decade. Winter and spring floods submerge the Mobile River Delta under restless waters that carve new channels and build new land. Yet amidst this chaos, ecosystems teem with species. On barrier islands, small plants thrive among shifting dunes and birds scrape out shallow nests in the sand. The flatwoods and grasslands resprout and bloom after wildfires, and the fertile sediments dropped by delta floodwaters sustain one of the wildest habitats on the continent. Alabama's coastal ecosystems—from barrier islands to river deltas—don't exist *despite* the frequent disturbances, they exist *because* of them. Before we explore this region, I must first introduce the system used throughout the rest of this book to classify ecosystems and natural landscapes.

Ecoregions and Ecological Systems

Biogeographers have developed several systems for classifying geographic regions and ecosystems in North America. These systems help us study biological patterns across large areas, but they also allow us to appreciate any particular ecosystem's or region's distinctiveness. Once an ecosystem's name is known, for example, one can easily find more published information about it. Without such nomenclature, it would be maddening to distinguish among the nearly 70 ecosystems and dozens of distinct ecological regions in Alabama.

Right: Level III ecoregions of Alabama. Numbers correspond to the US Environmental Protection Agency's classification system for North American ecoregions. Courtesy Ed Brands/University of Minnesota Morris.

Opposite: Level IV ecoregions of Alabama. Numbers correspond to the US Environmental Protection Agency's classification system for North American ecoregions. Courtesy Ed Brands/University of Minnesota Morris.

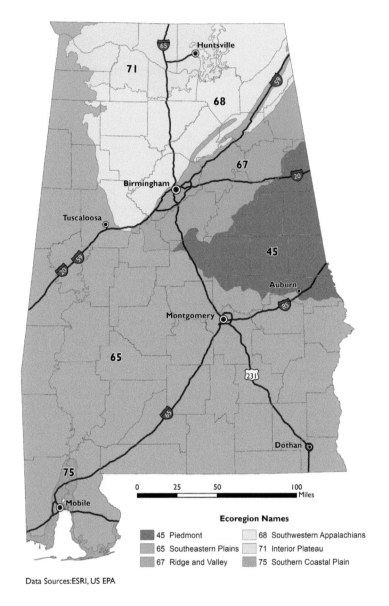

Ecoregion Names

45 Piedmont	68 Southwestern Appalachians
65 Southeastern Plains	71 Interior Plateau
67 Ridge and Valley	75 Southern Coastal Plain

Data Sources: ESRI, US EPA

For ecosystems, I've adopted NatureServe's index of ecological systems, defined as "recurring groups of biological communities that are found in similar physical environments and are influenced by similar dynamic ecological processes, such as fire or flooding."[1] This system is the most carefully constructed classification system for ecosystems.

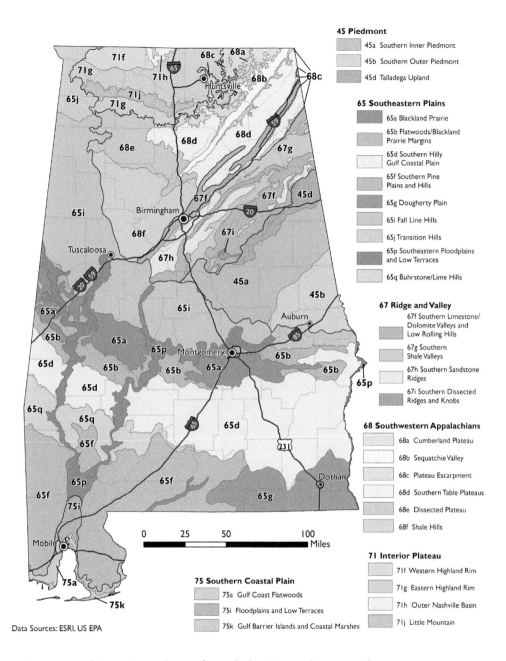

45 Piedmont

- 45a Southern Inner Piedmont
- 45b Southern Outer Piedmont
- 45d Talladega Upland

65 Southeastern Plains

- 65a Blackland Prairie
- 65b Flatwoods/Blackland Prairie Margins
- 65d Southern Hilly Gulf Coastal Plain
- 65f Southern Pine Plains and Hills
- 65g Dougherty Plain
- 65i Fall Line Hills
- 65j Transition Hills
- 65p Southeastern Floodplains and Low Terraces
- 65q Buhrstone/Lime Hills

67 Ridge and Valley

- 67f Southern Limestone/ Dolomite Valleys and Low Rolling Hills
- 67g Southern Shale Valleys
- 67h Southern Sandstone Ridges
- 67i Southern Dissected Ridges and Knobs

68 Southwestern Appalachians

- 68a Cumberland Plateau
- 68b Sequatchie Valley
- 68c Plateau Escarpment
- 68d Southern Table Plateaus
- 68e Dissected Plateau
- 68f Shale Hills

71 Interior Plateau

- 71f Western Highland Rim
- 71g Eastern Highland Rim
- 71h Outer Nashville Basin
- 71j Little Mountain

75 Southern Coastal Plain

- 75a Gulf Coast Flatwoods
- 75i Floodplains and Low Terraces
- 75k Gulf Barrier Islands and Coastal Marshes

Data Sources: ESRI, US EPA

For geographic regions, I've embraced the Omernik system developed by the US Environmental Protection Agency and the Commission for Environmental Cooperation and named in honor of its principal architect, James M. Omernik. The Omernik system classifies lands based on climate, geology, hydrology, soils, land use, physiography, vegetation, and wildlife. Though relatively new—you won't find it in older literature—it is the most important system in use today. It has a hierarchical structure with four levels, each designated with a Roman numeral. The approach divides North America into 15 level I ecoregions, placing Alabama and most of the eastern United States in the Eastern Temperate Forest. Three level II, 6 level III, and 29 level IV ecoregions subdivide Alabama, the latter two levels being the most useful for studying Alabama's biogeography. The state's level III ecoregions largely correspond to the physiographic provinces of the state with one significant exception: the Coastal Plain physiographic province is divided into two level III ecoregions, one of which is the Southern Coastal Plain.

PLOVERS AND STORMS

With its brilliant white sands merging into the teal waters of the Gulf of Mexico, many consider the barrier system to be the Southeast's most alluring landscape. Technically known as the Gulf Barrier Islands and Coastal Marshes level IV ecoregion, this strip of islands, peninsulas, and mainland Gulf beaches stretching across Alabama's 50 miles (80 km) of outer coast comprises the state's most dynamic set of landforms and ecosystems. Though its loose sands are continually being reworked by winds, waves, tides, and storms, the barrier system is a haven for thousands of species coping with the chaos just fine.

One of them is a shy and imperiled little bird known as the Snowy Plover, and each year it hurries to raise its young before the cyclones come. These small, squat, stubby-beaked birds are mostly white and pale gray but are handsomely trimmed with black markings around the face. Breeding starts as early as April and is complicated: mates and nesting sites must be chosen; eggs must be laid and incubated; and young must be fed and self-sufficient within a month of hatching. The process takes many weeks, and since cyclone season begins in midsummer, the threat of catastrophe ratchets upward each day.

Nests are shallow depressions scraped in the sand and lined with bits of debris. Though easily constructed, they are difficult to defend against predators. Thus, the plovers nest in flat, open expanses where

Snowy Plover, shy resident of the barrier system. Courtesy of David Sparks.

they can see trouble coming and either distract approaching predators away from the nest or use their camouflage to hunker down. But these tactics don't work against cyclones. Their nests are so close to sea level, even a minor storm can send killing waves through nesting areas.

As serious a threat as they are, cyclones are a friend to the Snowy Plovers. Here's how: barrier systems endure the brunt of the storms and waves from the Gulf, thereby protecting mainland coastal settlements and ecosystems. Cyclones bring hungry waves and unusually high tides, known as storm surges. Where barrier islands are narrow, storm surges can break across dunes and flatten the landscape. While dune plant and beach mouse populations suffer, the open, flat expanses created by the storm surge become prime nesting habitats for Snowy Plovers and other birds. What's more, storm surges will sometimes cut a channel across a peninsula or barrier island to create smaller islands that, if free of mammalian predators, become a paradise for breeding birds.

This happened on Dauphin Island when hurricanes Ivan and Katrina struck in 2004 and 2005, respectively. Dauphin Island is Alabama's largest barrier island and is the easternmost link in the Mississippi–Alabama barrier island chain. The cyclones overwashed the narrow western end of the island, stripped away vegetation, and left behind an expanse of flattened sand just inches above sea level. Additionally, the storms severed off the westernmost 7 miles of the island. The new island and its overwash zone is now a haven for ground-nesting aquatic birds, including sandpipers, terns, gulls, skimmers, and Snowy Plovers.[2]

With the threat of cyclones and the encroachment of coastal development, why don't these species breed in different habitats in the region? They have evolved bodies and behaviors making them well suited for breeding on the barrier system, but nowhere else. Nests are simple depressions just yards away from abundant food. Nesting is often in loose colonies to ensure more eyes and angry beaks are available should predators approach. Breeding is nearly simultaneous and timed to match peak food availability, just before the onset of cyclone season. And once a breeding area is colonized by vegetation and dunes, these versatile coastal birds will choose the best nesting site from what is available.[3] So, despite millions of years of sea level change, climate swings, and cyclones, the benefits of the barrier island have always outweighed the costs for the Snowy Plover and other species.

Barrier Ecology and Unsolved Mysteries
NatureServe describes four ecological systems for the barrier system, and there are places on the Alabama coast where you can see all from a single point. There's the Florida Panhandle Beach Vegetation, the narrow zone of sparsely vegetated habitat from the primary dunes to the shoreline. Its plants are hardy and fast growing, like Gulf Bluestem and Sea Oats, two grasses that play a crucial role in building and maintaining dunes after cyclones. Adjacent to this is the East Gulf Coastal Plain Dune and Coastal Grasslands, known also as the secondary and tertiary dunes of the barrier system. Its habitats include lightly vegetated sand expanses, meadows of grasses and wildflowers, and scrub dunes with stunted trees sheltered in their lee. This ecosystem's plants employ a great range of strategies to cope with limited freshwater, nutrient-poor soils, periodic exposure to salt spray, and erosion from storms (see chapter 2). Behind the scrub dunes is the East Gulf Coastal Plain Maritime Forest. These forests are safe from most storm surges and consequently support more species than any of the other barrier system habitats. For the most part, they grow on old dunes and swales.

Sand Pine, Live Oak, and Sand Live Oak top the high sandy areas, while Slash Pine and a thick understory of Saw Palmetto and shrubs fill in the low zones.

The Gulf Coast's maritime forests are globally famous among ornithologists and birdwatchers. Each fall and spring, dozens of land bird species cross the Gulf as part of a seasonal migration between tropical overwintering habitats and North American breeding grounds. In fall, maritime forests are the last stop for many of the migrants before they embark on a 600-mile (966 km) nonstop overnight flight. In spring, birds arriving after crossing the Gulf usually bypass the maritime forests and head for the lush floodplain forests farther inland. Occasionally, spring migrants encounter rain or strong winds during the crossing and become exhausted. Upon arrival, desperate multitudes drop from the sky and into the maritime forests. During these events, known as fallouts, migrants rest briefly then begin vigorously feeding to replenish their strength. Birdwatchers, or birders, yearn to witness fallouts, for one can see several thousand birds of over a hundred migrant land bird species in a day, including orange and black orioles, golden warblers, shining blue buntings, and dazzlingly red tanagers. The birds are often so hungry they allow people to observe them from just a few feet away. The maritime forests of Dauphin Island and the tip of the Fort Morgan Peninsula are legendary hot spots for observing migrating land birds and fallouts.

Because it is hot, dry, and surrounded by salt water, it seems odd that the barrier system hosts an abundance of freshwater wetlands. These wetlands vary in configuration but all belong to the Southeastern Coastal Plain Interdunal Wetland ecological system. The larger ones, like Shelby Lake within Gulf State Park, provide habitat for freshwater plants and animals, including ducks, fishes, turtles, and even American Alligators. It's generally accepted Shelby and similar lakes are former streams flooded as sea levels rose after the last ice age and sealed off from the Gulf by drifting sands. With the Gulf just a few hundred feet away, storm surges can infuse lakes with salt water and temporarily create brackish conditions.

Other areas of the barrier system contain a geological and ecological marvel—great expanses of narrow dune ridges and wetland swales

Painted Bunting, one of many land birds that rest and refuel on the barrier system during migration. Courtesy of Jack Rogers.

aligned in long alternating strips. Ridges can be several yards taller than adjacent swales, and the width of swales and ridges ranges between a few feet and several yards. Some prominent ridges run for miles while others merge into one another. Known to geologists as strandplains, the formations are strikingly visible in aerial photographs of Orange Beach and the Fort Morgan Peninsula. Grasses, wildflowers, and trees populate the ridges while wetland shrubs, sedges, and grasses occupy the swales. Nowhere else in the Southeast can you find vastly different ecosystems—desertlike dunes and soggy wetlands—in alternating, linear strips.

The strandplains' origins are a scientific mystery. Most geologists agree they are primary dune ridges that formed successively as sea levels dropped from their most recent high-water mark. However, there is energetic debate over when they appeared. The conventional view is they developed tens of thousands of years ago as sea levels dropped during the onset of the last ice age. Other geologists believe sea level after the last ice age attained a position 6–7 feet (2 m) higher than today before retreating to its current level. This Holocene highstand, as geologists call it, would have eroded away all landforms below the high-water mark. If true, then all lands and ecosystems of the Southern Coastal Plain within this zone, including the strandplains, have developed in the past few thousand years. Geologists in both camps cite evidence that includes interpretations of coastal topography and the age, position, and composition of sediments from sample cores from along the Gulf Coast. This mystery will probably be unresolved for years to come.

SAND STARVED

Though barrier system habitats and species have lost ground to coastal development, a more profound threat is that portions of the barrier system may completely vanish. Over the past 160 years, 35 percent of the Mississippi–Alabama barrier island chain has disappeared. This loss is deeply troubling given the importance of these islands to biodiversity, culture, and the protection of estuaries and the mainland.

Alabama's barrier systems survive because large southeastern rivers deliver sand—fragments of the ancient Southern Appalachian Mountains—to the coast. Most was deposited long ago, but some still arrives in the present era. This sand moves back and forth between the islands and shallow offshore areas during seasonal weather events like winter storms and cyclones. More influential is the westward drift of sand along the northeastern Gulf Coast in the nearshore current. This current has delivered the sand maintaining the modern barrier system since its formation several thousand years ago. Much of this sand originates from an extensive offshore deposit in the northeastern Gulf.

However, the situation is changing. In this new era of anthropogenic climate change, sea levels are rising and cyclone intensity (not frequency) has increased. It's a nasty combination for the barrier system, as seen in the damage Dauphin Island incurred during the first decade of the 21st century. Yet, sea level rise and cyclones are not the primary cause of the shrinking islands.

Right: Strandplain dunes and swale at Bon Secour National Wildlife Refuge, Baldwin County. Photo by R. Scot Duncan.

Below: Stark evidence of land loss along the Fort Morgan Peninsula, Baldwin County. Photo by R. Scot Duncan.

Instead, the islands are being starved of sand. Starting in the late 1800s, shipping channels were dredged and maintained between the islands, including the Mobile Ship Channel between Dauphin Island and Fort Morgan. These channels trap drifting sands, and when re-dredged, the spoil is hauled to disposal areas or released far offshore. As a result, sand eroded from the barrier system is not being replaced. Were it not for these channel-dredging practices, the barrier islands might be enduring rising sea levels and stronger cyclones just fine.

What will happen to Dauphin Island and other lands of the barrier system in the next century is unclear. The armoring of the island's shoreline with seawalls and jetties has minimized erosion on its eastern end. Perhaps more influentially, two small, offshore islands (Pelican and Sand) have recently migrated north and merged with the eastern end of Dauphin Island. This addition shelters the island's eastern end from cyclones and supplies it with much-needed sand. Dauphin Island's western end, however, struggles to survive.

Nursery Waters

North of Dauphin Island and adjacent to the mainland is Grand Bay, one of several shallow estuaries of the Mississippi Sound. Along the bay's shore is an expanse of wetland wilderness with tremendous biological productivity. More than 5 uninterrupted miles (8 km) of salt marsh extend from the Point aux Pins peninsula to the Mississippi state line, and another 5 miles continue beyond to Pascagoula, Mississippi. This coastline is true wilderness, if wilderness can be defined by the certainty of finding human solitude. Though nearby coastlines are intensively developed, one can roam this area for days without seeing another human.

The marsh is part of the Mississippi Sound Salt and Brackish Tidal Marsh ecosystem, another component of the Gulf Barrier Islands and Coastal Marshes ecoregion. Compared with the surf-pounded storm-torn barrier system, however, the salt marsh ecosystem is much calmer. The barrier system protects the marsh from the Gulf's waves and high salinity. These less-turbulent waters allow fine sediments—brought to the coast by large rivers—to settle out and build marshland. Salt marshes along the Alabama coast are also distributed along the pro-

The vast salt marsh of Grand Bay, Mobile County. Courtesy of Hunter Nichols/Hunter Nichols Productions.

tected northern margins of the barrier system and in sheltered portions of Perdido Bay and lower Mobile Bay.

Though appearing serene, salt marsh ecosystems are in constant motion due to tidal influence. Powered by several forces, especially variation in the gravitational pull of the moon, the tides flood and drain Alabama's marshes once a day on average. When the tide is high, saline water and fine sediments arrive, invading the marsh via an intricate network of convoluted tidal creeks. These channels begin several hundred feet across, but split repeatedly until each branch is a narrow cleft between the rushes. When the tide is low, the salt water retreats and the channels collect freshwater seeping from the mainland. Marsh species arrange their lives, from feeding to spawning, around this rhythm. During the spring tides that occur with a new or full moon, most of the marsh floods with salt water and then is completely emptied 12 hours later. During the weaker neap tides occurring when the moon is half full, lower elevations remain flooded while upper elevations stay dry.

These patterns of tidal flooding create obvious vegetation zones.

The marsh's upper margin is bounded by an irregular edge of pine woodland. Near this boundary grow a dozen or more marsh plants. This is the least stressful of the marsh zones since less salt water arrives, and there is continual access to freshwater draining off the land. Interspersed in the high marsh are barren areas known as salt pans. These are level and sparsely vegetated areas flooded only during spring tides. Water trapped here evaporates and leaves behind a thin layer of salt. Only halophytes—plants tolerant of extreme salt exposure—live in the salt pans. Halophytes have tricks to discard the salt taken up with soil water. Some have glands to pump out excess salt. Succulents, including Chickenclaws and Saltwort, store rainwater in swollen leaves and stems to dilute the salt absorbed during dry weather.

At lower elevations Black Needlerush dominates. Needlerush tolerates a wide range of tidal flooding and occupies most of the marsh. Its tall, dark-green and sharply pointed stems give the salt marshes of the northern Gulf Coast their characteristic hue and texture. Smooth Cordgrass, a hardy plant that can withstand incoming waves, high salinities, and daily immersion, populates the seaward and lowest margins of the marsh. The bright green blades and short height of cordgrass contrast markedly with the needlerush.

Offshore from the salt marsh is the Northern Gulf of Mexico Seagrass Bed ecosystem. Shallow meadows of Shoalgrass, Turtle Grass, and Widgeongrass often grow just beyond the cordgrass margin, but only where water clarity permits photosynthesis. None are true grasses, but all have grasslike stalks and leaves, and expanses of them resemble underwater meadows. Like marshes, seagrass meadows are critical habitats for many marine creatures. Some live there permanently, including the Chain Pipefish, which disguises itself as a blade of seagrass. Others are salt marsh creatures retreating into the seagrass during low tide.

Though trees are absent and herbaceous plants reign, salt marshes are among the most productive ecosystems on the planet. Fed daily by nutrient- and sediment-rich brackish water, sheltered by the barrier system, and bathed in sunlight, needlerush and cordgrass grow exceptionally fast. Their stems die quickly but are soon replaced, which is why the marsh's high productivity isn't apparent. Microbes and small invertebrates like young shrimp rapidly digest the dead stems of marsh plants. As in river ecosystems, these animals and their waste become food for other marsh creatures. Other animals graze on photosynthetic microbes and algae whose populations surge during high tide.

Were we to visit the Grand Bay salt marsh at low tide, we'd find wildlife hastily feeding before the tidal flooding begins. Among the stalks of needlerush and cordgrass, snails known as Marsh Periwinkles, sometimes numbering several dozen per square yard, patrol the mud and stalk bases, harvesting small plants stranded by the retreating tide. Small flocks of sandpipers probe expanses of exposed mud, or mud-flats, for invertebrates, while overwintering ducks dive after minnows trapped in the deepest channels. Graceful herons stalk minnows in narrow tidal rivulets, but the Reddish Egret dashes about with extended wings to corral and catch small fishes. Low tide is our best chance to see a Clapper Rail, one of the few birds living only in salt marshes. Shaped like small chickens plumaged in rust and brown, the rails have long legs and huge feet for traversing the soft mud, and long, decurved bills for pulling small creatures from their burrows. Rails are exceptionally shy and usually remain hidden, but some clappers sneak onto mudflats to feed at low tide. Raccoons capture crabs, fishes, and other tasty snacks on mudflats and in the shallows. Theirs are always the most abundant prints on the game trails weaving through the marsh.

Along larger tidal channels and outer margins of the marsh are exposed, jagged clusters of Eastern Oysters. These are small versions of the oyster reefs in deeper waters of coastal estuaries. Reefs only develop where salinity is low and there is a firm substrate for shell attachment. If salinity is too high, oysters become prey to American Oyster Drills, a predatory snail. Encouraged by increased salinities, the drills have decimated reefs along the Alabama coast in recent years. This has been a loss for biodiversity (and the seafood industry), for the reefs provide habitat and shelter for dozens of aquatic creatures. Oysters have few terrestrial predators, but among them is one of the most extravagant of coastal birds, the American Oystercatcher. With a bronzy-brown back, black head and neck, and white belly, the feathering of this large shorebird is striking, but not gaudy. Rather, it's the oystercatcher's long brilliant-red bill and legs that dazzle. Oystercatchers wield their bills in various ways. Some stab into oysters whose valves were left slightly open. Others hammer through weak points where the valves come to-gether. Still others roam the mudflats and pull clams and worms from their burrows.

The mudflats are the parade ground for armies of tiny fiddler crabs, whose exoskeleton armor is decorated with purples, golds, and greens. During low tide they emerge from burrows to march across the mud-flats by the thousands. They sift through the topmost mud for stranded plankton and leave behind a trail of processed mud pellets. There is also fraternization. Males spend much of their time waving a grossly en-larged claw to attract females. When it's time to breed, females choose a mate from the displaying males, then hide in his burrow until it's time to release the fertilized eggs.

The rising tide brings dissolved oxygen, silt, nutrients, and hoards of marine creatures eager to feed or take refuge in the marshland. Per-iwinkles migrate up plant stalks hoping to stay beyond the reach of hungry Blue Crabs, while fiddler crabs retreat into their foxholes be-fore big fishes arrive. For other marsh creatures, high tide is chow time. Worms, shrimps, crabs, and clams hidden in waterlogged burrows during low tide resume feeding. Small fishes are among the first tidal migrants to arrive, rushing up the channels several hundred at a time. Armed with nets, one could easily capture a dozen species or more. They hurry because there is limited time to feed before the tide ebbs and because predators like Speckled Trout, Redfish (Red Drum), and Southern Flounder are not far behind them. During high tide, they will hide within the cordgrass and needlerush while the predators pa-trol the channels.

These larger fishes attract apex predators. Small sharks enter larger tidal channels during high tide. Ospreys, hawks that dive into the water to capture large fishes, will hunt the channels, where their prey is con-centrated. Bald Eagles patrol the marshes for fishes, too, but resident Ospreys often chase them away. River Otters hunt fishes in larger tidal creeks, and Bobcats will slink from the flatwoods to hunt rails and rab-bits.

Perhaps the rarest species of the salt marsh is the Mississippi Dia-mondback Terrapin, the only southeastern turtle thriving in brackish water. Terrapins are part-time scavengers—dead fish are irresistible—

Diamondback Terrapin, perhaps the rarest species of the salt marsh. Courtesy of Andrew T. Coleman/Institute for Marine Mammal Studies.

but mainly use their powerful jaws to crush armored prey like periwinkles and crabs. Southeastern populations declined precipitously due to habitat loss and overhunting. Newer threats include drowning in crab traps and high rates of nest predation.

Predictable tidal change governs the daily and seasonal ecology of the salt marsh, but over longer time periods marshes are a messy compromise between destruction and creation. Rising sea levels disrupt ecological zones in the marsh, and waves erode the marsh's outer perimeter, especially when the barrier system is breached and Gulf swells churn across the sound. Rivers come to the rescue by supplying fine sediments that help marshes rebuild. Typically, marsh ecosystems survive changing sea levels because their species can quickly adjust.

Grand Bay's salt marshes are retreating, and they have been for a long time. Until sometime in the past few hundred or thousand years (it's unknown), the Escatawpa River emptied into Grand Bay and supplied its marshes with sediment. Ever since the larger Pascagoula River system captured the Escatawpa, the Grand Bay marshes have been losing ground. Furthermore, the fragmentation of Dauphin Island in recent centuries now allows Gulf swells to batter the marsh's outer perimeter.[4] Recent sea level rise triggered by anthropogenic climate change is also forcing erosion. Though geologists have carefully mapped these changes, the evidence is starkly apparent to any visitor. Pine islands scattered throughout the marsh are drowning. Trees along the seaward margin are dying from salt exposure, while sun-bleached trunks of dead pines rise above the needlerush to mark the graves of former islands.

Archaeologists have uncovered evidence of marsh loss dating back several hundred years. Sprinkled throughout the Grand Bay marshes are mounds of oyster shell. Topped with oaks, cedars, and other upland plants, they protrude awkwardly from the flat marshlands. With the oyster shells are broken conchs and whelks; bones of fishes, mammals, and turtles; charcoal; and pottery shards. These mounds are the middens of Native American settlements from the recent Protohistoric to the older Archaic cultural periods. Archaeologists were surprised to find beneath some mounds a basal layer of shells from a clam known as the Common Rangia. Rangia need water with salinity levels much lower than waters inhabited by oysters. Some interpret this as evidence these encampments were initially on inland freshwater creeks where rangia were abundant. As the coast subsided or sea level rose, or both, salt water and marsh species invaded the creeks and converted swamps and pine flatwoods to salt marsh. Later generations of inhabitants shifted their diet to keep up with these changes.

Just like these Native Americans, modern residents and visitors to the northern Gulf Coast enjoy the seafood species nurtured in oyster reefs, seagrass meadows, and salt marshes. These ecosystems are essential to sustaining the coastal economy and way of life. Salt marshes also protect people, property, and the economy by slowing and absorbing storm surges brought by cyclones. The critical question is whether the marshes can keep up with the sea level increases predicted for this century and beyond. While these ecosystems have migrated successfully in the past, today there are roads, farmlands, and settlements in their path. No one knows for sure what will happen next.

Fire, Floods, and Flesh-Eating Plants

Beyond the upper rim of the Grand Bay salt marshes and most other estuarine habitats along the Alabama coast is a level IV ecoregion of fire and flooded soils known as the Gulf Coast Flatwoods. Its ecosystems occupy a nearly level plain of clay and sand deposited in the late Tertiary and Quaternary. Like the other ecosystems of the Southern

Coastal Plain, these ecosystems must cope with dramatic environmental change. Although cyclones flatten large tracts of flatwood trees and inundate them with storm surges, more than anything else, seasonal wildfires and flooding shape these ecosystems and sustain their rich biodiversity.

As the name suggests, the prevailing ecosystem is the flatwoods. Technically known as the East Gulf Coastal Plain Near-Coast Pine Flatwoods, these are woodlands (lightly wooded habitats with abundant openings between tree canopies) founded on acidic sandy soils underlain by a claypan—a compact clay layer preventing groundwater from infiltrating deeply.[5] The claypan creates a perched water table that saturates or floods flatwood soils much of the year. Some of the storms bringing this water also bring fire. Violent summer storms pound the ecoregion with rain and lightning, the latter sometimes igniting wildfires. Historically, wildfires burned pine flatwoods every one to four years.

Few trees can endure a fire regime this arduous, so only two fire-tolerant pine species make up most of the flatwood's trees. Slash Pine prevails on the wettest soils, and Longleaf Pine dominates higher, better-drained areas. Both species have traits helping them survive fire (e.g., fire-resistant insulating bark), but wildfires are so frequent that only a few do. Thus, adult pines are widely spaced and much of the understory is below open sky.

Fire is friendly to flatwood flora. Abundant light, water, and warm temperatures promote a thick, diverse layer of grasses and wildflowers that readily burn. Left alone, grasses will monopolize sunlight, space, and nutrients at the expense of wildflowers. However, accumulations of dried grass stems are the main fuel for wildfires. After a fire, competitively weaker wildflowers quickly sprout, mature, and reproduce before grasses regain control. Ultimately, fire is also a friend to grasses. Young shrubs from nearby wetlands where fires are rare continually attempt to colonize flatwoods. Wildfires kill invading shrubs, keeping the flatwoods open. When wildfires are scarce, impenetrable shrub thickets form and shade out the understory within a few years. These thickets do not readily burn. Only intense fires during warm-season droughts can clear them and allow herbaceous plants to reestablish. In the current era, wildfires are rare due to habitat fragmentation and fire suppression by humans. Flatwoods and other fire-dependent ecosystems now depend on managed burns.

Alongside the pine flatwoods are East Gulf Coastal Plain Savannas and Wet Prairies. Savannas are grasslands with scattered trees, while prairies are grasslands lacking trees. These ecosystems and flatwoods intermingle and it's often difficult to draw clear lines of demarcation. In many ways, including soil, flooding, fire frequency, and flora, the flatwoods and grasslands are very similar. Pines are the common savanna trees, but wetter savannas can support Pond Cypress, another fire-tolerant conifer. It isn't understood why trees are scarce in the grasslands. Perhaps fire frequency is too high for many pines to establish. Subtle differences in soils or flooding patterns may play a role. Maybe grasslands are former flatwoods that lost their trees to hurricanes or logging (old roadbeds and ditches scar these landscapes).

Savanna and wet prairie in Forever Wild's Grand Bay Savanna Tract, Mobile County. Photo by R. Scot Duncan.

In select places, the flatwoods ecoregion is home to flesh-eating plants, especially in Southern Coastal Plain Herbaceous Seeps and Bogs. These wetlands are embedded within the mosaic of flatwoods, savannas, and prairies and share many ecological characteristics. What's different is the greater volume of freshwater seeping in from adjacent higher elevations. A claypan traps freshwater here for most of the year, keeping the soils waterlogged or submersed. These acidic waters are lacking in essential nutrients like nitrogen. Several plant taxa have evolved strategies for capturing and digesting animals to obtain these nutrients.

Pitcher plants—the most famous and spectacular of these carnivorous plants—are indicator species for this ecosystem. Pitcher plants have tubular leaves that can be tall, squat, or reclined. The leaf grows a hood protecting the tube from rain and secretes nectar to attract flower-seeking insects. The hood and upper tube can be red, yellow, or white to attract flower-seeking insects to their doom. As naïve insects explore the hood, some wander into the tube's interior, where they encounter stiff, downward-pointing hairs guiding them lower and preventing escape. Waiting below is a pool of water and digestive enzymes into which trapped insects eventually fall. Unable to scramble out, the hapless insects drown. The plant absorbs nutrients released as the insects disintegrate.

The abundant water and dense groundcover of the flatwoods ecoregion support dozens of amphibian and reptilian species. Most of them, plus the ecoregion's mammals and birds, also live in the Southeastern Plains level III ecoregion to the north. One exception is the Mississippi subspecies of the Sandhill Crane. This majestic bird stands about 3.5 feet tall (1.1 m) and resembles a grayish-brown heron with a brilliant red cap. Small flocks once stalked small animals in Alabama's coastal prairies, savannas, and bogs, but habitat destruction has them mostly confined to the Mississippi Sandhill Crane National Wildlife Refuge. Perhaps Alabama will support a population in the near future. Alabama's Forever Wild program and the Alabama Chapter of the Nature Conservancy (TNC) have preserved large tracts of the Gulf Coast Flatwoods ecoregion in the areas above Grand and Perdido bays. The tracts harbor tremendous species diversity but desperately need the return of frequent fires. With careful habitat restoration, the Mississippi Sandhill Crane could once again hunt these ecosystems of fire and flood.

Bays within Bays

With an area of 413 square miles (1,070 km²), Mobile Bay is the fourth-largest estuary in the United States. The bay ferries freshwater, sediment, and nutrients from the Mobile River Basin to the Mississippi Sound and Gulf of Mexico. The bay is perpetually brown because its waves, tides, and currents prevent fine sediments from settling. Sands can settle, however, which is why Mobile Bay's eastern and western shores are lined with beaches.

These beaches are interrupted where creeks and small rivers reach the bay. Miniature bays form here, complete with scaled-down deltas and salt marshes. The largest is Weeks Bay, home of the Weeks Bay Reserve, whose interpretive center enables visitors to explore salt marshes, pine flatwoods, and a seepage bog. If we boated up one of these side bays, we'd observe the bay's waning saltwater influence, like the ceding of Black Needlerush to Jamaica Swamp Sawgrass and other freshwater grasses. Farther up we'd encounter a new ecosystem, the East Gulf Coastal Plain Tidal Wooded Swamp. Populated by Bald Cypress, tupelo, and Sweetbay, the swamps endure an unusual phenomenon for freshwater wetlands, the daily rise and fall of tides. The headwaters are so close to Mobile Bay that during a rising tide the creeks reverse course. High in the headwaters there are freshwater tides, but downstream the tidal swamps are briefly exposed to saline water. Were it not for the plentiful rains keeping the creeks flowing, salt exposure would turn these swamps into salt marsh.

These tidal creeks are habitat for West Indian Manatees. Manatees are huge aquatic mammals with large rounded tails and paddlelike forelimbs. Typical adults weigh 1,000 pounds (454 kg) and measure 9 feet (2.7 m) in length. They are calm, slow-moving, and gentle animals that munch on aquatic plants in nearshore waters. Until recently, scientists knew little about Alabama's manatees. Past sightings were assumed to be strays from Florida, where the main US population resides. However, through the Mobile Manatees Sightings Network, an effort led by the Dauphin Island Sea Lab, dozens of sightings are logged yearly.

West Indian Manatee, now known as a regular inhabitant of Alabama's coastal ecosystems. Courtesy of Alan Cressler.

This and other studies have confirmed that Alabama hosts a small population favoring the small rivers along Mobile Bay's shoreline and the lower Mobile River Delta. Manatees need warm waters, so during the colder months, most of Alabama's manatees migrate to Florida. How long manatees have populated Alabama is unknown, but it is clear manatees have profound difficulty coping with coastal development. Boat collisions are the greatest source of injury and death—the backs of far too many bear horrendous scars from boat propellers. With only about 3,000 in the United States, manatees are yet another endangered coastal species.

JUBILEES: GOOD OR BAD?

Several times a year on hot summer nights, lethargic schools of flounders, catfish, eels, shrimps, crabs, and stingrays swarm the shallowest waters of eastern Mobile Bay. In one square yard a dozen or more creatures, including prized seafood species, will crowd together. Locals watch for these events and quickly spread the word—but only to friends—when they occur. People arrive with gigs, dip nets, and cast nets and fill their coolers with their catch. These famed but mysterious events are called "jubilees" in reference to the excitement associated with their occurrence. Mobile Bay is one of the few places in the world where jubilees occur. But what causes this strange phenomenon?

After years of study, scientists have some answers, though aspects of jubilees remain unexplained. Most jubilees occur on a rising tide in the predawn hours of August and September as animals attempt to escape hypoxic (poorly oxygenated) water. Hypoxic water accumulates in the bay's deepest depths where water has no atmospheric contact, but high tides push these waters near the shallows. An eastern wind associated with many jubilees seems to push surface water westward and pull hypoxic water up against the eastern shore. Elsewhere, such a breeze would oxygenate the surface waters, but the high bluffs lining the eastern shore cause this wind to bypass the bay's eastern margin. Jubilees probably happen late at night due to phytoplankton inactivity. During the day phytoplankton oxygenate water while photosynthesizing. During the night, when photosynthesis stops, oxygen levels steadily drop and bottom out in the predawn hours. Finally, jubilees are a summer phenomenon because warm water cannot hold oxygen well.

During jubilees, many species migrate to other portions of the bay, but slow-moving bottom dwellers become trapped against the shoreline. Jubilees are short-lived, for a shifting wind or turning tide will reoxygenate nearshore waters and allow the refugees to return to deeper waters. Jubilees are another example of the chaotic nature of the Southern Coastal Plain, but they rarely cause widespread mortality.

Are jubilees natural? Or are they symptoms of an unhealthy bay? The answer may be "yes." The first published account of a jubilee in 1867 refers to it as an annual event. Because this was long before major alterations to Mobile Bay and pollution became epidemic, it's generally accepted that jubilees are natural. However, dredging of deepwater shipping channels created unnaturally deep pockets in Mobile Bay where hypoxic waters accumulate, especially in summer. Additionally, excessive nutrients in river water (from fertilizer runoff, livestock waste, and treated sewage) entering the bay trigger summertime algae and plankton blooms that eventually create large hypoxic areas in the bay known as dead zones. Hypoxic waters have increased in recent decades and are almost certainly contributing to the frequency and strength of jubilees. No official system is in place to document jubilees, but reports of them to the local media have increased in recent years.

BUILD IT AND THEY WILL COME

There are dramatic seasonal changes in the number and species of shorebirds, seabirds, and waterfowl in Mobile Bay. During winter, many bird species populate all corners of the bay and adjacent waters. Nevertheless, by late spring most—including loons, ducks, and cormorants—have left to breed in the north. Their departure might seem ill-timed since the estuaries vibrate with baitfish and invertebrate prey during the summer. But, as with Snowy Plovers, finding safe nesting along the northern Gulf Coast is difficult.

This scarcity of breeding habitat had long been recognized when Wilson Gaillard, a Mobile dentist, proposed an island be built in Mobile Bay for breeding birds using spoil created from digging the Theodore Ship Canal along Mobile Bay's western shore. The proposal was controversial since the canal and the island would destroy habitats. Nevertheless, Gaillard's idea took flight and the 1,350-acre (5.5 km²) island was completed by 1982.

At the time, many coastal bird species were struggling after devastating population crashes caused by DDT, though the pesticide's use in the United States had been banned 10 years earlier. Brown Pelicans were among the hardest hit and were on the US Endangered Species List. In 1983, for the first time in recorded Alabama history, four pairs of Brown Pelicans nested on Gaillard Island. More came the next year, and the next, and the next. By the mid-1990s, over 10,000 pelicans bred on the island each summer. The rookery's success contributed to the recovery of pelicans in the eastern Gulf, and in 1985 the species was delisted in Alabama and Florida (it is still endangered in Mississippi, Louisiana, and Texas). The nesting pelicans have plenty of company. Thousands of Laughing Gulls and Royal Terns and over 10 other aquatic bird species nest on the island. For several, Gaillard Island is their only breeding colony on the Alabama coast. Perhaps this is because, unlike the storm-ravaged barrier islands, Gaillard Island is a stable destination off-limits to people. In any case, the island is a chief reason that coastal Alabama has more aquatic birds now than at any time in the past half century.

RESTLESS REFUGE

Our coastal wanderings now bring us to the top of Mobile Bay, where we meet the second-largest wetland in the United States, the Mobile River Delta (or Mobile-Tensaw Delta). Maintained by the processes described in chapter 7, the delta provides 250,000 acres (1,012 km²) of rivers, creeks, bays, bayous, sloughs, swamps, oxbow lakes, islands, marshes, and bottomland forests.[6] The delta begins where the Alabama and Tombigbee rivers join to form the Mobile River 30 miles (48 km) north of Mobile Bay. Within 5 miles, the Mobile loses control of its payload and spins off the Tensaw River to the east. Both split into other rivers—the Spanish, Apalachee, and Blakeley—just miles before reaching the bay. The delta's wildlife is abundant and over a thousand species find refuge here. With more than 100,000 acres (405 km²) of protected public land, the delta offers lifetimes of adventure for fishers, hunters, and naturalists.

The marshland at the delta's bottom is a battlefield between the rivers and the bay. This example of a Mississippi Sound Fresh and Oligohaline Tidal Marsh (*oligohaline* means low salinity) is a 6-by-6-mile (9.7 by 9.7 km) labyrinth of bays, channels, and freshwater marsh. The ecosystem is marvelously productive despite and because of strong seasonal changes. In winter and early spring, when river water overwhelms the delta, terrestrial animals flee to the uplands, and freshwater animals seek shelter in backwaters. These floods bring fresh sediments to the marshes. By summer, the floods subside and marsh plants grow profusely. Marshland expands as plants colonize the new sediments. The abundance of nutrients and freshwater supports a flora—including irises, water lilies, bulrushes, and arums—far more diverse than any salt marsh.

By midsummer the delta sheds its excess freshwater and salt-tainted tides invade the marshland farther each day. Terrestrial animals return to drier areas, while marine animals ride tides into the marshes and shallow bays to feed, seek shelter, and reproduce. The drought of late summer and early fall allows salinities to climb so high that many marsh plants established just months before die back. This, plus wave erosion driven by the summer sea breezes, returns some of the new marshland to the bay. After autumn's first cold snap, marine crabs, shrimps, and fishes head into the bay to wait out the coming floods. Freshwater spe-

Bald Cypress at sundown in the Mobile River Delta. Courtesy of Hunter Nichols/Hunter Nichols Productions.

cies then arrive from higher in the delta to occupy vacated habitats. Whether the marsh or the bay wins more area varies from year to year and depends upon the volume and duration of the seasonal flooding.

Farther up the delta the marshland transitions into dense tidal swamps like those up the small rivers perforating the bay's margins. Some of the swamp-lined channels are sloughs, former river channels with slow currents. Some sloughs lead to oxbow lakes—former river bends left behind when a channel changed course. Swamp and Water Tupelo and Bald Cypress populate the swamps surrounding sloughs and lakes. Bald Cypress is the archetypal tree of the southern swamp. It has a wide-buttressed base and a platform of spreading roots for stability on soft sediments. Cypress roots sprout clusters of low upright knobs emerging above the waterline. Known as cypress knees, their function is unknown to scientists.

The delta's cypress population is having a rough time. Today, cypress stands make up less than 20 percent of the delta's forests, but a century ago their extent was several times this. Settlers valued cypress' soft, rot-resistant wood, and they logged southeastern swamps heavily in the decades after the Civil War. Historical photographs and measurements of stumps revealed during droughts show the original trees were massive. Bases greater than 9 feet (2.7 m) in diameter were common and many were over a thousand years old. Unfortunately, slow-growing cypress often failed to reestablish. Instead, fast-growing tupelos now dominate the swamp canopy. Young cypress need full sun and several seasons of low water to establish, but tupelo capture most light and water levels are kept artificially high during the dry season due to release of reservoir water upstream.

At higher elevations, swamps give way to forests inundated for weeks at a time in winter and spring. These are the East Gulf Coastal Plain Large River Floodplain Forests. The roster of tree species varies from place to place, dependent on elevation and flooding frequency. Each flood drops another layer of nutritious sediment that maintains the floodplain and promotes rapid tree growth. The natural levees lining many channels have some of the greatest tree diversity. Tree limbs overhanging smaller channels are often draped with Spanish Moss, Resurrection Fern, Green Fly Orchid, and other epiphytes hinting of the tropical influence on the region's flora. Impassable thickets of Blue Stem Palmetto in the gloomy interior augment the jungle theme. Here and there, extra sediment has piled up and supports a Southern Coastal Plain Hydric Hammock populated by oaks, magnolias, elms, and other vegetation preferring less flooding. The high productivity attracts wildlife, and as water levels drop, floodplain forests become crisscrossed with game trails furrowed with the prints of White-tailed Deer, feral hog, Northern Raccoon, and Virginia Opossum. Within this fertile ecosystem, Native Americans established the Bottle Creek settlement and farmed corn (see chapter 5).

In recent centuries, the delta has provided safe harbor for people and wildlife. Since the demise of Bottle Creek, few peoples other than logging crews have used the delta on a regular basis. The unstable soils, flooding, intimidating wildlife, and hordes of biting and stinging insects have discouraged settlement in and along the margins of this and other large southern swamps. Even today's hardened hunters, anglers, and other adventurers give pause before entering the delta, for the tangled network of waterways writhing in unpredictable directions makes futile the use of traditional navigational aids. Southern folklore tells of how swamps have hidden hunted and marginalized peoples including rebel armies, pirates, bandits, smugglers, runaway slaves, moonshiners, hermits, and Native Americans. Biologists tell stories of how these swamps provide sanctuary for hunted and marginalized wildlife species escaping plows, guns, cars, boats, and bulldozers. One of these refugees is the American Black Bear.

Black Bears were widespread in the Southeast, but overhunting in the 19th century thinned their ranks and habitat loss in the 20th century hindered a rebound. The Mobile River Delta is one of the few places in the Southeast where they found refuge. Black Bears are large animals—adults weigh up to 350 pounds (159 kg)—requiring big territories.[7] Females need about 2,600 acres (11 km^2) of habitat while males can require over 15,000 acres (61 km^2). Bears prefer forests with shrub thickets in which they hide and forage for berries, nuts, roots, grasses, and the occasional small animal. Despite its large size, the delta supports only 50–100 bears. A major limitation on population growth may be the availability of large hollow trees for winter dens. Logging crews usually remove such trees to make room for younger trees that produce better lumber.

The delta is also refuge for reptiles whose populations have declined elsewhere due to hunting and habitat loss. The American Alligator is among the delta's 70 species of reptiles and is the ecosystem's apex predator. The delta was a stronghold for the species during its decline in the past century. Snakes are rare nearly everywhere due to the widely held and grossly misguided conviction that "the only good snake is a dead snake." However, snakes flourish in the delta. Such beauties as the Rainbow and Florida Redbelly snakes, and the overly feared Western Cottonmouth, are a few of the delta's most common serpents.

Among the delta's turtles are species, including snapping turtles and softshells, that have been hunted for centuries. The delta is refuge to them and other turtles including species of slider, cooter, musk, mud, and map turtle. Turtles are the delta's most frequently seen reptiles, owing to their habit of basking on logs and roots above the waterline. Some are herbivores, others eat snails and mussels, while others capture fishes. One, the Southern Black-knobbed Map Turtle, lives only in the delta and has a shell armored with black spikes.

The delta's most famous turtle is the Alabama Red-bellied Turtle, the official reptile of Alabama and one of its most endangered. It has a low carapace and black skin with yellow pinstripes and its underside is a gaudy red or reddish-orange. The species inhabits the bays and channels of the lower Mobile and Pascagoula river deltas. Its most immediate threat is access to and security of nesting sites. Like all aquatic turtles, the red-bellied scrambles ashore to bury eggs in nests that are afterwards unattended. Prime nesting habitats are beaches along river channels, but these are also popular with boaters and campers whose activities can disturb nesting females and destroy nests. Egg and hatchling predation by Fish Crows, Raccoons, feral hogs, and the Red Imported Fire Ant take a huge toll, as do vehicle collisions with females attempting to nest near roadways that cross the delta.[8] State agencies and environmental groups have installed low fences to keep nesting turtles off roadways.

Many birds with diminished populations find refuge in the delta. In the lower delta, every low tide reveals a mudflat cafeteria to herons, egrets, sandpipers, and plovers. These birds are particularly abundant during spring and fall when stopping for a quick snack during migration. In winter, other species take up temporary residence when aquatic habitats in the north are frozen over. Over 17 species of duck and other waterfowl overwinter in flocks numbering in the thousands. Visitors often spot White Pelicans, enormous birds with 9.5-foot (2.9 m) wingspans, flying in formation over the marshes or cooperatively herding minnows in the bays.

The most exquisite bird finding sanctuary in the delta is the Swallow-tailed Kite. Kites are sleek birds of prey known as agile and graceful flyers. The Swallow-tailed Kite is gleaming white with iridescent black wing feathers and a forked tail. They breed during the summer, spending their days diving and wheeling in pursuit of dragonflies or snatching small vertebrates from the forest canopy. Unlike most raptors, Swallow-tailed Kites nest and roost in loose colonies. After the young have fledged, they feed in large flocks and roost communally in the same trees year after year. By fall, kite flocks have left for the trop-

Swallow-tailed Kite, agile hunter of the delta's skies. Courtesy of Jack Rogers.

ics. Habitat loss has caused their extirpation throughout much of the United States, but they continue to breed in Alabama.

For the Florida Panther, Red Wolf, and Ivory-billed Woodpecker, the delta's size and wildness wasn't refuge enough. The extirpation of these species was one of the early signs the delta is struggling. River water is choked with excessive sediments, nutrients, and toxins. Floods cannot peak and wane naturally because water levels are regulated with upstream dams. And then there is the Highway 90 Causeway, an earthen dike across the lower delta at the head of the bay built with dredged bay sediments. The completion of the Causeway in 1927 ushered in a new ecological era for the lower delta. The Causeway became an important cultural resource for Mobile, hosting fish camps, restaurants, bars, and parks, but its construction destroyed hundreds of marsh acres. What's worse, it blocks the daily and seasonal exchange of water and migrating animals—including shrimp, crab, and fish larvae—between the bay and the delta.

Our relationship with the delta may be improving. There is more awareness of the delta as a valuable cultural and natural resource, and more people than ever are involved in its study and conservation. Logging is declining and the state's Forever Wild program protects huge tracts. There are even discussions about replacing portions of the Causeway with bridges to restore connectivity between the bay and delta. The delta's ecosystems and most of its species are adaptable and attempts to help them should be impressively successful.

THE LOST RIVER

Tucked away on the eastern border of Baldwin County is the Perdido River and its bay. *Perdido* translates to "lost" from Spanish. Local folklore claims the bay acquired its name because it was difficult to spot the bay's entrance from the Gulf. Today it is only lost by being overshadowed by its neighboring bays, Mobile and Pensacola, which are eight and four times larger, respectively. The Alabama–Florida border bisects most of the bay, then ascends the Perdido River and follows it to the very northwestern corner of Florida.

Mobile and Perdido bays share a similar sequence of ecosystems from the Gulf to their deltas, but similarities stop there. Their size and shape differ due to the size of the rivers feeding them. Mobile Bay is a straight shot to the Gulf largely because its outflow overpowers the Gulf's nearshore current. Perdido Bay receives much less freshwater and it reclines eastward because its mouth has been pushed westward over millennia by the nearshore current. Whereas the Mobile River is a brownwater river, so named for the suspended sediment it always carries, the Perdido River is a blackwater river typical of the coastal drainages whose formation and ecology were described in chapter 7. Blackwater streams drain sandy terrains, carry few sediments and nutrients, and are acidic and stained a dark golden-brown. Consequently, Perdido Bay has stronger tidal mixing with the Gulf and it is clearer and more saline than Mobile Bay.[9]

The Perdido River lacks large floodplains. Instead, it has steep, white-sand banks occasionally interrupted with patches of the Southern Coastal Plain Blackwater River Floodplain Forest ecosystem. Common trees of this forest are Bald Cypress, Water Tupelo, and Atlantic White Cedar, the latter being rare in other southeastern ecosys-

White-sand banks and tea-colored waters of the Perdido River, Baldwin County. Courtesy of Hunter Nichols/Hunter Nichols Productions.

tems and characteristic of blackwater rivers. Beyond the riverbanks and floodplain forests is pine woodland we'll explore in the next chapter.

The small size of Perdido Bay and its river has protected their ecosystems. Because of its narrow outlet to the Gulf, shallow waters, and small river, early settlers largely overlooked Perdido Bay. Though logging, forestry, and pollution from paper mills have taken their toll, the Perdido is one of the healthier river basins along the northern Gulf Coast. Alabama's Forever Wild program and the Florida Chapter of TNC have begun safeguarding large tracts of marsh, swamp, and pine woodlands on both sides of the river. The history of Perdido Bay and its river illustrates a point conservationists know well: sometimes being lost is a good thing.

AHEAD: LANDLOCKED COASTLINES
A chaotic mix of cyclones, floods, fires, and tides continually reconfigures the Southern Coastal Plain's ecosystems. By provoking destruction

and renewal, these forces maintain biodiversity unique to the Southeast. How rising sea levels will affect these ecosystems and their human settlements is not entirely clear. We can be encouraged that most of the Southern Coastal Plain's species are resilient. They have been subjected to the coast's myriad disturbances and great climate swings in the past few million years. If we help coastal species respond to rising sea levels, we'll ensure that future human generations can enjoy their company and the natural resources they provide.

Next, we'll journey across a series of stranded coastlines known as the Southeastern Plains level III ecoregion. The lands comprising this terrain were once subjected to many of the same chaotic forces described above.

9
Southeastern Plains

REDISCOVERED FRONTIER

To the uninitiated, the Southeastern Plains might seem to be a dull, monotonous expanse. This, the largest level III ecoregion in Alabama, lacks the dramatic landscapes of Alabama's mountains and coast and instead hosts flat expanses of pine plantations, clearcuts, and croplands. Parks and nature preserves are scant, and most tourists hastily cross the region to destinations beyond. It is understandable that one might conclude the Southeastern Plains have little to offer students of biodiversity.

Nothing could be less true. This ecoregion is a biological Eden, unsurpassed in Alabama for its ecological and species diversity. What's more, it is one of the most biodiverse ecoregions in North America. There are dry tallgrass prairies, bogs brimming with carnivorous plants, and vast swamps. The ecoregion hosts the core range of the Longleaf Pine ecosystem, the most species-rich forest on the continent outside the tropics. Its diversity of wetlands and uplands supports one of the greatest assemblages of herbaceous plants, reptiles, and amphibians in North America. How is it possible that a landscape with no mountains, no coast, and a long history of intensive land use is one of the most biologically diverse regions of the temperate world?

The answer is provided by the fundamentals of geology, climate, and evolution. The Southeastern Plains is a series of ancient seafloors wrapping around the Southern Appalachian Mountains. The oldest were deposited in the Cretaceous when the Gulf of Mexico reached at least as far inland as the Fall Line, while the youngest were deposited in the Tertiary as the Gulf receded. These marine sediments became the various soil formations at the surface today.[1] Shallow reefs became limestone, barrier islands became sandhills, and deepwater deposits of microorganisms became chalk. Because each deposit has different nutrients, chemistry, drainage, and erodibility, each supports a unique assemblage of ecosystems and species. The whole ecoregion enjoys a near-tropical climate in which sunlight and rainfall are plentiful and winters are mild. The resulting high primary productivity sustains thriving ecosystems and multitudes of species. Finally, there were several episodes in deep time that promoted speciation and endemism in this ecoregion, including climate and sea level changes in the Tertiary and the absence of Pleistocene glaciation.

Consequently, in terms of biodiversity the Southeastern Plains ecoregion is anything but dull. Furthermore, it is a biological frontier still harboring mystery and surprise. Though it was one of the first regions explored by early naturalists, modern biologists are still deciphering its ecology and cataloging its species. As did its first explorers, we'll begin our trek through the Southeastern Plains near the coast and proceed inland until we reach Alabama's highlands.

Plains of Fire and Pine

Much of the story about the Southeastern Plains is a story about Longleaf Pine. Longleaf woodlands, technically named the East Gulf Coastal Plain Interior Upland Longleaf Pine Woodland, covered much of the lower Southeastern Plains, including its two southernmost level IV ecoregions, the Dougherty Plain and the Southern Pine Plains and Hills. Both regions are flat to slightly rolling plains of dry, sandy or clayey soils interrupted by wetlands. Like the pine flatwoods of the Southern Coastal Plain to the south, lightning-ignited fires often started in late spring would roam the landscape for weeks until extinguished by rain. Frequent burns kept the woodlands clear of young trees and shrubs but nurtured Longleaf Pine and a diverse understory flora. Dominant within the latter was Pineland Threeawn, commonly known as wiregrass, whose thin blades were a key fuel for wildfire. This grass was so abundant that the Dougherty Plain has been dubbed the "Wiregrass Region."

How can longleaf survive wildfires while other trees cannot? For starters, young longleaf don't grow up, they grow down. This may seem backwards for a tree needing direct sunlight throughout its life. For many years, young longleaf resemble a tuft of grass on the surface

Right: Longleaf Pine woodlands such as these once covered most of southern Alabama. Photo by R. Scot Duncan.

Below: Low-intensity wildfires keep Longleaf Pine woodlands open and brimming with species. Courtesy of Alan Cressler.

while their taproot drills downward in search of reliable groundwater. During this so-called grass stage, many young trees survive fire because their dense shield of long needles protects the tree's core from a passing fire's heat. Once they tap groundwater, young trees bolt skyward. They race against one another for a position in the canopy, but they also race against time. During this phase, longleaf lack the needle shield and don't yet have the thick bark that will insulate them from fire as adults.

Longleaf woodlands are constantly under threat of invasion. If fires become infrequent, the pines are crowded out by fast-growing, fire-intolerant trees. This is why longleaf help spread fire through the woodland by showering the understory each season with resin-filled needles and cones. Longleaf resin is a thick, tenacious, and highly flammable fluid produced in high volumes and stored in the trunk to repel attacking diseases and insects. Needles and cones contain resin to reduce pest damage, and once shed they become fuel for wildfires. Yet this strategy is risky. Should too much fuel accumulate, the next fire can be so hot it can kill adult trees. Even for longleaf, playing with fire is dangerous.

When Europeans arrived on the continent, longleaf woodlands covered 90 million acres (364,217 km²) of the Southeast, an area nearly three times the size of Alabama. However, this woodland wonderland was not to last. Longleaf's wood is strong, its resin slows decay, and its tall, straight trunks with minimal branching were easily milled. As southern towns grew up, pine woodlands went down. Loggers abandoned cutover land, and with no follow-up management to nurture the next crop, longleaf only regenerated where a few seed trees and fires survived. An entirely different ecosystem, the Southern Coastal Plain Dry Upland Hardwood Forest, replaced most longleaf woodlands.[2] These shady forests of oaks, especially Laurel, Post, and Southern Red, were previously restricted to areas where wildfires were less frequent.

Another factor in the downfall of the longleaf woodlands was turpentining. Cultures in Europe and Asia thousands of years ago discovered conifer resin could be processed into tar and pitch, two naval stores used to seal ship hulls and protect ropes and wood against corrosion and pests. By the early 1700s, longleaf naval stores extracted from felled trees were one of the Southeast's most lucrative exports. Later, resin was extracted from living trees at an industrial scale, then distilled into turpentine and rosin. Before petroleum was discovered, these extracts and dozens of their by-products were used in households, farms, and industries. Turpentiners tapped stands of longleaf until the weakened trees succumbed to pests or erupted in flame when a wildfire ignited their exposed resin.

Other longleaf woodlands were converted to agriculture or quietly faded away as fires became rare in the developed landscape. Eventually, the Southeast's longleaf nearly disappeared. Today, longleaf woodlands occupy a meager 3 million acres (12,140 km²). The decline of this ecosystem drove many species to the brink of extinction. Fortunately, concern for longleaf as a biological and cultural resource is on the rise. Longleaf woodlands are being protected, degraded stands are being restored, and longleaf is being replanted.

Some of the best longleaf woodlands in Alabama are in the Conecuh National Forest in south-central Alabama. The Conecuh and the adjacent Blue Spring Wildlife Management Area (WMA) to the east (hereafter, "the Conecuh") encompass over 80,000 acres (324 km²) and overlie the boundary between the Southern Pine Plains and Hills and Dougherty Plain ecoregions. The Conecuh is the northernmost piece of a huge assembly of protected lands extending to the Gulf. Most of the Conecuh had been clearcut when purchased by the US Forest Service in 1936. Though foresters preferred longleaf, its propagation was not understood and clearcuts were planted with other pines. Today, foresters are returning fire to the forest through controlled burns designed to minimize risk and maximize longleaf restoration. This policy change is largely due to the enforcement of the US Endangered Species Act to save a little woodpecker from extinction.

RCWs

If you want a taste of the primeval longleaf woodlands, seek out Red-cockaded Woodpeckers. These birds, often called RCWs, are endemic to the Southeast and highly endangered. It's no wonder, given their near-complete dependence on longleaf woodlands and their exceptionally complicated life history. RCWs reside in mature longleaf stands where fires are frequent, the trees are old, and the understory is a thick carpet of herbaceous plants. Such was the habitat I was hoping to find when early one recent April morning I pulled off of Covington County Road 4 onto a dirt track near the crossroads town of Wing in the Conecuh National Forest.

Near the turnoff were about a dozen Longleaf Pines with broad white bands painted around their trunks. I quickly parked, knowing these bands marked RCW nesting and roosting trees. Stepping out of

Red-cockaded Wood-
pecker at a nesting cavity
in Longleaf Pine wood-
land. Courtesy of Mark
A. Bailey/Conservation
Southeast, Inc.

the car, I was greeted by a spectacular sunrise and the dawn chorus of spring birdsong. Orchard Orioles, Blue Grosbeaks, and half a dozen other birds were busily feeding, singing, and settling the morning's disputes. Dozens of longleaf trunks sliced the rising sun into blinding slivers and for a few brief minutes the ground was transformed into bands of sparking dew-coated grass and chilly blue shadow. The forest had been burned several weeks earlier. Green understory plants were few, and a thin layer of blackened debris crunched underfoot. Tufts of wiregrass and young longleaf seedlings were resprouting. Near the marked pines I heard the excited squeaks of an RCW and soon spotted it high above flaking off pine bark while hunting small insects. RCWs are about 8.5 inches (22 cm) long and trimmed in black and white; a white cheek-patch is diagnostic. I saw another RCW with an insect in its bill fly to a nesting cavity in one of the marked trees, a sign the birds already had hungry chicks.

The ecology of RCWs is one of the greatest stories of North American ornithology. It all begins with the difficulty of being a woodpecker in a fire-prone ecosystem. Most woodpeckers excavate cavities in the soft wood of dead trees or rotting limbs, but resin-filled dead pines are highly flammable and do not linger in longleaf woodlands. Thus, RCWs must excavate cavities in living Longleaf Pines. Because live trees are bursting with sticky resin that would foul feathers, RCWs excavate cavities patiently. They chip away until the volume of resin oozing from the wound becomes hazardous. Excavation pauses several days or weeks until the resin stops flowing and hardens. Another complication is that longleaf wood is dense and difficult to excavate. However, older trees are often infected with the Red Heart of Pine fungus (often just "Red Heart") that softens the inner wood. Only trees 60–80 years old have had time to be infected and are large enough to house a nesting cavity. When construction is nearly finished, RCWs notch holes, called resin wells, in a perimeter around the cavity. The sticky cascade created discourages intruders, especially tree-climbing snakes that could trap birds and eggs in the cavity.

It can take years, but once RCWs have constructed a new cavity, it may be used for decades by multiple generations. However, because tree cavities in longleaf woodlands are rare, RCWs must defend them against more than two dozen vertebrate species also desiring cavities. Disputes can be intense and RCWs frequently lose. However, by providing this essential resource for other species, RCWs are an unwilling keystone species in their ecosystem.[3]

RCW social behavior is also complex and is tied to the scarcity of tree cavities. While young female offspring disperse to find a mate, one to four immature male offspring usually stay behind to help their parents raise newly hatched siblings. Helpers stick around because nesting cavities in the ecosystem are rare, vigorously defended, and difficult to excavate. First-year mortality rates of males leaving the family group, or clan, are high. Furthermore, helper males spread their genes indirectly in the population just as effectively as if they were breeding on their own. Any two full siblings share 50 percent of their genes, the same amount a parent shares with its offspring. Each helper is ultimately gambling he will win the competition to become the clan's next breeding male when the breeding male dies.

Loss of the longleaf woodlands decimated RCW populations. The species might be gone by now were it not for significant intervention by biologists. Signs of their handiwork were all around me that April morning near Wing. The recent fire had been a prescribed burn set to keep the woodlands open. Leaf litter and other fuels had been raked back from the base of cavity trees to reduce fire damage. Midsized pines had been removed because RCWs will abandon nesting trees if the midstory becomes crowded. Most importantly, I noticed several of the active RCW cavities were artificial cavity boxes inserted into the tree. Since natural cavities are so difficult to construct, inserts have greatly improved RCW population growth. In some areas, so many young woodpeckers hatch that biologists relocate some to augment smaller populations. While RCWs are still highly endangered, the careful management of remaining populations provides hope the species will survive.

THE LONGLEAF UNDERGROUND

Shelter is a premium in a landscape where any keen-eyed predator can see for a hundred yards or more, where summer heat bakes the landscape, and where roving fires can appear unexpectedly. One solution many creatures have evolved is to burrow and tunnel underground, especially where soils are sandy and easily dug. Among them is the Southeastern Pocket Gopher, a rodent with small ears, stout forelegs, and long claws for digging. Pocket gophers thrive underground nibbling on roots, tubers, and bulbs of herbaceous plants. They maintain tunnels a few inches below the surface, dumping excavated sand in small piles that are the only hint of their presence. Like RCW cavities, the tunnels provide shelter for other woodland creatures. Over 80 invertebrate species use their burrows—many have never been found elsewhere.

Another gopher digs into the sandy hills of the longleaf ecosystem, but this one is a large, lumbering tortoise. Gopher Tortoises, or "gophers," as many know them, grow to 15 inches (38 cm) in length. They have a broad head, powerful rear legs, and forelimbs designed for shoveling sand behind them when digging. In southern Alabama they live on open, sandy hilltops of Longleaf Pine and Turkey Oak—habitats known as sandhill.[4] Gophers are gentle herbivores that browse on herbaceous plants. They spend much of their life resting in their burrows that, at up to 9 feet (2.7 m) deep and more than 20 feet (6.1 m) long, provide escape from predators, temperature extremes, and fire. Ironically, burrowing tortoises need a lot of sunlight. As reptiles, gophers cannot generate their own body heat, and they control their body temperature through behavior, especially by basking in the sun. If fires become infrequent and trees fill the canopy, gophers seek new digs where sunlight is still plentiful for them and the understory plants they eat.

Gophers are famous as keystone species. Their burrows provide refuge to over 360 species of other animals. Many duck into burrows only when fire approaches, but a handful depend on them for most of their lives. One species frequently cohabitating with gopher tortoises is among the most feared animals of the Southeast, the Eastern Diamondback Rattlesnake. With its venomous bite and large size—some exceed 8 feet (2.4 m)—it demands the respect of anyone wandering the pinelands. Diamondbacks are well-camouflaged ambush predators whose markings include large dark diamonds edged with yellow. As with other rattlesnakes, a stack of pellet-filled hollow buttons (the rattle) at the tail's tip is shaken to make a loud whirring sound when threats approach. During threat displays, the snake assumes a coiled position, hisses loudly, and cocks its head in preparation to strike. This chilling spectacle can be heard dozens of yards away. Even so, diamondbacks do not pick fights and unprovoked bites to humans are rare and usually nonlethal.

Diamondbacks live in upland habitats in the Southeast, primarily where there is dense understory cover for their favored prey: mice, rats, squirrels, and rabbits. In warmer months, they find a spot where prey are likely to travel, assume a coiled position, and wait, sometimes for days, until a meal wanders past. During cooler months, rattlesnakes hunker down in stump holes or burrows, often ones used in past years. Tortoise burrows are favorite dens, and the two species cohabitate without dispute. However, when returning to a tortoise burrow, rattlesnakes need to be extremely cautious, for these are also favorite dens of their chief predator, the Eastern Indigo Snake.

Indigo snakes are gorgeous reptiles with a glossy blue-black color, a wash of red or brown across the face, and a calm demeanor. Their large size—up to 8.6 feet (2.6 m)—and muscular build help them subdue

their prey until they can swallow it alive. They'll capture any animal they can swallow, but in the Southeast, they seem to have a preference for rattlesnakes, whose fangs are no match for an indigo's thick scales, speed, and strength. Indigo snakes inhabit several southeastern ecosystems, but in Alabama they have always been found near Gopher Tortoise populations.

Populations of RCWs, Southeastern Pocket Gophers, Gopher Tortoises, Eastern Diamondback Rattlesnakes, Eastern Indigo Snakes, and other longleaf woodland species have plummeted over the past 150 years. Loss of longleaf habitat is primarily to blame, but several species have also been overhunted. Gopher Tortoises have been taken for food by the rural poor, and the beauty and docility of indigos made them popular pets. Gopher Tortoises and Eastern Indigo Snakes have been federally protected for many years, but diamondbacks are only now being considered for protection under the Endangered Species Act. Diamondbacks are usually killed on sight, and in some areas rounded up during rattlesnake rodeos. Rodeo participants find burrows, then use treble hooks on the end of long flexible tubes to yank snakes to the surface. Though the practice is now illegal, gasoline has been used to flush snakes from burrows, a practice that killed or injured other burrow occupants. Captured snakes are taken back to a festival, where they are put on display and then killed. Vendors sell fried rattlesnake meat and trinkets made from snakeskin and rattles. The Opp Rattlesnake Rodeo is the last of these annual events in Alabama. Given its popularity, it will likely continue until diamondbacks are legally protected, or until not enough can be slaughtered to justify the festival.

LONGLEAF OASES

Imagine you're hiking through the Dougherty Plain's longleaf woodlands on a summer day. You are hot, parched, and weary. As if anticipating your needs, the trail takes you downhill, through a thicket, and opens to a sparkling pool of clear water bubbling up from below. The emerging water is nearly transparent, tainted only by a bluish hue created when its dissolved minerals scatter the light. Whether you ease or dive in, the waters are breathtakingly cold and simultaneously uncomfortable and refreshing. Springs like this are one of the most exquisite natural treasures the Southeast offers, and they are distributed throughout the Dougherty Plain.

Beneath the plain's surface of sand and clay is a bed of Eocene and Oligocene limestone whose steady dissolution by acidic groundwater has left the ecoregion dimpled with springs, sinkholes, bogs, bays, swamps, ponds, and lakes. Springs, like Blue Spring in the Conecuh,

Blue Spring (Solon Dixon Forestry Education Center, Covington County) is one of many oases within the dry Longleaf Pine woodlands. Courtesy of Beth Maynor Young/Cahaba River Publishing.

are some of the few places where the limestone is exposed. One of the Southeast's most rewarding natural history experiences is to snorkel in a spring. The water is so clear that while drifting across the surface you may feel as though you are flying above the fishes, white sands, and spring mouth below. Diving down, you can study the sculpted limestone walls and find fossils of marine snails and clams whose modern descendants are just dozens of miles away on the coast. If you are a strong swimmer, you can maneuver to the entrance of the cave to feel its surging waters, peer into its darkness, and wonder about the marvels it hides. Though quite chilly in the summer—springs stay about 68° F (20° C) all year—during the winter a spring can seem invitingly warm.

Springs influence the ecology of only a small swath of adjacent land. In their natural state, smaller springs are often completely sheltered by a tree canopy of mesic forest species, while larger springs are open and ringed by cypress and tupelo. Beyond the waterline, the spring's valley harbors species typical of northern mesic forests like beech and various maples, while others are semitropical ferns, palms, and magnolias. None of these plants could survive in the dry, fire-prone woodlands nearby.

The Dougherty Plain also has sinkholes, craterlike depressions formed when an underground cavern collapses and the surface slumps downward. Water sinking into some sinkholes disappears into a network of crevices in the limestone below. In others a plug of impermeable clay allows water to collect seasonally or permanently. In either case, Southern Coastal Plain Sinkhole ecosystems are a haven for plants needing cool, moist environments.[5] Sinkholes vary in shape and vegetation based on the size, depth, steepness, and age of the depression. One of the more dramatic sinkholes of the Dougherty Plain in Alabama is Blue Pond at Auburn University's Solon Dixon Forestry Education Center. When I visited in 2009, this 80-foot-deep (24 m) sinkhole held a pond more than 350 feet wide (107 m). However, every few years the clay plug at the bottom of the sinkhole blows out and the entire pond drains in a few short hours.

LOVE IN THE LIMESINK

There are other oases in the dry woodlands of the Dougherty Plain. Isolated ponds with gently sloping banks known as depression ponds, or limesinks, are so common that some call this the Limesink Region. Some are permanent lakes, others are wooded swamps, yet others swing from being dry to flooded and support herbaceous plants (the latter are part of the East Gulf Coastal Plain Depression Pondshore ecosystem). In all cases, these ponds are magnets for wildlife. Herbivorous aquatic turtles are common and easily spotted when surfacing to breathe. Mammals visit at night for a drink or to hunt the shoreline for smaller pond visitors. American Alligators inhabit some ponds. It might seem strange for an alligator to live in the midst of an upland pine forest, but the ponds attract ample food for this ambush predator. As they do in other wetlands, alligators dig strategically placed holes in the shallows at the water's edge to maintain a pocket of deeper water. They wait in these so-called 'gator holes until they can seize some luckless mammal that ventures too near.

As glamorous as large vertebrates are, the most impressive biological spectacle of the limesinks is the amphibian mating frenzy occurring each spring and early summer in smaller ponds. Dozens of species, including toads, frogs, and salamanders, migrate to limesinks to court or be courted. Smaller ponds are preferred because their periodic drying kills fishes that would devour amphibian eggs and young; adult amphibians actually "taste" the water through their skin to determine whether fish are present.

These events are best witnessed by arriving at a limesink just at sunset during the peak of amphibian breeding and waiting quietly during the long summer dusk. Once the sun is down, nocturnal animals emerge. Chuck-will's-widows, a bird of the goatsucker family, shout their namesake song at extraordinary volumes through the woodlands. Common Nighthawks, another goatsucker, and various bat species fly erratically over the pond and adjacent woodlands pursuing insects. The understated hoots of a Great Horned Owl drift through the evening air, while a mother raccoon and her kits quietly scamper to the shore for a quick drink.

About the time stars and planets appear, the pond's tranquility is interrupted by the first calls of frogs and toads eager for romance. The impatient males breaking the silence are soon joined by more and more others until there is a deafening cacophony of trills, growls, booms, peeps, and chirps that can be heard a half mile away. Each species has a distinctive call its males use to woo females. Calls signal a male's whereabouts and inform females of his size, health, and vigor—information critical to choosing the best mate. During the day, frogs quickly leap to safety when approached, but at night they freeze in a spotlight's beam and can be closely studied. It seems impossible that females hear individual males during the raucous ensemble, but as the evening wears on, more and more males and females join in amplexus, a pseudocopulatory embrace in which females allow males onto their back so sperm and egg can be released simultaneously into the water. After a few hours, most females have made their choices and the chorus ends. Only a few lonely males will sporadically call during the rest of the night.

Gopher Frogs, perhaps the rarest southeastern frog, depend on a handful of limesinks. They grow to be 4 inches (10 cm) long and are light brown or gray with numerous dark blotches and spots. The species has an improbable life history. Like all amphibians, the frogs are highly vulnerable to desiccation, yet they spend most of their time in the hot, dry sandhills of the longleaf woodland. The frogs hide by day in burrows dug by other animals, especially the gentle landlord of the sandhill community, the Gopher Tortoise. At night, when the sandhills have cooled, the frogs emerge to hunt small invertebrates near the mouth of their home burrows. In spring, adults migrate up to a mile (1.6 km) to spend a few rowdy nights breeding in depressional ponds. With their reliance on a rare ecosystem and an imperiled tortoise, plus migrations across roads and other hazards, it is no wonder Gopher Frogs face extinction.

Scientists have long sought to understand limesink ecosystems. They are perplexingly diverse in their size, hydroperiod (how long and when they hold water), and ecology. It's generally accepted they form when limestone below the sandy surface dissolves and the surface slumps. Acidic, nutrient-poor groundwater and surface water collects in depressions, but only accumulates when a claypan beneath the lake's sandy bottom prevents leakage into the bedrock. If the slumping is deep and widespread, the claypan is robust, and water is plentiful, a large lake forms and can support fishes and other fully aquatic organisms. The largest of these in Alabama is Lake Jackson, which at 500 acres (2.0 km²) and over a mile in length is Alabama's largest natural lake. Sometimes, however, a lake's claypan fails, and it drains into the aquifer within a few hours, if not minutes.

The hydroperiod and ecology of midsized and small ponds seem to be controlled by exposure to wildfire and the claypan's depth. In

Gopher Frogs, perhaps the rarest southeastern frog, in amplexus. Courtesy of Kevin Messenger/Alabama A&M University.

general, when the claypan is deep, sands above it absorb much of the water it receives, and the pond is more likely to dry regularly. Smaller ponds will dry each year, while larger ones will dry during prolonged droughts. In either case, when water levels are low, grasses, wildflowers, shrubs, and trees colonize the sandy soils revealed. If the pond is adjacent to frequently burned woodlands, dry season fires sweep through the colonizing vegetation and kill fire-intolerant plants. Wet season inundation kills flood-intolerant plants. Together, flooding and fire help maintain the open character and dominance of herbaceous plants in many limesinks. In contrast, a shallow claypan keeps water close to the surface, and water is likely to be present all year. Such ponds are often filled with Pond Cypress and tupelos if depths are not excessive, though their tree seedlings only establish during long intervals of low water and no fire.

The Forgotten Citronelle Ponds

Few are aware the Southern Pine Plains and Hills ecoregion has its own unique wetland, the Citronelle pond. Limesinks and Citronelle ponds are similar. Both have shallow banks, acidic waters, and a claypan. Both vary in size, with Citronelle ponds varying from 1 to over 200 acres (0.8 km²). Both have been dubbed grady ponds, a wetland named after a soil type they and similar southeastern wetlands develop. However, unlike limesinks, the origin of the Citronelle ponds is a geologic mystery.

The Southern Pine Plains and Hills lack limestone. Instead, a soft and easily eroded Miocene-aged formation, known simply as the Miocene Series, underlies most of the ecoregion. Millions of years of erosion have turned this former seafloor into a hilly terrain. Longleaf Pine woodlands once prevailed, but today the ecoregion mainly hosts plantations of other pines. Near the coast, much of the Miocene Series lies beneath a younger, Pliocene-Pleistocene-aged soil of sands, clays, and gravels known as the Citronelle Formation. These deposits have resisted erosion more than the Miocene soils. Consequently, the terrain is more level and row agriculture is common. The Citronelle ponds formed on these plains, but given the stability of the underlying sediments, there is no clear mechanism for their formation. Still, their resemblance to limesinks suggests to scientists that dissolution of deep soil layers is involved.

There is also mystery about the original ecology of the Citronelle ponds. Today, most support isolated swamps of Pond Cypress and/ or Swamp Tupelo, an ecosystem known as the Southern Coastal Plain Nonriverine Cypress Dome. *Dome* refers to the typical pattern of taller trees at the center being ringed by smaller trees. Though a few ponds are open or shrub-dominated, the predominance of swamp trees has prompted the suggestion that the original ecosystem was swamp. However, prior to human settlement, these wetlands were embedded within longleaf woodlands, and ponds with shorter hydroperiods probably burned frequently and supported herbaceous cover like that of similar limesinks of the Dougherty Plain.

Most Citronelle ponds today are wetland islands in a sea of agriculture. Pond densities in some areas were once greater than 25 ponds per square mile (10 per km²), but most small ponds have been plowed over. Surviving ponds provide refuge for many wetland species despite suffering periodic logging and agrochemical contamination. However, they have largely been overlooked by conservationists. As best as I can tell, none are on public lands in Alabama, and Florida and Mississippi protect only a scant few.

SWAMP JUNGLES
Baygalls, bay heads, bay flats, titi flats, and green-heads—these are local names for junglelike swamps of the lower Southeastern Plains known

to ecologists as Southern Coastal Plain Seepage Swamps and Baygalls (hereafter, "baygalls"). Baygalls are choked with an impassable tangle of trees (especially Sweetbay, Swamp Tupelo, and Buckwheat Tree), shrubs, ferns, sphagnum moss, and briers. Baygalls usually occupy the valleys between hills, collect groundwater, and channel it downhill to feed the region's blackwater streams. Their moist soil, shady understory, and absence of dry ground fuels protect baygalls from fire even though historically they were usually surrounded by fire-prone woodland.

Baygalls are dwarfed by the Southern Coastal Plain Nonriverine Basin Swamps of the lower Southeastern Plains. These densely vegetated wetlands are so large they are also known as bays. Some scientists suggest the basins developed on former river oxbows or stranded coastal embayments, but the Dougherty Plain's basins may have originated via limestone dissolution over a broad area.

Basin swamps collect water seeping from the surrounding landscape and stay flooded at least half the year thanks to a claypan beneath. The flooding helps swamps repulse most wildfires. A thick layer of peat—partially decayed plant material—accumulates below the surface and acids leaching from it stain the water black. Shallow basin swamps are prone to drying and burning, so Bald Cypress, Slash Pine, and other plants tolerant of inundation and surface fire prevail. Deeper basin swamps support broadleaf trees less tolerant of fire, including Swamp Tupelo, Sweetbay, and Red Maple. Only cypress and tupelo survive in the deepest basin swamps having near-continuous inundation. All basin swamps eventually burn; subsurface peat fires can smolder for days, baking roots and killing trees and shrubs. Such areas will harbor open water habitat until shrubs and trees reestablish during subsequent droughts. Being unsuitable for development, basin swamp ecosystems remain largely intact in the lower Southeastern Plains.

Hungry Plants

Southern Alabama is home to some of the strangest plants in the world. Nested within the longleaf woodlands of the Dougherty Plain and Southern Pine Plains and Hills are bogs populated by plants that lure, swallow, drown, trap, and digest small animals. These ecosystems, the East Gulf Coastal Plain Interior Shrub Bogs, are world renowned for their carnivorous plants. Botanists have found 27 species in southern Alabama, making the area a global center for carnivorous plant bio-diversity.[6] Other herbaceous plants also thrive here, and in some spots a phenomenal 30–50 species grow in 1 square meter of ground.

The bogs are sunny wetlands at the base of hills where a claypan forces water to the surface. During the wet season, groundwater emerges and flows across the surface in tiny rivulets. The adjacent basin swamps and baygalls receiving this water are a menacing presence. Swamp shrubs and trees endlessly attempt to colonize the bog. They move in quickly, form thickets, and capture the light needed by smaller bog plants. Fires come to the rescue from the longleaf woodlands bordering a bog's upper margins. Wildfires sweep through bogs, kill invading plants, and preserve open habitat.

Carnivorous plants derive all their energy from the sun. They extract from animals nutrients, especially nitrogen, rare in the bog's acidic waters and soils. Pitcher plants famously capture insects in a tube-like trap whose bottom holds a pool of water laced with digestive enzymes. Alabama's pitcher plants come in a variety of forms including the short and copper-colored Wherry's Pitcher Plant, the squat and pot-bellied Purple Pitcher Plant, and the reclined Parrot Pitcher Plant. The most flamboyant is the Crimson Pitcher Plant (or White-topped Pitcher Plant), whose tall snowy-white trumpets are decorated with a web of deep-red veins resembling blood vessels.

Three other plant groups complete the shrub bogs' carnivorous menagerie. Low-growing sundews display small, spoon-shaped leaves radiating from a central bud and held horizontally. Each leaf grows numerous short hairs secreting a sticky droplet resembling dew. To some insects the offering is irresistible and deadly. When an insect becomes ensnared, the leaf slowly curls inward toward the plant's center, where more sticky hairs adhere to ensure death. Leaves release enzymes to accelerate decay, then absorb the nutrients oozing from the carcass. Alabama's most conspicuous sundew, Tracy's Sundew, is unusual for displaying dew-laden hairs along stalks rising more than a foot into the air. Butterworts grow low and present yellowish-green leaves covered with fine sticky hairs that trap insects. Leaves curl inward from the margins to secure trapped insects and bathe them with enzymes. Bladderworts, tiny plants of the wettest bog areas, lack roots but grow a network of

White-topped Pitcher Plants at the Nature Conservancy's Splinter Hill Bog, Baldwin County. Courtesy of Sara Bright.

thin stems sporting nearly microscopic sacs, or bladders. When small swimming invertebrates trigger hairs near a sac's mouth, the sac quickly inflates, sucks in the unlucky animal, shuts the mouth, and begins digestion. Bladderworts are inconspicuous until they display small showy flowers that hover above the surface on thin stems.

With such extraordinary flora, it is easy to overlook the shrub bog's wildlife, which includes dozens of amphibian and reptile species, the rarest being the Pine Barrens Treefrog. This handsome small frog is green above with a chocolate-brown or purple mask. Bright yellow-orange spots on its brown flanks and legs add a splash of color revealed only when jumping. These treefrogs breed in the shrub bogs, but little else is known about their ecology. Curiously, there are three widely separated population centers in the United States: southern Alabama–northwestern Florida, New Jersey, and the Carolinas.

Another rare animal is the Henslow's Sparrow, a northern breeder overwintering in the bogs and wet savannahs of the Southern Coastal Plain and lower Southeastern Plains. With its olive-green head and reddish-brown wings, it is one of the more colorful sparrows. Henslow's Sparrows are exceptionally shy, prefer thick groundcover, and scamper through dense herbaceous vegetation rather than fly.

Only a few shrub bogs have survived in Alabama, the most spectacular of which is the Splinter Hill Bog near the headwaters of the Perdido River. The Alabama Chapter of the Nature Conservancy (TNC) and the state's Forever Wild program have purchased over a thousand acres of this vast, uniquely situated bog and adjacent longleaf woodland. Such measures are dearly needed, for populations of carnivorous plants, Pine Barrens Treefrogs, Henslow's Sparrows, and other bog species are declining.

THE RED HILLS

The Red Hills region of southwestern Alabama is a postmodern frontier for students of biodiversity.[7] Despite the unusual character and rich biodiversity of this region, more formally known as the Buhrstone/Lime Hills level IV ecoregion, relatively few naturalists and biologists explore or study it. Tall hills arise from the Coastal Plain with a geologic complexity rivaling the most tortured mountains in the state. Woodlands of pine and oak once graced the ecoregion's hilltops before agriculturalists arrived, but steep ravines still conceal a handful of rare and endemic species, plants and animals whose core distributions are far to the north, and new species never before described. Hill streams resemble Alabama's mountain creeks in form and fauna. Where large rivers cut through the Red Hills, steep bluffs loom over the river and provide magnificent vistas. Erosion along streams and rivers reveals ancient biodiversity in the form of fossil gardens of oysters, scallops, snails, and corals.

The ecoregion has one of the most sparse human populations in Alabama, though it has supported over 150 years of agriculture, ranching, and timbering. Many farms and ranches have closed due to soil depletion and the industrialization of agriculture. Where towns once thrived, highways meet at empty crossroads populated by abandoned homesteads and collapsing storefronts overgrown with Kudzu and Mimosa. Decades of intensive land use spared little of the region's natural ecosystems. All we know of its original ecology are a few tantalizing notes written by the first visiting naturalists.

Marine deposits from the Eocene and Oligocene, including claystone, sandstone, and limestone, are the foundation for the Red Hills. These durable sediments became hills as softer sediments of the surrounding regions eroded more quickly. Six distinct mineral soil deposits are present across the ecoregion, and several can be exposed on a single hillside. The hills attain elevations of 550 feet (167 m) above sea level and rise up to 400 feet (122 m) above the local terrain. Though quite rare below the Fall Line, limestone dissolution has created a few caves. The southern portion of the ecoregion includes the Hatchetigbee Dome, a portion of the earth's crust shoved upwards by a deep pillar of rising salt. The dome forces water with salt, sulfur, magnesium, and calcium to the surface through mineral springs and artesian wells and exposes rock otherwise located at the surface a dozen miles to the north (salt domes are explained in chapter 14).

On the ecoregion's hilltops, rocky, dry areas originally supported Longleaf Pine woodlands with a diverse herbaceous flora. Where soils were moister and fires less frequent, broadleaf trees dominated, principally Southern Red, Black, Blackjack, and Post Oak, plus Pignut and Mockernut hickory.

The hill slopes are cut with dim ravines sheltering displaced species. During the late Tertiary and Pleistocene, the Southeast became drier and colder, and the moist ravines were refuge for the region's flora. Later, during the Holocene's warming, the ravines provided refuge for species needing cool, moist conditions and escape from increasingly frequent wildfires. In recent times, the rugged terrain provided escape from the agriculture and timbering that swept the Southeastern Plains. Surviving refugees from these episodes are part of today's Southern Coastal Plain Mesic Slope Forest. These ecosystems develop wherever topography in the Southeastern Plains provides cool, moist conditions. Their extent and biodiversity is greatest in the Red Hills.

A dim ravine within Forever Wild's Red Hills Tract, Monroe County. Courtesy of Hunter Nichols/Hunter Nichols Productions.

Until recently, the only public access to mesic slope forests of the Red Hills has been Haines Island Park.[8] This US Army Corps of Engineers' park is on the Alabama River west of the small town of Franklin and northwest of the smaller town of Scratch Ankle. The road into the park cuts through overgrown longleaf woodland, then descends across a steep bluff to the river bottomland, dropping 247 feet (75 m) in its last half mile (0.8 km). At the bottom, towering trees of the East Gulf Coastal Plain Large River Floodplain Forest surround campsites, a boat launch, and a shaky-looking single-car ferry.

Trails lead into a primordial forest, where trees of impressive girth tower above. Their ranks include American Beech, Tulip-poplar, and Swamp Chestnut Oak. The Spruce Pines are among the tallest and most distinctive trees. They are elegant conifers, having an open canopy with short needles and a tight gray bark. The species is more shade-tolerant than any other southeastern pine, which helps it survive in this shady forest. Above any new seedling are multiple canopy layers of every green imaginable. Enough light filters through to also support ferns and shrubs including buckeyes, Anisetree, Blue Stem Palmetto, and Needle Palm.

Two magnolias also help define these forests. One is Southern Magnolia, a belligerent competitor whose dark leathery leaves absorb light so effectively that no other plants grow beneath. The species grows incrementally, wedging upward toward the canopy. In contrast, the Bigleaf Magnolia has enormous thin leaves measuring over 30 inches (75 cm) long and they, excluding the palms, are the largest leaves of any North American plant. Bigleaf Magnolias rarely reach the canopy and instead harvest sunflecks and indirect light captured by its huge leaves. While young Southern Magnolias keep their limbs tightly reined in, the limbs of young Bigleaf Magnolias twist in all directions searching for light patches.

The rarest animal of the region is the Red Hills Salamander, Alabama's official state amphibian. It is a peculiar animal found nowhere else but in Alabama's Buhrstone/Lime Hills. This brown or gray unmarked salamander can grow to 10 inches (25 cm) and is unusual for having a prehensile tail comprising half its length. Like Tolkien's Hobbits, these salamanders rarely wander, staying loyal to their burrows within steep slopes where erosion has exposed siltstone and clay. By

The endemic Red Hills Salamander, Alabama's official state amphibian. Courtesy of Danté Fenolio.

night they wait at the burrow entrance to ambush invertebrates wandering past. By day they retire to their burrow's humid depths. Despite a superficially solitary lifestyle, Red Hills Salamanders usually occupy burrows where others reside a few inches away.

Much is still unknown about the species, but researchers suspect populations expand slowly. Mature mesic slope forests and their humid understory provide the best habitats; logging will dry out the understory to the salamander's detriment. Given these vulnerabilities plus extensive habitat loss throughout the ecoregion, the US Fish and Wildlife Service listed the species as threatened in 1976. Recently, the salamander received a boost in protection when TNC and the state cooperated to buy 1,800 acres (7.3 km²) of mesic slope forest a few miles east of Haines Island Park. The Forever Wild property is open for hiking, hunting, and nature observation, and there is hope this initiates a new wave of exploration by scientists and ecotourists.

CUESTAS AND THE POCOSIN IMPOSTER
North of the Buhrstone/Lime Hills is the Southern Hilly Gulf Coastal Plain. Stretching completely across Alabama, it's the largest level IV ecoregion in the state. In the west it is often under 20 miles (32 km) across, but in the east it averages 50 miles in width (81 km). Despite its size, the ecoregion has less protected habitat than any of the state's other ecoregions.

The Southern Hilly Gulf Coastal Plain resembles the Red Hills, but the topography is less extreme and its ecosystems are less diverse. The ecoregion shares many sediment types with the Red Hills, including Cretaceous and early Tertiary sands, clays, and a lime-clay mixture termed *marl,* but it encompasses different physiographic regions, including the Buhrstone Hills Subdistrict, Southern Red Hills District, and the Chunnenuggee Hills District east of the Alabama River.

Cuestas are a topographic feature unique to this ecoregion. Pronounced "KWAY-stas," these are elongated hills of erosion-resistant mineral soils. Cuestas formed when the earth's crust beneath the Southeastern Plains was pulled down and southward as the Gulf of Mexico grew in weight. In western Alabama, crustal deformation tilted westward as river sediments accrued in the lower Mississippi River Basin and adjacent offshore areas. Consequently, cuestas in eastern Alabama run east–west, while those of western Alabama run northwest–southeast. The southern face of a cuesta is broad and gently sloped and is composed of the cuesta's youngest soils. By facing southward, these slopes support drier ecosystems like pine woodlands. The northern face of a cuesta is usually a steep and foreshortened cross-sectional cut through several soil types. Sheltered from direct sun, northern slopes sustain mesic slope forests like the Red Hills. The cuestas in western Alabama are usually taller than those to the east, and their ecology is more similar to that of the Red Hills.

The Southern Hilly Gulf Coastal Plain's mix of cuestas, unorganized hills, and low flatlands has a long history of agriculture, ranching, and timbering; little remains of its original ecosystems. From early accounts and deductions based on its current vegetation, the ecoregion supported broad expanses of pine woodland. Drier, sandier upland soils hosted longleaf woodlands, while moist, clayey soils sustained East Gulf Coastal Plain Interior Shortleaf Pine-Oak Forest (hereafter, "Shortleaf Pine woodland"). Shortleaf Pine woodlands require recurring low-intensity wildfires, but the pine prefers cooler conditions where fires are less frequent than in longleaf woodlands. At low elevations where fires were infrequent, pine woodlands transitioned into broadleaf forests.

The Pike County Pocosin is an ecological mystery in the Southern Hilly Gulf Coastal Plain that has puzzled scientists and naturalists for 150 years. It is an area of steep ravines on coarse, sandy soils just east of Troy. Nothing quite like it exists in the Southeast. The flat uplands support an open, dry community sparsely populated by scrubby trees and shrubs (e.g., Arkansas, Turkey, and Running oak; Devilwood; Mexican Plum), Eastern Pricklypear, and lichen mats. Historically, wildfires were unable to penetrate this sparse habitat and reach the pocosin's deep ravines. The pocosin's cool, sheltered slopes protect species typical of mesic slope forests but unusual in southeastern Alabama. The two communities harbor rare, threatened, and endangered plants including Sessileleaf Pinelandcress, Elliott's Gentian, and Nodding Nixie. Sadly, in the 1960s and 1970s the pocosin's uplands were converted to pine plantation, and many ravines were filled with debris. While much of its original character is lost, Troy State University and the Forever Wild program now protect much of the pocosin and are trying to repair its ecosystems.

There is no accepted explanation of how this ecosystem formed or how its ravines became populated by species so far from their normal ranges. Furthermore, no one knows how this ecosystem became known to locals as a pocosin. Pocosins are Atlantic Coastal Plain wetlands with high water tables, a pine overstory, frequent fires, and peat deposits—clearly, the Pike County Pocosin is an imposter.

Alabama's Tallgrass Prairie
Let's go back in time . . .

It is an early 17th-century May morning in a tallgrass prairie several miles due west of the future site of Selma. For the past hour we have been tracking from downwind a herd of 14 bison. We've crept to about 100 yards (91 m) away and are watching them from a patch of short oaks. The bison are feasting on the fresh grasses and herbs sprouting thickly from the rich soil. Mostly we just see the tall shoulders and backs of the adults. Occasionally one ambles to a new position, and we see its massive blocky head and stout horns. We keep a careful eye on the bull that has, at least for now, won the rights to this group of females. He is feeding on the far side of the herd but keeps gazing away from us and his herd to the southwest.

The bison have shed much of their thick winter coats, but several still have matted tangles of darker fur dangling from their shoulders and necks. We notice worn bark on the trees concealing us and clumps of fur on the heavily trampled ground. Realizing this is a favored back-scratching haunt for shedding bison, we glance behind us for any bison coming here with an itch. Bison are ornery and might not take kindly to our presence.

Several mothers calved about a month ago, and where the grasses are shorter, we catch glimpses of their young. They stay close to their mothers, and at one point we watch a calf nuzzle her mom for a quick and vigorous round of nursing. Two older calves initiate a brief, playful chase. They dart with surprising speed around their watchful mothers, sometimes kicking their hind legs into the air and bleating with excitement. Otherwise, the herd is quiet, save for an occasional grunt.

The peaceful scene is interrupted when a loud, distant roar from a bull in a herd to the southwest demands everyone's attention. Our herd's cows glance in that direction for a few seconds before returning to feeding, but our bull stares intently for nearly a minute. It is early enough in the season that bulls are still fighting to capture or defend access to females, and we dearly hope to see males sparring during our day of observation. It seems we may get our wish.

This scene is not entirely fantasy. As evidenced from historical accounts, archaeological finds, and fossils, herds of bison roamed Alabama until several hundred years ago. This is often a surprise to those assuming pre-Columbian Alabama was heavily forested. However, prairie, savanna, and woodland were the norm across most of Alabama throughout the Holocene. The largest swath of non-forest ecosystem in the Southeastern Plains is the Blackland Prairie level IV ecoregion.[9] In Alabama it stretches from the border with Mississippi nearly to Georgia, averaging 25–30 miles (40–48 km) wide. In Mississippi, the ecoregion bends northward and barely crosses the Tennessee border. Its grasslands are part of the Southern Coastal Plain Blackland Prairie and Woodland ecosystem and are the easternmost extent into the Southeast of the tallgrass prairies once blanketing the Midwest.

In its primeval state, the blackland prairie must have been exquisite, for many earlier visitors wrote enthusiastically of its allure. It was a complicated patchwork of prairie and woodland. There were a few large prairies occasionally interrupted by patches of scrubby oaks or shrubs, but most prairies were midsized or small due to being separated by streams and their riparian forests. The original prairie flora is thought to have been very similar to that of the midwestern tallgrass prairies. Grasses included Sideoats Grama, Indian Grass, and various bluestem species. Wildflowers included White Prairie Clover and other legumes, plus members of the sunflower family including Blazing Star, Compass Plant, and Prairie-dock.

The blackland prairie is among the last remnants of the arid ecosystems widespread in the Southeast during the Pleistocene's ice ages. When the climate warmed to its present state, the prairie resisted the expanding forest largely because of a thick deposit of Cretaceous chalk beneath. Chalk formed from the microscopic shells of marine creatures. Soils on this calcium-rich deposit are black and roughly

Blackland prairie wildflowers at Forever Wild's Old Cahawba Prairie Tract, Dallas County. *Clockwise from top left:* Butterfly Milkweed, Lemon Beebalm, Gray-headed Coneflower, Common Rose-pink. Photos by R. Scot Duncan.

neutral in acidity, the former attribute inspiring various names for the region. Chalk is notoriously dense and impervious to roots and water, and trees colonizing the prairie had no groundwater to help them survive drought. Furthermore, the soils hindered establishment of the many southeastern trees preferring acidic soils. Another deterrence was damage to tree roots as clayey prairie soils swelled and cracked during droughts. Despite these challenges, several hardy trees and shrubs like Eastern Red Cedar thrive on blackland prairie soils. However, they could not overtake the prairie because they are fire intolerant. Frequent wildfires helped maintain blackland prairie, just as they do in midwestern tallgrass prairies.

Cotton and cattle killed the Alabama prairie. Cotton agriculture arrived in the mid-19th century and any prairie with tillable soil was converted. In the early 20th century, ranching spread to where topsoil had been depleted and farming was unprofitable. Cattle could be raised on worn-out fields and in shallow-soiled prairies that had escaped the plow. Today's blackland prairies occupy less than 1 percent of their original distribution, and most remnants are on soils so degraded that even ranching is unprofitable.

Prairie can be restored. Such projects are becoming common in the Midwest, and a few are springing up here in Alabama. Ecologists at the University of West Alabama are studying how to restore tallgrass prairie, while TNC and the Forever Wild program are restoring prairie in the Old Cahawba Prairie Tract in Dallas County. Fire is used to thin encroaching trees and shrubs and each spring the tract is now ablaze in the yellows, pinks, and oranges of wildflowers such as Pinnate Prairie Coneflower, Old Cahaba Rosinweed, and Purple Prairie Clover. Perhaps one day we'll have restored enough prairie to reintroduce bison to Alabama's blackland prairie.

More Chalkland Legacies

One of the most stunning geologic formations in Alabama is found where hilly portions of the Blackland Prairie ecoregion are crossed by large rivers. The rivers have carved gleaming-white chalk bluffs sparsely garnished with cedars and other hardy species that can survive on small ledges. This unique ecosystem, the East Gulf Coastal Plain Dry Chalk Bluff, is one of the other two ecosystems the Cretaceous chalks of the upper Southeastern Plains have helped shape.

Intermingling with the Blackland Prairie ecoregion is the Flatwoods/Blackland Prairie Margins level IV ecoregion. This is a region of level, deep soils that sustained dry woodlands of Post Oak, Shortleaf Pine, and Loblolly Pine. Recurring fires prevented transition to full forest and nurtured a rich herbaceous understory, and dense clays prevented groundwater storage for periods of drought. Hilly areas are also present, especially where the ecoregion incorporates portions of

Blackland prairie and goldenrod at sunset, Forever Wild's Old Cahawba Prairie Tract, Dallas County. Courtesy of Emily Horton Reilly.

the Chunnenuggee Hills physiographic district west of the Alabama River. Steeper hillsides and ravines support mesic slope forests similar to those of the Red Hills to the south.

OVERLOOKED PRAIRIES

Given the prominence of the Blackland Prairie ecoregion in the state's ecological history, it's easy to overlook the smaller regions of blackland prairie in Alabama. They are in level IV ecoregions far south of the Blackland Prairie ecoregion. Most of what is known about them is from descriptions by Alabama's first naturalists.

One set of prairies is in northeastern Wilcox County where Tertiary-aged limestone nears the surface and weathers into calcium-rich clays on which grasses thrived. Apparently, the original vegetation of the Wilcox County prairies was similar to that of the blackland prairies. Land use patterns were also similar and settlers converted deeper prairie soils to cropland and hay production and thinner soils to pasture. Ranchland with remnants of the prairie's flora and structure can be seen along roadsides in the area between the towns of Pine Apple and Snow Hill.

Another set of prairies existed in northern Washington and southern Choctaw counties on clays, marls, and sands of a soil formation known as the Jackson Group. Ecologically, this area is the southeastern tip of the Jackson Prairie level IV ecoregion of southeastern Mississippi. Alabama's Jackson Prairie is so small it was excluded from Alabama's ecoregion map. Early explorers in Mississippi wrote of the beauty of the Jackson Prairie, but its fertile soils were quickly put to work. Little is known of its original flora except for what has survived in prairie fragments along roadsides, in cemeteries, and in small patches within Mississippi's national forests.

THE OLDEST HILLS

Just below the Fall Line boundary between the Southeastern Plains and Alabama's uplands, a range of hills rises up from the plains. These are the Fall Line Hills and Transition Hills level IV ecoregions. Their ecology blends characteristics of the Southeastern Plains and the highlands beyond. They are also the oldest hills of Alabama's Southeastern Plains, sculpted from clay, sand, and gravel deposited in the Cretaceous when the oceans had penetrated deeply into the continent.

The Fall Line Hills begin at the Georgia border, wrap around the terminus of the Southern Appalachian Mountains, and spill into Mississippi and a tiny corner of southwestern Tennessee. These tall hills range to 1,000 feet (305 m) above sea level and 200–400 feet (61–122 m) above the local terrain. Longleaf Pine woodlands cover much of the southern Fall Line Hills. As with their counterparts farther south, frequent fires historically maintained a landscape of open woodland and rich herbaceous understory. Yet there are differences. Little Bluestem replaces wiregrass as the dominant grass, and hard sediments prevent Gopher Tortoises from digging burrows and living here.

Longleaf woodlands of the Fall Line Hills also differ by mixing with other ecosystems. Shortleaf Pine woodlands prevail in the ecoregion's north, where conditions are cooler and moister. Oak forests prevail in valleys and on some north-facing slopes sheltered from wildfire and the sun's heat. White, Laurel, Post, Southern Red, and Chestnut oak are the most common. Two of Alabama's rare oaks, Arkansas and Georgia oak, inhabit the ecoregion's south.[10]

Today's Fall Line Hills are mostly managed stands of Loblolly Pine grown for lumber and paper. However, Longleaf Pine has a stronghold in the Oakmulgee Ranger District of the Talladega National Forest, which straddles six counties of the Fall Line Hills. For decades, the US Forest Service mainly planted Loblolly Pine, but longleaf woodland restoration is now a top priority and the district hosts the largest population of Red-cockaded Woodpeckers in Alabama.

The district's valley swamps and wetlands harbor the endangered Mitchell's Satyr. Before its chance discovery here in 2000, this butterfly's nearest known population was in North Carolina. A recent discovery of a population in Mississippi suggests the satyr isn't as rare in the Southeast as once thought.

Another natural treasure of the Fall Line Hills is the Alabama Canebrake Pitcher Plant, a species endemic to the lower Coosa River watershed in Elmore, Chilton, and Autauga counties. These are stately, golden-green plants with crimson venation and flowers. They thrive in interior shrub bogs wetted by seeps at the base of tall hills. Like other pitcher plants, they require saturated acidic soils and abundant sunlight. Water for the seeps falls as rain to the northeast, sinks deep below ground, flows for dozens of miles downhill across thick clay layers, and reemerges where erosion has exposed the clay and allows the groundwater to escape. Unfortunately, most known Alabama Canebrake Pitcher Plant populations have been wiped out by agriculture and competition with woody plants in areas subjected to fire suppression. About a dozen populations survive. None are publically owned, but one of the largest is secured within TNC's Roberta Case Pine Hills Preserve.

The Transition Hills ecoregion occupies a small area centered where the borders of Tennessee, Mississippi, and Alabama meet. Despite being bisected by the Tennessee River, the ecoregion has hills reaching to 980 feet (299 m) above sea level and 400 feet (122 m) above the local terrain. The ecoregion's geology and ecology are a blend of the Southeastern Plains and the Interior Plateau level III ecoregion across the Fall Line. Upland habitats are on Cretaceous sediments typical of

Alabama Canebrake Pitcher Plant, endemic to Alabama's Fall Line Hills ecoregion, Autauga County. Courtesy of Alan Cressler.

the former, while streams cutting into the Paleozoic-age bedrock resemble the rocky streams of the latter. The Transition Hills are beyond the range of Longleaf Pine and instead support East Gulf Coastal Plain Northern Dry Upland Hardwood Forests dominated by White and Chestnut oak.

There are two WMAs in Alabama's Transition Hills. The Freedom Hills WMA south of the Tennessee River is underlain by the Hartselle Sandstone. Small seep and bog ecosystems throughout the hills persist where the sandstone forces groundwater to the surface. North of the Tennessee River is the Lauderdale WMA. Here, the Transition Hills rest on a foundation of Fort Payne Chert. The section of the Tennessee River between the two WMAs is Pickwick Lake, a large impoundment famous among birders. In winter, the reservoir's high fish densities and mudflats attract dozens of bird species. Bald Eagles gather in large numbers to pick off fishes, especially those weakened by the cold. Resident eagles nest along the reservoir's secluded shorelines and slopes. In summer, visiting birders seek the Warbling Vireo, an accomplished songster that breeds along the lakeshore but is rarely seen elsewhere in Alabama.

Crossing the Line

Shaped by sunlight, water, and fire, the ancient seafloors that are now the Southeastern Plains harbor an extravagant diversity of ecosystems and species. Though its lands have been worked for many human generations, the region is still a biological frontier, full of ecological mysteries to solve and species to study. Next, our explorations take us over the Fall Line and into the plateaus, mountains, and valleys of Alabama's highlands. Though we leave behind a biological Eden, the biodiversity awaiting us is every part as intriguing.

10

Ridge and Valley

Appalachian Origins

The Ridge and Valley level III ecoregion rises from the serene flatlands and rolling hills of Alabama's Coastal Plain like an ancient leviathan rising from the sea. Where the beast first breaches at the Fall Line, there is a hill-studded hide of thick clays stretched across a rocky skeleton. Farther inland, hills coalesce into spinelike ridges, some with exposed rock, running for miles along the backbone. Valleys between rocky ridges collect water into streams flowing with an energy lacking on the plains to the south. It's a new landscape to explore, one alive with biodiversity and beauty unlike anything we have thus far witnessed.

Central Alabama is where the great Appalachian Mountains begin (or end). The four level III ecoregions comprising the Southern Appalachians are the Ridge and Valley, the Southwestern Appalachians, the Piedmont, and the Blue Ridge; the latter isn't in Alabama. The Ridge and Valley is the longest and most consistently mountainous. As its name hints, long low ridges and narrow valleys are characteristic. You can enter the mountains where they begin near Centerville, then journey 900 miles (1,448 km) to southeastern New York without ever leaving the ecoregion's sheltered valleys.

The tortured topography of the Ridge and Valley and the rest of the Appalachians is the legacy of a dramatic geologic upheaval. Three hundred million years ago (during the Pennsylvanian Subperiod) the Gondwana supercontinent began ramming the eastern margin of North America to eventually form an even larger supercontinent, Pangaea. The collision zone was rocked by earthquakes and punctured by volcanoes. The lands becoming the Ridge and Valley crumpled under the strain. Crustal slabs were shoved up and onto land farther away from the collision, a process known as thrust faulting. Elsewhere, sidelong pressures compressed the crust into a series of elongated domes

Long ridges and valleys typical of the Ridge and Valley ecoregion, Ruffner Mountain Nature Preserve, Birmingham. Courtesy of Bob Farley/F8PHOTO.

and valleys. The domes fractured and rain, wind, and gravity began eroding the rubble. Softer rocks, including limestone, shale, and siltstone, yielded first; harder rock, like sandstone, weathered more slowly. Mountain domes, potentially thousands of feet higher than today's mountains, were reduced to valleys whose adjacent ridges are remnants of the domes' bases.[1] Thrust fault ridges were worn down to nubs of the most resistant rock. River systems carried the eroded sediments to the nearby oceans, where they became seafloor sediments and eventually rock. Though evidence is accumulating of a less dramatic and more recent mountain-building period, these older events established the fundamental geology and topography of the Southern Appalachians.

Because of this violent birth, the Ridge and Valley's geology and topography are complex. Geologists have subdivided Alabama's portion of the region, which they call the Tennessee Section of the Valley and Ridge Province, into seven physiographic districts. Biogeographers have used a simpler approach, partitioning the area into two valley and two ridge level IV ecoregions.[2] Due to the region's fine-scale complexity, neither mapping system is perfect; valley units contain ridges, and ridge units contain valleys.

Our ramble through the Ridge and Valley begins in the upland ecoregions. We'll study a bit more geology, then hike across one of the ecoregion's prominent mountains to observe how the rugged terrain shapes ecosystems. Then we'll explore the ecoregion's lowlands, including two recently discovered ecosystems, and finish with a tour of the ecoregion's nature preserves.

HOGBACKS, WINDGAPS, AND KNOBS

The Southern Sandstone Ridges level IV ecoregion ranges throughout the Southern Appalachians and in Alabama includes the Ridge and Valley's most prominent mountains. These elongated ridges run southwest to northeast and are topped with a thick and durable sandstone cap whose origins can date to 500 million years ago. Ridgeline peaks can be 2,030 feet (619 m) above sea level and 1,000 feet (305 m) above the local terrain. Some ridges are just a few feet across and are called hogback ridges for resembling the raised spine of a feral pig. Broader ridges support ecosystems similar to the table plateaus of the Southwestern Appalachians level III ecoregion to the north.

A single mountain can run for miles, but notches in the ridgeline, known as windgaps, are frequent. Faults and subsequent erosion created some, while others are the remnant paths of streams crossing the landscape before the most recent phase of crustal uplift. Larger windgaps provided opportunities to thread highways and railways through the mountains and were important travel routes for Native Americans and white settlers. During the Pleistocene, windgaps were probably important corridors for seasonal migrations of bison, mammoths, and other megafauna.

The other upland ecoregion of the Ridge and Valley is the Southern Dissected Ridges and Knobs level IV ecoregion. It is a small collection of low knobs, hills, and short ridges just west of Sylacauga. The ecoregion also occupies a small area of Georgia and is widespread in eastern Tennessee. In Alabama, these mountains rise 400–700 feet (122–213 m) above the local terrain and are partially formed of metamorphic rocks, including slate and quartzite, created by tectonic pressure as the Appalachians grew. These rocks resist erosion more than those of the surrounding terrain.

The Ridge and Valley's mountains and tall hills are mostly forested because their steep slopes and rocky soils have shunned agriculture. From afar, these forests appear as a monotonous carpet of green. On closer inspection, there is ecological variation governed by rapid transitions in bedrock type, aspect, elevation, and slope position. To understand this complex ecology, we'll visit Oak Mountain State Park near Pelham and hike up and over Double Oak Mountain.

OAK FOREST INCLINE

We'll begin midmorning early in the summer along the banks of Hogpen Creek in the foothills north of the mountain. Compared with where we're headed, streamside life here is easy. Soils are thick and fertile, the terrain is level, and water is abundant. The common trees of this South-Central Interior Small Stream and Riparian ecosystem are competitive species suited to benevolent conditions, including Tulip-poplar, Red Maple, and Winged Elm.

We leave the tranquil streamside and bushwhack up the nearest slope toward the ridgeline. It is hot and the grade is steep, so after several minutes we pause to rest and study the forest. The forest is dramatically different from the riparian forest below. Because of gravity's insistence, the soils are thinner and drier. Accordingly, trees are shorter, the forest is more open, and oaks tolerant of these conditions prevail, especially Northern Red, Southern Red, and Black oak. This is the Allegheny-Cumberland Dry Oak Forest and Woodland, one of the most widespread and variable ecosystems in Alabama's mountains.

Also shaping this forest is the slope's aspect. This northwestern flank of the mountain receives sunlight at an oblique angle.[3] While there is less light for photosynthesis, it is cooler and the soils dry out slowly. Consequently, the growing season is longer on these northwestern slopes than elsewhere on the mountain but is shorter than in the valley below. These conditions cause wildfires to be infrequent and favor the survival of many plants and animals more typical of regions farther north.

Near where we rest is an oak downed during a recent spring storm. Its exposed roots still grip chunks of sandstone ripped from the bedrock. Rooting on steep slopes is a challenge for large trees. The sandstones and shales of these slopes weather into slightly acidic soils, but other mountains in the Ridge and Valley expose dolomite and limestone, rocks producing alkaline soils supporting Southern Ridge and Valley/Cumberland Dry Calcareous Forest and trees rare on Double Oak Mountain.

We see signs this forest is changing. Logging crews swept through most upland forests of the Ridge and Valley several times before the end of the past century, including the park's forests. Some of the pioneer trees that colonized logged forests still survive in the canopy above us, including Virginia Pine, Black Cherry, Blackgum, and Persimmon. These species cannot tolerate shade and have mostly yielded to the oaks. However, the oak-filled forests inspiring the mountain's name are also temporary. Highly competitive, shade-tolerant trees, especially Pignut and Mockernut hickory, American Beech, and Southern Sugar Maple, have established and will replace many oaks over the next century.

We resume our climb, but the steepening slope slows our progress. It also affects the forest. The soils are even thinner, rockier, and drier here than below and the trees are shorter. Chestnut Oak is the dominant tree. The stress, plus frequent exposure to strong winds, has made the older trees stocky and gnarled.

Eventually we near the sandstone ledge along the ridgeline. The short trees here sometimes yield to shrub thickets. Vulnerable to browsing herbivores, these shorter plants defend themselves with stiff twigs and thorns that scratch and tear as we push past. With the ledge in sight, we pick our way carefully up a field of lichen-encrusted boulders. This talus reminds us that mountains are impermanent, even those capped with a sandstone slab. Just beneath the ledge the grade steepens and we grasp at boulders, roots, and tree trunks to pull ourselves higher.

When we reach the ledge we notice its own micro-ecosystem. The gray, weathered rock is encrusted with lichen patches. Mosses and ferns absent on the forest floor grow where soil has accumulated in cracks and small ledges. These are tough organisms, but they wouldn't survive in the shade of the forest far below. Some ledge fissures are several feet deep. In one, we find a freshly shed snakeskin, warning us to be vigilant—outcrops are a favorite hangout for venomous snakes.

We scramble through a cleft to the top of the ledge and are rewarded with a sweeping view of the Cahaba River Valley and adjacent ridges and hills. A rising breeze cools us and carries the song of a Scarlet Tanager, a gaudy bird of the oak forests below. A Turkey Vulture sails by at eye level. Soaring birds often coast great distances on winds driven upward by mountain slopes. After resting and enjoying the vista, we turn to the low trees and shrubs surrounding the outcrop. These hardy plants are stunted for all the reasons we noted earlier and include Sand Hickory, Scarlet Oak, Mexican Plum, and several blueberries. This community is another variation of the Allegheny-Cumberland Dry Oak Forest and Woodland ecosystem.

Like every mountain, Double Oak Mountain has a quirk. Instead of one, it has two ridgelines (hence the "Double") and a shallow valley between them. Formed by thrust faulting, the mountain is two thick slabs of crust stacked upon one another. Shackleford Gap Fault, the fissure between the slabs, lies beneath the perched valley between the ridges. The sandstone of the northern slab is where we are resting and the southern slab forms the mountain's southeastern flank.

We head toward the southern ridge through a short woodland of hickories, oaks, and shrubs. With its southeastern aspect, the slope intercepts more sunlight than the mountain's northwestern slopes. The drier conditions and the near-surface bedrock keep the canopy low. At the valley bottom runs Peavine Branch, the stream draining this portion of the mountain. Near the stream we enter a Cumberland Seepage Forest whose boggy soils harbor moisture-loving plants including Tulip-poplars, Royal and Cinnamon Ferns, Mountain Laurel, and the rare Mountain Camellia. This ecosystem receives abundant water leaking from the surrounding sandstone slopes. It seems far too lush and tranquil to be resting on a deep geologic fault. Though the shade and cool stream are welcoming, we don't linger—we are eager to reach the southern ridge, where we'll find two new ecosystems and Alabama's only endemic tree.

PINE WOODLAND DECLINE

Reaching the southern ridgeline of Double Oak Mountain, we find a view of a forested valley and the neighboring mountain. Below us is a sunny woodland dominated by Longleaf Pine and attended by Black-jack Oak. This is the mountain longleaf ecosystem, a version of the Southeastern Interior Longleaf Pine Woodland.

To those who associate longleaf woodlands with ecoregions below the Fall Line, this ecosystem in the mountains may seem wildly out of place. Yet there are fundamental similarities between these ecosystems. Sandstone lies beneath this slope all the way to its base; its rocky soils are thin, sandy, and nutrient-poor; and with its southeastern aspect, this slope receives a hard dose of sunlight each clear day. These factors impose droughtlike conditions on this mountainside much of the year, conditions like those in many environments where Longleaf Pines reign to the south. Moreover, this ecosystem needs fire to survive. While the dry, thin soils slow the invasion of pioneers, fire is needed to repulse the aggressive trees, especially Red Maple, Sweetgum, and

High-elevation Allegheny-Cumberland Dry Oak Forest and Woodland, Mountain Longleaf National Wildlife Refuge, Calhoun County. Photo by R. Scot Duncan.

Above: Mountain longleaf woodland, Oak Mountain State Park, Shelby County. Photo by R. Scot Duncan.

Right: Boynton's Oak, Alabama's only endemic tree species. Photo by R. Scot Duncan.

Blackgum, that are the vanguard for an invasion that would replace this rare woodland with a shady forest of species common in the region.

Once common in the Ridge and Valley and Piedmont ecoregions of Alabama and northwestern Georgia, most mountain longleaf ecosystems were wiped out by logging and urbanization. Fire suppression is snuffing out many of the remnants. Previously, lightning-sparked fires roamed the ecoregion unhindered, revisiting mountain slopes every 5–10 years and maintaining the open character of longleaf woodlands and other highland ecosystems. Today, wildfires are rare and are quickly extinguished to protect people and property. Unburned woodlands are becoming shady, cool forests with less unique biodiversity.

One of Alabama's flagship endemic plants is threatened by the decline of the mountain longleaf woodland. Boynton's Oak is the state's only endemic tree, as well as one of the only trees endemic to a single state.[4] It is small and easily overlooked, but here along the ridge we find it everywhere. It forms scrubby thickets, but occasionally one shares the canopy with other woodland trees. While scientists haven't yet carefully studied it, Boynton's Oak seems to need abundant sunlight and depends on thin soils and fire to prevent more competitive species from crowding it out. Like other scrub oaks, it survives fire and resprouts from the roots.

Hiking down the southeastern slope, we notice signs that Boynton's Oak and the mountain longleaf woodlands of the park are in peril. Fires are now rare, shrub thickets have formed, the canopy is closing, and fire-intolerant trees are invading. The transformation is most apparent in side-valleys where soils are deepest.

We weave through scratchy shrub thickets and small meadows and scamper down small ledges until the woodland abruptly opens to an expanse of bedrock, open sky, and sparse vegetation. This exquisite ecosystem is a sandstone glade. Countless seasons of wind and rain have worn the bedrock smooth. Small, gnarled trees and weathered boulders interrupt the surface. Crisscrossing fissures and their shallow, gritty soils sustain small hardy plants including Little Bluestem, Nuttall's Rayless Goldenrod, Talinum, Pineweed, Smallhead Blazing Star, and Eastern Pricklypear. They enjoy plentiful sunlight, but must endure desertlike conditions.

Sandstone glade and scattered wildflowers near Gadsden, Etowah County. Courtesy of B. William Garland.

Glades are open ecosystems where bedrock exposure and extreme soil conditions keep trees away. They are rare in the Southeast, but Alabama has glades of sandstone, limestone, granite, and dolomite. All have thin, dry soils and provide little rooting space and few nutrients. These ecosystems harbor unusual plants that handle brutal extremes but cannot survive alongside more competitive plants dominating milder environments. Their adaptations for survival are manifold. Many have small or narrowed leaves that efficiently retain water and avoid overheating. Some have hairlike trichomes covering their leaves and stems to reflect sunlight and reduce water loss. A few succulents store water in their tissues for periods of drought. Stress in the glades peaks in summer, so many species complete growth, flowering, and seeding in spring.

There are a few short trees in the glade. One is the Georgia Oak, a rare species found only in a few counties in Alabama, Georgia, and South Carolina. Most of the other trees are Longleaf Pines. They photosynthesize during the few days each year when water is available. The stress and slow growth has wrought them into marvelously gnarled forms that made them undesirable to loggers. Many are legacy trees over 200 years old. Rarely, a larger pine has a long-abandoned cavity of the Red-cockaded Woodpecker. RCWs maintain a handful of populations in the Ridge and Valley, but they abandoned Double Oak Mountain last century (last sighted in the 1970s) when the forest grew too dense for the bird's liking.

Ridge and Valley sandstone glades are abundant on several of the south-facing slopes of the Southern Sandstone Ridge's taller mountains. Those on Double Oak Mountain extend for miles along the mountain's length. Ecologists have largely overlooked this ecosystem. Ridge and Valley sandstone glades lack their own NatureServe ecological system designation and are rarely mentioned in the scientific literature. There are sandstone glades of the Southwestern Appalachians level III ecoregion, but those differ greatly by having shallow slopes and a different flora.

We've no more new ecosystems to explore on this mountain, so we head back. Our hike revealed how rapid topographic and geological change creates a remarkable range of ecosystems in the Ridge and Valley's uplands. We'll next explore the ecoregion's valley lowlands and a similarly diverse range of ecosystems and unusual species.

VERDANT VALLEYS

Below the Ridge and Valley's rocky uplands are the floodplains and gentle hills of its valley ecoregions. These lush landscapes provided early agriculturalists with an ideal combination of resources: abundant water, level terrain, and rich soils. Settlers converted nearly every corner to agriculture and timberland in the 19th and early 20th centuries. The valleys are so amenable to agriculture that most areas are still in agriculture and low-lying hills are in timber, pasture, or hay production.

Valley terrains also fostered the growth of cities. Birmingham, Alabama's largest city, was founded in Jones Valley to tap the area's mineral wealth—all ingredients for making iron were accessible within a few miles of the city's center. The valleys' level terrain eased the building of factories, commercial districts, and the all-important railroads. Clear running creeks, springs, and rivers supplied water for people and industry and conveyed their wastes downstream.

Despite nearly wholesale transformation, a handful of unusual eco-systems and species survive in the Southern Shale Valleys (hereafter, "Shale Valleys") and the Southern Limestone/Dolomite Valleys and Low Rolling Hills (hereafter, "Limestone Valleys") level IV ecoregions. Their stories provide us with glimpses of the region's original biodiversity. Once again, these stories begin with geology.

The Shale Valleys are mostly underlain by shale, a sedimentary rock that resists water infiltration and limits groundwater supply. Other sedimentary rocks, including sandstone, mudstone, limestone, and dolomite, are also present. The soils weathered from shale, sandstone, and mudstone are slightly acidic, while those from limestone or dolomite are alkaline, and the flora of the Shale Valleys varies accordingly. A smattering of karst features, including springs, sinkholes, and caves, exist where alkaline rocks prevail.

The Limestone Valleys are less variable, but with abundant limestone and dolomite, its karst features are more developed. Most traveling these valleys are unaware that below their feet is a three-dimensional catacomb of caverns and underground rivers. Cave entrances are in upland areas where groundwater does not saturate the bedrock, while springs emerge where subterranean streams breach the surface. These springs harbor some of the Southeast's rarest animals.

THE SPRING EFFECT

The Watercress Darter is one of these rare animals. Its survival story is a nail-biter, for the entire distribution of this endangered fish is within the Birmingham metropolitan area. Watercress Darters prefer submerged vegetation, including its namesake plant, Watercress, a Eurasian crop species brought to the Southeast early last century. This fish is small and adept at hiding, so most who visit its home springs will never see it. This is regrettable, for during the mating season the male's flanks are a dazzling orange, its dorsal fins are banded blue and red, and other fins are turquoise.

The darter is tightly dependent on springs. Its colors are useful in courtship and its favorite aquatic plants flourish only because spring water is reliably clear. Spring water is also stable in temperature, chemistry, and flow rate, which enables the plants of the spring to thrive

during the harshest weeks of winter and summer. These conditions form a unique ecosystem that in the Southeast's highlands has facilitated the evolution of endemic snails, crayfishes, and fishes like the Watercress Darter.

When Birmingham was founded in the late 19th century, the darter likely inhabited many of the area's abundant springs. Today, however, the darter inhabits only four springs within its original range.[5] Soon after its discovery in the 1960s, it was federally listed as endangered because of its limited distribution within a large metropolitan area. The US Fish and Wildlife Service (USFWS) then established the Watercress Darter National Wildlife Refuge in Bessemer to protect one of the populations.

Springs saved the species from extinction. In what ecologists have dubbed the "urban stream syndrome," urbanization of a watershed triggers the decline of stream ecosystems. Forests are converted into impervious surfaces that prevent rainwater from penetrating the ground, while stormwater systems shunt this rainwater to streams, causing flash flooding and in-stream erosion. During droughts, water levels plummet because groundwater reserves haven't recharged. Springs, however, have small surface watersheds and tap into limestone aquifers. Thus, while Birmingham grew, four of the darter's springs endured and saved the species.

An even rarer spring dweller is the Pygmy Sculpin, a strange little fish living only at the head of Anniston's Coldwater Spring. The sculpin has a name and a face over which only an ichthyologist could fawn. With its big head, upturned mouth, and large eyes, it resembles an ill-tempered underwater toad. As its name implies, it is very small (1.5 inches, 3.8 cm) relative to other sculpin. Its black-and-brown mottling provides camouflage to evade predators and ambush prey. The big head and wide mouth of this bottom dweller match a voracious appetite for prey much larger than what seems reasonable for a small fish. Individuals of similar sculpin species have died when trying to swallow fishes larger than themselves. One of the Pygmy Sculpin's foods is the Sculpin Snail, a species endemic to Coldwater Spring.

Male sculpin invest considerable effort in raising young. Males attract females by constructing and defending shallow gravel nests. Successfully courted females deposit eggs in the nest, which the male then fertilizes. He then guards the offspring until they hatch and are independent. Should a male lose his nest to a challenger, the victor will often cannibalize the loser's young.

The Pygmy Sculpin and the Sculpin Snail are highly vulnerable to extinction due to their restricted distribution. Though Coldwater Spring is protected, up to half of its daily output is withdrawn to supply water to residents in and near Anniston.

FLATWOODS SURVIVOR

Long lists of aquatic species are usually cited to demonstrate the importance of Alabama's rivers to biodiversity. Yet, a few terrestrial organisms, plants mostly, are strictly dependent on rivers for shaping the surrounding landscape. One of these is the Alabama Leather Flower, a small vine of the Coosa River Valley with a blue-violet, bell-shaped flower. The leather flower is a highly endangered plant known only from a few sites on the Coosa River's floodplains. All known populations survive along pasture edges and road and gas pipeline right-of-ways, habitats seemingly unrelated to the river. The explanation for this paradox involves the Coosa River's success at grinding down the Southern Appalachian Mountains.

The Coosa is powerful. Its headwaters have pulled countless layers off the Ridge and Valley's mountains. The river channel has worn through mountains and replaced them with broad floodplains where, prior to damming of the river, limestone and dolomite clays were deposited during annual floods. Because these dense alkaline clays resist water infiltration, Coosa River Valley floodplains remained under several inches of water during late winter and spring.

What developed on the floodplains is a forest known as the Coosa flatwoods, a version of the South-Central Interior Large Floodplain ecosystem. Trees tolerant of wet alkaline soils populated flatwoods, including Red and Silver Maple, Sycamore, and bottomland oaks including Willow, Water, Shumard, Overcup, and Cherrybark. Flatwoods could extend several miles from the river and into the floodplains of tributaries. Smaller versions of the flatwoods would have occupied the valley floodplains elsewhere in the Ridge and Valley.

Some botanists believe the Coosa flatwoods were the original habitat of the Alabama Leather Flower. Because the vine needs abundant sunlight, they suggest the species grew where trees had died and sunlight streamed into the understory. In the few remaining flatwoods, such openings house an impressive flora, including the highly endangered Green Pitcher Plant.[6]

The flatwood expanses were not to last. Most were converted to pastures, farms, and tree plantations. Others were swallowed by the river's reservoirs or settlements. Damming of the river and its tributaries altered the hydrology of the few remaining forests. Today, the Alabama Leather Flower persists in only five small locations. All are highly vulnerable spots regularly disturbed by human activities. There is good news, however. Two populations are now on protected lands and assessments suggest the species could easily dodge extinction if it can be established in enough protected and well-managed habitats.

LOST PRAIRIES OF THE COOSA RIVER VALLEY
In 1994, a 10-foot-tall (3 m) sunflower species was resurrected from extinction. Botanists found the Giant Whorled Sunflower during the exploration of a new ecosystem previously unknown to science, the Coosa River Valley prairie. These prairies had gone undetected since most were destroyed in the early 19th century. A handful of former prairie sites have been identified based on the presence of prairie wildflowers growing in pasture margins and roadside right-of-ways. However, with

so little prairie remaining, ecologists and botanists are struggling to understand how it formed, where it was distributed, and what it was like originally.

The prairies were on a foundation of limestone in the low hills of the Shale and Limestone Valleys ecoregions. They ranged from glade and gladelike environments on thin soils to small patch prairies on deeper soils (these are the Southern Ridge and Valley Calcareous Glade and Woodland, and Southern Ridge and Valley Patch Prairie ecosystems, respectively). These ecosystems (hereafter, the "Coosa Valley prairies") harbor a similar suite of grass and wildflower species adapted to alkaline soils. Many are also found in Alabama's blackland prairies and the Midwest's tallgrass prairies. A few, like the aforementioned sunflower, are rare species surviving only in a few southeastern locations.

Fortunately, a handful of intact Coosa Valley prairies survived in one location. Unfortunately for Alabama, they are just over the border in Floyd County, Georgia. Embedded within lands managed for commercial timber production, these relict ecosystems persisted on harsh soils where timber wouldn't grow. After their discovery, the owner (Temple-Inland) generously donated nearly a thousand acres via a conservation easement to the Georgia Chapter of the Nature Conservancy (TNC) to protect the prairies. They are small, but they are the most floristically diverse ecosystems in Georgia, sheltering such rarities as Mohr's Barbara-buttons, Prairie Rosinweed, Sideoats Grama, Smooth Blue Aster, and Wavyleaf Purple Coneflower.

The Giant Whorled Sunflower is probably the most endangered of the Coosa Valley prairie species. This stunningly tall species was discovered in 1892, but known populations were soon lost and for most of the past century the species was declared extinct. Then, 102 years later, it was rediscovered during surveys of the prairies and two nearby locations in Alabama. Remnant populations were precariously small, and the plant became a candidate species for the US Endangered Species List.[7] After several years of protection and nurturing, the number of sunflower stems counted during annual surveys was increasing. However, subsequent genetic analyses revealed that most new stems were sprouts from the same individuals. Thus, the number of genetically distinct survivors is vastly smaller than previously believed, and the need for the sunflower's study and protection has intensified.

The Coosa Valley prairies provide hints about the region's original ecology. All prairie plants need full sun, but prairies are continually under the threat of tree invasion. While the prairie's clayey, calcium-rich soils, especially the thinner soils, slow the invasion of most trees, wildfires must have regularly purged the prairies of trees like Eastern Red Cedar, which is tolerant of prairie soils. Because Coosa Valley prairies can be separated by dozens of miles, lightning-ignited fires were either numerous, sprawling, or both. Supporting this deduction, early accounts of the Coosa River Valley report that the slightly higher terrains adjacent to the Coosa flatwoods were woodlands of Longleaf Pine, Shortleaf Pine, and upland oaks. Such woodlands only persist where fires are regular. Thus, it seems the Coosa River Valley was a re-

Above left: Patrick Thompson (Auburn University) inspects a Giant Whorled Sunflower from a reserve population at the Davis Arboretum. Courtesy of Patrick Thompson/AU Davis Arboretum.

Above right: Close-up of Giant Whorled Sunflower. Courtesy of Alan Cressler.

gion that burned regularly and supported ecosystems quite similar to those found in the Southeastern Plains. Indeed, several biologists, from herpetologists to botanists, have written on the Coosa River Valley's likeness to ecoregions below the Fall Line.

BIBB COUNTY SURPRISE

As exciting as the discovery of the Coosa Valley prairies was, an even more dramatic finding occurred about the same time in Bibb County. In spring 1992, Jim Allison and friends were paddling down the Lower Little Cahaba River. Allison, an accomplished botanist, was conducting a streamside survey and noticed an open rocky habitat on a nearby hillside. Stopping to investigate, he discovered a previously unknown glade ecosystem populated by extremely rare plants and, more intriguingly, many wildflowers he did not recognize. After months of careful fieldwork and taxonomic study, Allison and colleague Timothy Stevens documented eight plants completely new to science and wholly endemic to the few dozen dolomite glades occupying this small area of the county.[8] In addition, he found 7 species previously unknown in Alabama and 60 rare and endangered plants in the glades and adjacent forests and rediscovered one species, Dwarf Horse-nettle, that had not been seen for over 150 years. Such spectacular discoveries are typically the lore of the 18th and early 19th centuries. Like the discovery of the Coosa Valley prairies, Allison's finding serves as a reminder that we still have much to learn about southeastern ecosystems.

The Ketona Dolomite glades (technically, the Alabama Ketona Glade and Woodland ecosystem) are similar to other glades: bedrock is at or near the surface; soils are thin, droughty, and nutrient-poor; trees are few or absent; and herbaceous vegetation is sparse and composed of hardy plants, most of which complete their life cycle before summer. The dolomite-weathered soils have extremely high magnesium concentrations that may prevent the invasion of the glade by some trees and other plants. Fires arriving from adjacent longleaf woodlands also prevent invasion, especially where soils are deeper.

Like most outcrop ecosystems, the glades occur in a narrow topographic zone with distinctive surface geology. In this case, the glades occupy a narrow 11.2-mile (18 km) tract within Bibb County along a boundary between the Limestone Valley and Southern Sandstone

Ridge ecoregions on outcroppings of the Cambrian-aged Ketona Dolomite. No other dolomite glades are known in the Southeast. Glades vary in size from a few square meters to 12 acres (0.05 km²). Some stand alone; others are in loose clusters.

The glades are an attractive landscape. Their slopes of fine white gravel and thin soil are interspersed with low ridges of weathered blue-gray bedrock. Where soil has accumulated, there are tangles of grasses and wildflowers, while areas with thin soils sustain the most drought-tolerant plants. Above the glades are longleaf woodlands, and below them are dense riparian forests. Several glades are perched on bluffs overlooking the nearby river, whose down-cutting over millions of years may have helped expose the dolomite. Most glades are privately owned, but TNC protects several in the Kathy Stiles Freeland Bibb County Glades Preserve.

Chris Oberholster (TNC) surveys a Ketona Dolomite glade at the Nature Conservancy's Kathy Stiles Freeland Bibb County Glades Preserve. Courtesy of Beth Maynor Young/Cahaba River Publishing.

New plants discovered in the Ketona Dolomite glades. *Clockwise from top left:* Sticky Rosinweed, Alabama Gentian Pinkroot, Cahaba Paintbrush, and Cahaba Daisy Fleabane. Courtesy of Sara Bright, Alan Cressler, B. William Garland, and Sara Bright.

PRESERVING THE RIDGE AND VALLEY

Fortunately, the Ridge and Valley's uplands are still heavily forested, and a constellation of nature preserves provides opportunities for exploration. The largest is the Talladega National Forest's Shoal Creek Ranger District in northeastern Alabama. Most of the district is within the Piedmont level III ecoregion, but its northernmost tier lies within the Ridge and Valley's Southern Sandstone Ridges. A mosaic of shortleaf, longleaf, and dry oak woodlands interspersed with seepage bogs once blanketed the district's steep slopes. A century of logging, fire suppression, and plantation forestry left the original ecosystems in disarray, but biodiversity conservation has become an important management priority and woodland restoration is under way.

Dugger Mountain Wilderness Area is within the Shoal Creek District's Ridge and Valley territory. Wilderness areas are federal lands on which management is limited to maintaining hiking trails. The mountain belongs to the Weisner Ridges physiographic district and boasts the second-highest peak in Alabama (2,140 feet, 652 m). Seven hundred and fifty plant species have been cataloged within its 9,200 acres (37 km²). This diversity survives largely due to the mountain's history of repelling logging on its steep slopes. There are stands of Virginia Pine that established when settlers cleared the lower elevations long ago, but higher elevations sustain shortleaf and longleaf woodlands and variations of dry oak forest. Unfortunately, an insufficiency of wildfire is allowing broadleaf forests to replace the woodlands.

We can thank the US Army for some of the finest mountain longleaf woodland left in the Southeast, and they saved it through regular bombardment. This woodland is on Choccolocco Mountain, the largest of the Weisner Ridges. Choccolocco begins just south of Piedmont, in the Shoal Creek District, and ends over 20 miles (32 km) to the southwest, just east of Anniston. The southern end of the mountain lies within the Mountain Longleaf National Wildlife Refuge, but previously it belonged to Fort McClellan. Beginning in 1917, the mountain's woods were the setting for combat simulations, and frequent wildfires produced by artillery practice maintained the longleaf woodlands and other fire-dependent ecosystems. Since it was unsafe to send soldiers where unexploded ordinance (UXO) was abundant, fires were allowed to burn. Some of the Longleaf Pines are legacy trees predating the arrival of white settlers. Tree-ring analysis of a pine felled in a storm revealed it was a seedling in 1713.

Permanent closure of the fort began in the 1990s and involved the transfer of 9,016 acres (36 km²) to the USFWS with a mandate to protect and restore mountain longleaf woodland. Since the refuge's establishment in 2003, removing the UXO still littering a third of the refuge has been the priority. The preservation and restoration of longleaf woodlands through thinning and prescribed fire has begun. The protection of other ecosystems is also a priority, for many other rare species dwell on the mountain, including the southernmost population in the eastern United States of Ground Juniper.

Through its Forever Wild program, the state of Alabama has also protected lands for biodiversity conservation and outdoor recreation in the Ridge and Valley. Tracts purchased on Indian Mountain in Cherokee County have secured 594 acres (2.4 km²) of dry oak forest and woodland to help link the Pinhoti hiking trail to the start of the Appalachian Trail in Georgia. A tract on Coldwater Mountain provides recreational opportunities and safeguards the watershed of Coldwater Spring. The preserve protects woodlands and forests typical of the ecoregion plus two sinks where surface streams disappear into the earth. A Forever Wild addition to Ruffner Mountain Nature Preserve in Birmingham expanded recreation and preserved mining ruins and dry calcareous forest on Red Mountain. Several miles to the north, the Turkey Creek Tract protects steep valley forests plus a mile (1.6 km) of Turkey Creek. The creek is home to the Vermilion Darter, a brilliantly colored fish occupying only 13 miles (21 km) of streams in the Turkey Creek watershed. Additional Forever Wild investments preserved the Cahaba River WMA and expanded the natural areas protected at Tannehill Ironworks Historic State Park.

The crown jewel of the Ridge and Valley's nature preserves is the Cahaba River National Wildlife Refuge in Bibb County. The refuge is one of the only places where the public can explore a free-flowing montane river ecosystem in Alabama. Established in 2002, the 3,800-acre (15.7 km²) refuge protects habitat for five federally listed (threatened or endangered) species and mountain longleaf woodland. The main attraction is the Fall Line shoals. Across this shallow expanse, hundreds of rivulets tumble between boulders and clusters of American Water-willow and Shoals Spider-lily. The lily, known in Alabama as the Cahaba Lily, produces exquisite white blooms, and their display on the shoals draws hundreds of visitors each spring. There are river beaches of tan sand strewn with artifacts telling of the ecoregion's history and biodiversity. Weathered driftwood comes from forests high in the watershed. Shiny coal fragments are a legacy sediment from historic mines. Polished river glass and ceramic shards are the litter from abandoned mining settlements. The river has sculpted Pennsylvanian-aged sandstone bedrock jutting from the sand into unexpected shapes. The rocks

are textured with arrangements of orange, brown, red, and purple minerals, details most noticeable in winter when more vibrant colors are dormant. Freshets cast up spent shells of native snails and mussels. Older mussel shells are bleached and worn, but fresher specimens have an interior luster of pearly white, cream, peach, or purple. Having survived the past century without damming, the Cahaba is one of the few southeastern rivers where one can find such molluscan diversity.

PLATEAUS TO COME

When we began exploring this ecoregion, I likened the Ridge and Valley to an ancient rocky leviathan rising from the sea-like plains. This excessively dramatic analogy highlights the remarkable topography the ecoregion brings to the Southeast. Then again, there is a softer dimension to the landscape. Sure, there are rugged forested ridges, but there are also gentle lowlands similar to the plains to the south. Perhaps juxtaposition of the rugged and the gentle best characterizes the Ridge and Valley. Regardless, this topographic diversity is why the ecoregion sustains such impressive biodiversity.

Next, we journey through the Southwestern Appalachians, another mountainous level III ecoregion. This is a tableland featuring sandstone plateaus, vertical escarpments, and underground caverns. Its ecological and species diversity is paramount within Alabama's highlands.

11

Southwestern Appalachians

From Seafloor to Plateau

The Southwestern Appalachians level III ecoregion is a naturalist's wonderland. Extending from Kentucky to Alabama, the ecoregion is a high sandstone plateau on which the elements have been carving for millions of years. Where sandstone endures, there are woodlands, bogs, glades, and barrens. Where it failed, water has sluiced through soft rock below and created cliffs, escarpments, canyons, and one of the world's most extensive cave systems. With topography this complex and a warm, wet climate, this ecoregion is among the most biodiverse on the continent and hosts dozens of endemic species.

It all began 354 million years ago in the Mississippian Subperiod when the rocks of the Southwestern Appalachians began forming. During this time, a slow collision of the earth's major continents was creating a mountainous debris pile on the fringe of North America. The mass of the young Southern Appalachian Mountains caused downwarping of the crust in the adjacent region. The shallow sea this created deepened as the mountains grew. The basin first supported a thriving marine reef environment, but in the Pennsylvanian it received great volumes of eroded sediment from the adjacent mountains. Over tens of millions of years, debris from shallow reefs became limestone, seafloor muds became shale, sand became sandstone, and coastal plain swamp deposits became coal.

This continued until about 285 million years ago when in the early Permian the intercontinental collision ended. The shifting stresses on the crust forced the basin to rise several thousand feet above sea level to form a large plateau region. The land mostly rose evenly, but uplift created several broad domes and valleys. The domes fractured extensively during uplift and were soon washed away. All remaining of them today are long anticlinal valleys. In a strange twist of geologic fate, the valley bottoms between those domes persist as the high plateaus of the Southwestern Appalachians.

The Southwestern Appalachians supply much of the biodiversity making the Southern Appalachian Mountains a global biodiversity hot spot. Though settlers transformed the plateau tops and valley bottoms into cropland, pasture, or water reservoirs, the rugged plateau margin, or escarpment, shelters many natural ecosystems. Because the escarpment also provides exciting topography and dramatic vistas, nature preserves are abundant and attract ecotourists. There are six state parks, two national wildlife refuges, a national forest, a wilderness area, a national monument, a national preserve, several holdings of the Alabama Chapter of the Nature Conservancy (TNC), a handful of Forever Wild preserves, and a few commercial preserves. We'll visit several protected areas in our tour of the Southwestern Appalachians, beginning on Lookout Mountain, where we'll find spectacular terrain and the best opportunities to study plateau ecology.

Lookout Mountain Anatomy

Lookout Mountain is a long, narrow plateau beginning near Gadsden and ending at Chattanooga, Tennessee. The mountain has a long history of attracting humans. Native Americans, gold-thirsty conquistadors, and detachments of warring armies have all used it as a refuge and a resource. Settlers last century transformed the plateau into farming communities and, more recently, vacation destinations.

Many ecotourists ascend the mountain to visit DeSoto State Park and Little River Canyon National Preserve. Most come to see the gorges, waterfalls, and forested canyons carved by Little River, one of the few mountaintop rivers in the United States. This unsullied river is the centerpiece of the park and preserve. It is designated an Alabama Wild and Scenic River and is one of the few Outstanding National Resource Waters in the state. The river's challenging whitewater attracts canoeists and kayakers from throughout the country. Regardless of why they come, most visitors would agree that Little River Canyon is one of the most stunning landscapes in Alabama.

The level uplands of Lookout Mountain are part of the Southern Table Plateaus level IV ecoregion. Sometimes referred to as the table-

Winter in Little River Canyon National Preserve, DeKalb and Cherokee counties. Courtesy of Alan Cressler.

lands, this is the largest ecoregion of the Southwestern Appalachians, encompassing the tops of Lookout, Sand, Blount, and Brindley mountains. These relatively flat-topped plateaus have survived for millions of years because a thick layer of the Pottsville Sandstone protects softer rock layers below. Because the soils weathered from this sandstone are acidic, sandy, and nutrient-poor, agriculture conquered the tops of these mountains only after fertilizing and other modern agricultural techniques became widely available.[1]

Still, the sandstone cap is not invulnerable. Water is persistent, powerful, and efficient at finding a rock's weaknesses. Over millions of years, erosion created precipitous slopes along Little River and the margins of Lookout Mountain. These slopes belong to the Plateau Escarpment level IV ecoregion. Their slopes are so steep that they are often managed for timber or just left alone.

RETURNING WOODLANDS

We begin our tour of Lookout Mountain's ecosystems driving across the tableland toward Little River Canyon National Preserve. As we cross the gently rolling countryside, we pass forests with a closed canopy of oak and pine and a dense understory of shrubs and young trees. Are these at all like the presettlement forests? Not likely. Based on early accounts and inferences from the plateau's current ecology, much of the plateau was a woodland of oak and pine with an understory of low shrubs, grasses, and wildflowers. Lightning-sparked fires maintained the open nature of the woodlands. Wildfires wandered the tableland for miles and weeks, driven by fuel availability and prevailing winds, and were stopped only by stream valleys, the escarpment margin, or heavy rain. The fire return interval was probably about five years, longer than intervals in southern Alabama due to the tableland's higher elevation, cooler temperatures, and fewer thunderstorms. Where fires were most frequent, Shortleaf Pine woodlands, a variant of the Southern Appalachian Low-Elevation Pine Forest, prevailed. Areas that burned less frequently sustained mixed woodlands of Shortleaf Pine and fire-tolerant oaks including Southern Red, Blackjack, and Post. Where topography blocked all but the occasional fire, pines ceded to oaks and the Allegheny-Cumberland Dry Oak Forest and Woodland prevailed.

Arriving at the north end of the preserve, we stop at an overlook above Little River Falls, where the canyon begins. Along its 17-mile (27 km) journey through the canyon, adjacent cliffs and steep slopes tower an average of 400 feet (122 m) over the river. Most visitors come for the 11-mile (18 km) scenic drive along the canyon's western rim and its many overlooks and trails. Established in 1992, the preserve is maintained by the National Park Service to conserve the natural landscape, improve its accessibility, and protect several endangered species. The preserve is still growing in scope and size. In early 2009, the Little River Canyon Center was established as an interpretive and educational center, and several weeks later the US Congress approved plans for expansion.

We turn from the river to examine the adjacent forest, which is the best example of the tableland's original woodland. Unlike the forests we saw on our drive, the canopy is more open, pines are common, and we can see several hundred yards into the forest. The pines' lower trunks are blackened from the prescribed fires used to open the forest. It is still part forest and part woodland, but where broadleaf trees have been thinned by fire, grasses and wildflowers flourish. The forests will need continued burning to once again become a thriving shortleaf woodland.

We head into the pines seeking the highly endangered Green Pitcher Plant. In the tablelands, these bright-green, yellow-topped carnivorous plants live at the base of gentle slopes where near-surface sandstone forces groundwater to saturate, and sometimes flood, the soil. While most of Alabama's carnivorous plants reside below the Fall Line, the Green Pitcher Plant has found a niche in the sunny seepage bogs of the mountains. These bogs, varieties of the Cumberland Seepage Forest ecosystem, host species of wetland grasses, wildflowers, and ferns that thrive when frequent fires prevent trees and shrubs from usurping available sunlight.

Arriving at one of the bogs, we are greeted by the purple, orange, and yellow blooms of wildflowers basking in the late-spring sun. The pitcher plants stand out with their bright-green color and tubular leaves known as trumpets. We notice they grow in clusters, with each plant sprouting several of the trumpets used to trap invertebrates. As in the bogs of southern Alabama, the acidic, waterlogged soils here lack key nutrients that captured and digested animals can provide. Strangely, the huge, greenish-yellow, nodding flowers must successfully attract invertebrates, especially bumblebees, for pollination while just inches away the trumpets lure other invertebrates to their death.

After decades of human enterprise, seepage bogs in Alabama's highlands are exceedingly rare and the Green Pitcher Plant faces ex-

tinction. The species survives in only 35 colonies, most of which are in Alabama's Southwestern Appalachians and Ridge and Valley ecoregions. Over half of the remaining colonies have fewer than 50 plants and extirpation is imminent. Most populations are on private lands, which complicates conservation efforts. A few small colonies survive in DeSoto State Park, Little River Canyon National Preserve, and several TNC holdings. For now, the careful attention given to the species by conservation organizations, state and federal agencies, and a few concerned landowners prevents these elegant plants and their ecosystems from disappearing altogether.

We continue hiking and near the top of a gentle rise find another version of the Southern Appalachian Low-Elevation Pine Forest common in the Southern Table Plateaus. Small herds of deer moss (a lichen, actually) and patches of grasses, sedges, and native blueberry species thrive beneath a full canopy of Virginia Pine. The sandstone is near the surface and soils are only a few inches deep. Mostly, soils are dry, but a few spots collect water and sustain hummocks of sphagnum moss. When winter rains are plentiful, soils are moist, but water is scarce in drier months because there is scant groundwater storage. The existence of these forests is something of a mystery. Virginia Pine cannot tolerate fire, yet its saplings grow only in full sun. It seems likely these are replacement forests that colonized barrens (see next section), pasture, clearcuts of Shortleaf Pine woodland, or former stands of Virginia Pine wiped out by catastrophic fire.

Soil depth drives much of the Southern Table Plateaus' ecology. Forest and woodland prevail where soils are deeper; bogs occupy shallow, saturated soils; and Virginia Pine forests grow where soils are shallow and dry. We'll next visit a plateau ecosystem where the thinnest of sandstone soils push organisms to the extreme.

Sandstone Glades

Just down the road at Lynn Overlook we find another view of the deepening canyon and a treeless exposure of the Pottsville Sandstone with intermittent patches of small, colorful plants. This is a sandstone glade, an example of the Cumberland Sandstone Glades and Barrens ecosystem. Both are treeless, but glades have extremely thin soils or none whatsoever, while barrens have just enough soil to sustain herbaceous plants, usually grasses, on at least half their area. Barrens need occasional wildfire to kill invading trees, while glades seem to be maintained by erosion.

Very few sandstone glades survive in Alabama; most are in the

Elf Orpine and other sandstone glade plants, Little River Canyon National Preserve, DeKalb County. Photo by R. Scot Duncan.

Southern Table Plateaus, and a few are known from the Dissected Plateau level IV ecoregion of the Southwestern Appalachians. The best example of a barrens ecosystem is Chitwood Barrens Preserve on Sand Mountain, maintained by TNC. Other sandstone glades exist in the Ridge and Valley ecoregion, but they have steeper slopes than the Cumberland glades and barrens. The shallow slopes of the Cumberland Plateau's glades may be what allows them to support a more diverse flora.

We notice this immediately when we examine the glade at Lynn Overlook. Water seeps from the woodland margin at the top of the glade and slowly trickles across its surface. The water pools in scattered rounded depressions etched into the bedrock over millennia by the acidic groundwater. Such pooling is possible only because the slope is gentle. In the water are semiaquatic plants that grow for only a few weeks each spring. The most conspicuous is Elf Orpine, a small, bright-red succulent forming thick mats.[2] Thin layers of saturated, gritty soil sustain mosses, grasses, and wildflowers. Among the latter, Oneflower Stitchwort's delicate white flowers hover above the rock on needle-thin stems; gaudy Yellow Sunnybells display panicles of bright yellow flowers; and the Little River Canyon Onion, a species endemic to the canyon, spices up the flatrock with accents of pink. Dry patches of sandstone support black, yellow, and grayish-green species of lichen. Cactuses grow in dry areas where enough soil has collected.

The glades are one of the most austere southeastern ecosystems, but their tough plants have all the light they need. However, growth and reproduction are challenging. Though water is abundant in early spring, glades are bone-dry by midsummer. Thus, most species complete their annual growth in just a few short weeks in spring.

Acidic Cliff Dwellers

Let's descend into the rugged terrain of the canyon's Plateau Escarpment ecoregion. Though the escarpment and tableland meet at the edge of the canyon rim, their ecologies are miles apart. To begin, we head upstream to DeSoto State Park and its network of escarpment trails. The park is named after Hernando DeSoto, the first European to explore the Southeast's interior in his unsuccessful and bloody search for gold. Two of his officers explored Lookout Mountain and Little River, hence the inspiration for the park's name. Nearly 400 years later, in 1939, the Civilian Conservation Corps built the infrastructure for the park. The Corps' enduring handiwork includes many of the trails wandering across the escarpment and its forest. Though the escarpment here lacks the depth of the canyon downstream, its gentler slopes allow us to easily sample several escarpment ecosystems.

From Azalea Cascade trailhead, we explore the heath bluff forests lining the slopes above the river. The trails radiating from here are famous among native plant enthusiasts for the spring flowering of rare and showy plants. Beneath a closed canopy of trees that tolerate cool conditions and moist acidic soils—especially American Beech, Red Maple, and Shortleaf Pine—a tangled understory of acid-tolerant shrubs is the main attraction. Most are members of the family Erica-ceae, whose plants are often referred to as "heath." The family's repre-sentatives on these slopes include Native Azalea, Catawba Rhododen-dron, Mountain Laurel, Sourwood, and several blueberries. Wandering the bluffs, we find these shrubs in full bloom, brightening the shady slopes with bursts of pink, rose, yellow, and orange. Where the shrub thicket relents, patches of herbaceous plants thrive, including Liverleaf, Beetleweed, Fairywand, and native gingers.

Heath bluff forests are one of many variations of the South-Central Interior Mesophytic Forest, an ecosystem found throughout the Ap-palachians in moist, sheltered conditions. Acidic groundwater shed by the adjacent tableland keeps heath bluffs wetted. In spring, water satu-rates the slopes and pours from the ground in seeps and small cascades. In summer, the bluffs' sheltered position keeps them moist and cool. Along the Azalea Cascade trail, the steep hillsides above Laurel Creek allow the heath community to extend from the bluffs to a position on the escarpment that elsewhere would be dominated by oak forest.

Within the heath bluffs and other forests of the escarpment are small cliff ecosystems known as the Cumberland Acidic Cliff and Rockhouse. They form where erosion exposed the plateau's sandstone caprock. These short cliffs line many of the bluffs along Little River but are also found at higher elevations where the river ran before it had cut so deeply into the caprock. Most cliff faces are a few yards tall and sheltered beneath a forest canopy. Small fissures collect minuscule volumes of soil where small plants have rooted. Larger clefts may be colonized by Mountain Laurel or rhododendron. In the wet season, small playful waterfalls drop off some cliffs, but groundwater also drib-bles in thin sheets down the rock faces, sustaining dense micro-gardens of ferns, mosses, and liverworts. This display of sandstone, water, and scrambling flora beneath a canopy of mature forest is among the most quintessential of Appalachian compositions.

On warm nights, salamanders emerge from their lairs in the cliff's moist fissures to patrol for insects, rivals, and mates. The most exquisite of them is the Green Salamander. Growing to nearly 6 inches (15 cm) in length, this secretive amphibian inhabits sandstone and limestone cliff faces, rocky outcrops, and moist caves of the Southwestern Appa-lachians. Crisp greens and yellows bejewel its black back in an intricate

Above left: Mountain Laurel, Little River Canyon National Preserve, DeKalb County. Photo by R. Scot Duncan.

Above right: Catawba Rhododendron, Little River Canyon National Preserve, DeKalb County. Photo by R. Scot Duncan.

pattern resembling the cliff's mosses and lichens. Its flattened profile aids with wriggling into deep protective crevices to hide from daylight, the winter chill, or voyeuristic naturalists. Cliff life presents challenges for a salamander. Most amphibians lay their eggs in water, but the Green Salamander might live hundreds of yards away from a stream or pool. So, it affixes its eggs to the roof of its crevice home, and hatchlings then crawl to nearby moss beds for prey, moisture, and cover. Alabama is home to the southernmost population of this and several other Appalachian salamanders, but habitat degradation due to logging, urban sprawl, and home construction in mountainous areas threatens the Green Salamander's survival. The species is now protected in Alabama and many other states.

Along the margins of the caprock, we discover places where weaker rock beneath has fallen away to create shallow, isolated caves known as rockhouses. Some are only a shallow overhang, while others once housed Native American encampments. Rockhouses do not support true cave species but will attract shelter-seeking surface dwellers. In a

Rockhouse, Little River Canyon National Preserve, DeKalb County. Photo by R. Scot Duncan.

few places, fraying of the caprock has created bridges and rock towers, or chimneys. Where undercutting has been extensive, large caprock slabs have collapsed onto the slope below and fractured into a jumble of massive stone blocks known informally as rock towns or boulder fields. The rockhouses, bridges, chimneys, and boulder fields of the upper escarpment attract visitors who delight in admiring and scrambling over, under, and through the unraveling caprock.

The acidic cliff and rockhouse ecosystems don't compare in terms of grandeur to the tall cliffs found downriver in the canyon. Our next stop is back in the preserve, where we'll investigate these ecosystems and hike to the canyon floor to meet the river face-to-face.

Green Salamander, inhabitant of wet acidic cliff and rockhouse ecosystems. Courtesy of Kevin Messenger/Alabama A&M University.

CLIFF HANGERS

From several of the preserve's overlooks, we notice dense oak forests covering most of the canyon's slopes. Heath bluff forests occupy some steeper slopes, but mostly in side-canyons and north-facing canyon slopes receiving less sunlight. We also see a river that has grown in size, speed, and ambition. The river is a furious white wherever it is forced to tumble over and between the massive hunks of the sandstone cap thrown in its way from the rim. Where its path is unobstructed, the river calms and is an alluring blue. Driftwood tree trunks litter the rocky banks, warning of the river's no-nonsense attitude during flooding. It is easy to see why some call this "The Little Grand Canyon" or "The Grand Canyon of the South," though at least one other gorge in the Southern Appalachian Mountains vies for this title. Regardless, it is one of the deepest canyons in the East.

Along the drive there are many overlooks from which we can examine tall cliffs. Dry cliffs and the jumbled boulders beneath belong to the Southern Appalachian Montane Cliff and Talus ecosystem. The cliff face may be the most unforgiving of Alabama's ecosystems, for soils are absent and water is scarce. Here and there, a nervous, stunted Virginia Pine has colonized a narrow ledge. Otherwise, only lichens seem capable of inhabiting the dry cliffs. The rough-and-tumble boulder piles at the cliffs' base support a few trees, but the recesses between the boulders provide shelter for wildlife. Occasionally a plateau stream plummets into the canyon. The waterfall bathes the cliff in moisture and nutrients, thereby enabling mosses, algae, ferns, and liverworts to establish a Southern Appalachian Spray Cliff ecosystem.

A few trails lead from the canyon rim to the river below. Some are marked, others not, and all are steep. They are walked by the fishers, hikers, kayakers, and students of biodiversity who find the canyon and its wilderness irresistible. We pick a trail at random and head down through the forested escarpment. After several minutes, we near the canyon floor and feel a slight but welcomed drop in air temperature caused by the river's cold waters. Had we been here in the morning, we'd have felt much cooler air that had sunk into the canyon overnight. Where the canyon's walls block the rising sun, this chill can linger well into the morning. These cooling influences help the lower canyon slopes retain sufficient moisture all year to support a tall, rich forest.

Recent rains have the river frothing with agitation, and the roar from the rapids dominates the canyon's acoustics. Still, the river is subdued compared to one of its floods. When the region receives abundant rain, the impervious caprock quickly shunts the stormwater to the canyon. Floodwaters move boulders that haven't budged for months or years previously. Trees are ripped from the riverbank and sent racing downstream. The trunks of survivors along the riverbank bear scars from decades of battery during floods. Tangles of fine debris in their lower branches and a bank swept clear of leaves reveal that the river flooded recently. Were we here in summer after the seasonal rains have slackened, the canyon floor would be a serene, contemplative place. The stream would flow leisurely and its murmurings would mingle with the songs of the canyon's bird life.

The US Fish and Wildlife Service lists several threatened and endangered species that *may* occur on the floor of the canyon. The uncertainty of their listing reveals how little we know of the canyon's ecology and the difficulty of exploring this rugged terrain. Known endangered species of the canyon include the Green Pitcher Plant, Harperella (a wildflower), Blue Shiner (a fish), and a poorly known aquatic plant, Kral's Water Plantain.

If the waters were lower, we might spot the plantain. It prefers clear waters, where it spends its life rooted among rocks submerged no more than 3 feet (0.9 m) deep. Its narrow leaves grow sickle-shaped in swift currents and straight where the current is weak. The species mainly reproduces asexually, but where sunlight is plentiful plants sprout emergent flower stalks to attract pollinators. When listed as threatened under the Endangered Species Act in 1990, the plantain was known only from Little River, but biologists have discovered it in several nearby localities since.

As we scramble upward on the return hike, we pause frequently to catch our breath and admire the purple shadows emerging from their midafternoon alcoves and retaking the canyon for the night. They restore a depth and texture to the landscape that was hidden at midday and complement the warm tones on cliffs still catching light from the reddened sun. We've completed our whirlwind tour of Lookout Mountain's ecosystems, but we have just scratched the surface of the Plateau Escarpment ecoregion. At elevations below that of Little River Canyon, there is an escarpment realm harboring the most luxuriant and mysterious ecosystems in the Southeast, and we're headed there next.

Karst Escarpment

Two places on Sand Mountain where the public can hike down the full escarpment, from the plateau rim to the valley margin, are Buck's Pocket and Lake Guntersville State Park. On such a hike, there is a point at which you would leave behind the acidic, sandstone slopes of the upper escarpment and enter the alkaline, Mississippian-aged karst slopes of the lower escarpment. Spotting the transition is easy if bedrock is exposed. Otherwise, the trees will reveal the transition. Above the boundary, look for Virginia Pine, Chestnut and Southern Red oak, and Winged Elm. Below the boundary, look for their counterparts: Eastern Red Cedar, Chinkapin and Post Oak, and Slippery Elm. Not all indicator species are plants—large land snails are rare on acidic soils but are often conspicuous in karst forests. Their spiraled shells contain the same calcium atoms as those in the shells of marine snails and other reef organisms that became the lower escarpment's limestone over 320 million years earlier.

The karst forests of the lower escarpment vary greatly. Forests of drier slopes are Southern Ridge and Valley/Cumberland Dry Calcareous Forests. These ecosystems prevail in the Southwestern Appalachians ecoregion on south- and west-facing limestone slopes bearing the summer sun's full force. The limestone's porosity prevents groundwater accumulation, which, combined with the heat, keeps the forest dry. Consequently, trees don't grow very tall, the canopy is broken, and forests may resemble woodlands. Cedars, being tolerant of dry conditions, are a good marker for these forests.

Moist karst slopes sustain forests with more tree species than any other forest in Alabama. Over 30 tree species may be tallied in some spots. These variants of the South-Central Interior Mesophytic Forest occupy the middle and lower escarpment where water is abundant, especially coves, canyons, and north-facing slopes. Common trees include Shagbark Hickory, Basswood, Ironwood, Eastern Hophornbeam, and Black Walnut. In summer, the understory is dim, but in spring before the forest canopy thickens, a carpet of wildflowers reawakens and blooms.

The lower escarpment is prominent north of the Tennessee River in Jackson and Madison counties, in an area geologists call the Jackson County Mountains. This landscape is like no other in Alabama. These densely forested mountains form branching networks of ridges and small isolated hills. They are remnants of a large sandstone plateau once similar to Sand Mountain but whose caprock is nearly gone. Great expanses of limestone are now fully exposed and, in a geologic time frame, are rapidly eroding. For now, these mountains harbor some of the most lush and diverse forests in the Southeast.

Remnants of the Jackson County Mountains' sandstone caprock comprise another of the state's level IV ecoregions, the Cumberland Plateau. This ecoregion makes up the main body of the Southwestern Appalachians in Tennessee, but in Alabama it survives only in the highest elevations of Jackson and Marshall counties. The Cumberland Plateau differs from the Southern Table Plateaus by being founded on the Pennington Formation and having a higher elevation where the climate is measurably cooler. This colder climate stymies agriculture, so forests are more prevalent on the Cumberland Plateau than on the Southern Table Plateaus.

Below the branching ridgelines of the Jackson County Mountains are valleys carved by large creeks and small rivers, including the Paint Rock and Flint. These valleys are incursions from the west of the Interior Plateau level III ecoregion. The convergence of the two ecoregions produces an exquisite landscape of bucolic valleys and steep forested mountains. One such valley contains one of Alabama's most revered natural wonders, the Walls of Jericho.

WALLS OF JERICHO

The Walls of Jericho is a Forever Wild preserve nestled within the Jackson County Mountains along the Tennessee border. The "walls" are steep limestone cliffs lining a narrow canyon. With significant help from TNC, the preserve was added to the state's portfolio of Forever Wild properties in 2004. The hike to the canyon's entrance begins on the Cumberland Plateau, then drops through lush mesophytic forest into the valley of Hurricane Creek, a major tributary to the Paint Rock River.

Turkey Creek, a tributary to Hurricane Creek, deserves the praise for carving the canyon and the Walls.[3] The canyon's trail follows the creek and its sequence of small waterfalls and clear, limestone-floored pools. Lush forests on the canyon's steep slopes maintain rare trees including Smoketree, Kentucky Coffeetree, and Kentucky Yellowwood. Higher in the canyon, the Walls begin where the slopes yield to sheer limestone cliffs.[4] Subject to frequent high-velocity floods, the canyon floor and lower walls are nearly bare. Above the floodline, the walls are draped with ferns, wildflowers, and firm-rooted trees and shrubs. The creek has carved wide basins, narrow chasms, and deep plunge holes directly into the bedrock. At one point the creek plummets into an underground cavern and emerges downstream from a breach in a cliff. The cliffs above belong to the Southern Interior Calcareous Cliff ecosystem and sustain sparse and precarious populations of herbs, shrubs, and stunted cedars.

The mosaic of mesophytic forest and neighboring ecosystems at the Walls supports a tremendous diversity of wildlife. At the mouth of the canyon, the valley offers abandoned pastures, forest margins, springs, and beaver ponds. The habits of summering birds illustrate how these

A narrowing within the Walls of Jericho (managed by Forever Wild and Tennessee's Department of Environment and Conservation). Courtesy of Bob Farley/F8PHOTO.

habitats enrich species diversity. Blue Grosbeaks and Eastern Kingbirds prefer the pastures; Indigo Buntings and Blue-winged Warblers work the forest edge; Kentucky and Worm-eating Warblers lurk in the forest understory; and Scarlet Tanagers and Parula Warblers forage in the treetops. These birds feed themselves and their young on the bounty of insects available in the summer, then return to the tropics for the winter when ecosystem productivity in northern Alabama grinds to a halt.

The Walls of Jericho and mature mesophytic forests of the Jackson County Mountains are home to several rare montane species. One is the Cerulean Warbler. This summer visitor is a favorite among birders for the male's sky-blue color and the female's subtle blue-greens. This warbler was one of the most common in the eastern United States, but a marked population decline last century has biologists concerned about its future. In Alabama, the cerulean prefers fully mature mesophytic forests with occasional gaps in the canopy where large trees have died. Most such forests have been clearcut and replaced by second-growth forests that have not yet developed the structure the cerulean requires.

The Ruffed Grouse is another rarity in the Jackson County Mountains and is the species most likely to trigger a heart attack. Distantly related to the Wild Turkey, the grouse is a stout bird measuring up to 19 inches (48 cm) in length and 1.5 pounds in weight (0.7 kg). Like the Cerulean Warblers, the grouse of northeastern Alabama are the southernmost population of the species. Grouse have brown streaking and barring for camouflage; long legs for running through dense forest; rounded wings for powerful bursts of flight; and a crested head for good looks. However, most people encountering the grouse will see few of these features. Grouse sit perfectly still when danger nears. Should the threat approach too closely, the bird will explode into flight at short range with heart-stopping speed and noise, then quickly disappear. Only briefly in spring will the grouse advertise its presence. Perching on a log or boulder, males pound their wings to make a loud drumming to attract females and ward off would-be competitors. Even then, this shy bird is seldom seen and most dedicated birders have never seen one in Alabama.

Rodents with a penchant for interior decorating also inhabit the escarpments of northeastern Alabama and mountainous areas farther north. The rare Allegheny Woodrat resembles the familiar rats of the urban environment, but its thickly furred tail and unusual behaviors set it firmly apart. In densely forested escarpments, woodrats build cozy nests within the crevices of caves, cliffs, and tall outcrops. They decorate nests with seemingly random objects including bones, feathers, bottle caps, coins, and shell casings. Woodrats nest in small colonies and stay active all winter. Thus, survival depends on how many seeds and nuts are stored in the nest each autumn. Should you search for a woodrat colony, use your nose—the colony uses conspicuous, communal latrines. They are shy animals, and this may be all you get to see.

The arduous hike in and out of the Walls of Jericho is not for everyone, but fortunately there are other places where the public can explore the lower escarpment, including the Skyline Wildlife Management Area, TNC's Keel Mountain and Sharp and Bingham Mountain preserves, and Monte Sano State Park. A few other preserves protect portals to caverns, the most unusual and poorly studied ecosystem in the state.

The Underworld

With over 2,300 caves, the Jackson County Mountains are the most cave-rich region of North America and are a global hot spot for cave biodiversity. Within Jackson, Madison, and Marshall counties, there are almost 90 species of troglobites, animals completely adapted to life underground, including spiders, shrimps, crayfishes, pseudoscorpions, flatworms, fishes, and a salamander. Rare and endemic plants cluster around their entrances. Already, the region is famous as the third most biodiverse cave region in the temperate world. However, biologists have inventoried merely 10 percent of the Jackson County Mountains' caves and more caves are discovered each year. They anticipate discovering dozens of new species in the coming decades.

The Jackson County Mountains are riddled with openings into a network of caves, channels, and chasms. Sinks in perched valleys swallow mountain streams and sinkholes appear where surface rocks have collapsed into an underground void. The moist walls of these shafts are variants of the Southern Interior Sinkhole Wall ecosystem and support rarities like Green Salamanders, Bulblet Bladderferns, and Walking Ferns. The cavern networks were carved by subterranean streams and rivers long ago. Some house continuously running streams. Oth-

Spelunker descends into Falling Cave, Jackson County. Courtesy of Alan Cressler.

ers are usually dry but become torrents during heavy rains. Still others are completely dry now but once hosted streams before tectonic uplift pushed them beyond the reach of groundwater.

Cavern ecosystems can challenge our understanding of how ecosystems function. After all, the energy supply for most ecosystems is sunlight. Nevertheless, the sun's power reaches deep below the surface into the absolute darkness. Many cave dwellers are aquatic species, and the energy and nutrients they use arrive as organic compounds in groundwater seeping down from the surface. Aquatic microbes harnessing these compounds are the base of the food chain. Because energy supply is limited, cave animals and their populations are usually small. A few caves receive high volumes of nutrients and energy from bat guano. Such caves sustain a thriving detritivore community of roaches, crickets, and millipedes, plus the centipedes and spiders that hunt them.

Should you be tempted to explore these caverns, be forewarned they are treacherous places with slick mud, steep drop-offs, confusing labyrinths, and the continual threat of absolute darkness. Underprepared adventurers can get lost, injured, or worse. Furthermore, planning and expertise are needed to avoid damaging cave ecosystems and their fragile formations. Despite these challenges, many cannot resist the thrill of subterranean adventure and the chance to marvel at rarely seen and dazzling geologic formations, so the Jackson County Mountains are popular with recreational spelunkers.

For those wanting a tamer experience, several commercial and public caves in the ecoregion provide safe underground adventures. Cathedral Caverns State Park offers the most spectacular of the public caves. Guides lead visitors on a well-lit pathway through a 126-foot-wide (38 m) entrance and deep into the 2-mile-long (3.2 km) cavern. Such a tour is a must for those wanting to fully experience the ecoregion's ecology. However, don't expect to see many cave animals on the tours. These sensitive species are usually absent in frequented caves. Besides, most tours spotlight flow formations and folklore, not critters. If it's cave fauna you want, wait at the entrance to a maternal roosting cave of the Gray Bat at sundown in summer and you will witness one of the most astounding wildlife spectacles in Alabama.

ENDANGERED SPECTACLE

In northeastern Alabama you can stand beneath swarms of Gray Bats—several hundred thousand strong—emerging from a cave at dusk for the nightly hunt. By day, they sleep and tend young in tightly packed crowds hanging from a cavern ceiling. By night, they feast on insects hatching from nearby waterways. Gray Bats—1 of 16 chiropteran species in Alabama—inhabit the south-central United States, preferring karst topography with nearby bodies of water. Yet despite a seeming abundance of habitat and impressive emergences at favored roosts, Gray Bats are one of our most endangered mammals. Their highly social nature and need for specific roosting conditions put them at risk.

In general, bats roost in a variety of situations, including caves, hollowed trees, and attics. Some roost singly, others roost in small groups. The Gray Bat is exceptionally gregarious and prefers roosting with as many other Gray Bats as possible. However, not every span of cavern ceiling can suitably host thousands of dangling bats. Summer caves must have sufficient air flow, and the bats favor rooms with domed

Emerging Gray Bats at Sauta Cave National Wildlife Refuge, Jackson County. Courtesy of Emily Horton Reilly.

ceilings that trap their body heat. In the summer, females and males roost in separate caves. In winter, the sexes cohabitate in deep caves where cold air accumulates and helps them quickly slow their metabolism and enter hibernation. These strategies worked well for millions of years, then a new species began venturing into southeastern caves.

A single visit by human explorers can kill hundreds of bats. Hibernating bats rely on fat reserves to get them through the winter. Bats unnaturally awakened during hibernation can burn their fat reserves too soon. This can force them to end hibernation early, emerge into a winter landscape devoid of food, and starve. In summer, disturbances to maternal caves cause mothers to panic and drop their flightless pups to the ground, where they die. Too many such disturbances last century caused major population declines of Gray Bats and other species, as did the flooding of important caves when the Tennessee and other rivers were impounded. Thanks to federal protection, conservation measures, and bat-friendly spelunking practices, Gray Bat populations have been recovering.

At several locales in northeastern Alabama, you can watch the mass emergence of bats. The caves are specially gated to keep people out but to allow bats to pass freely. A popular spot for bat watching is Sauta

Cave National Wildlife Refuge near Scottsboro, where 200,000–300,000 females roost and raise pups each summer. It's best to arrive just before twilight. At first, just a few eager bats trickle out of the cave. Then the trickle becomes a stream, and the stream becomes a torrent. Thousands pour through the exit every minute, visible as blurred streaks against the dimming sky. Those with keen hearing can detect their high-frequency chirps, emitted as they echolocate through the flock and the forest canopy. Sometimes hawks or owls ambush bats as they emerge, and Gray Rat Snakes will hang above the cave's entrance and pick off hapless bats.

Though Gray Bat populations have been rebounding, a new menace has appeared. In 2006, biologists in New York State detected a fungus-caused bat disease known as white-nose syndrome (WNS). The disease attacks multiple species and is spreading rapidly through the eastern United States, killing up to 90–100 percent of bats in affected colonies.[5] The fungus itself is not fatal, but infected bats frequently awaken from winter hibernation and deplete their fat reserves prematurely. WNS just recently reached Alabama, and it is still unclear how badly the state's bats will be affected. Once again, the Gray Bat's gregarious nature makes them vulnerable—diseases spread easily through large, dense populations like roosting and hibernating bat colonies.

An Inland Sea

Nestled between Sand Mountain and the Jackson County Mountains is the Sequatchie Valley level IV ecoregion, a long anticlinal valley extending from north-central Alabama into eastern Tennessee. The valley hosts Guntersville Lake, where each fall tens of thousands of aquatic birds arrive to overwinter. Among them are freshwater species from the Midwest and Canada and marine species from the North Atlantic Ocean. So many terns, gulls, loons, cormorants, grebes, coots, and ducks descend on the lake that on most winter weekends there are birders scanning the lake through binoculars and telescopes. Bald Eagles will also be scanning the lake and its flocks. The lake is a stronghold for eagles that hunt the waterfowl.

Guntersville Lake is an impoundment of the Tennessee River and dominates about half of the Sequatchie Valley within Alabama. Settlers thoroughly converted the valley to agriculture by the late 19th century and the composition of the original ecosystems is unknown. Limestone underlies the valley floor, so presettlement ecosystems were probably similar to those of the Limestone Valleys and Low Rolling Hills of the Ridge and Valley. Flatwoods of broadleaf trees occupied the river's floodplains, while trees of the Southern Ridge and Valley/ Cumberland Dry Calcareous Forests would have populated higher elevations below the escarpments.

Surrounded by mountains and hosting the largest river in Alabama, the Sequatchie Valley was an obvious target in the 1930s for the Tennessee Valley Authority. Though the dams on the Tennessee River caused the extinction and extirpation of several dozen fishes and mollusks, the flooding created an artificial ecosystem that uniquely contributes to the state's biodiversity. Guntersville Lake is essentially a small inland sea comparable in size and biological productivity to bays of the northern Gulf Coast. Populations of deepwater river fishes readily took to the new environment, including Largemouth Bass, sunfishes, and catfishes. This spawned a thriving sports fishery and attracted overwintering aquatic birds, plus their predators and admirers.

Warrior Coal

The first of two remaining level IV ecoregions to explore is the Shale Hills, located at the southern end of the Southwestern Appalachians. Most of the ecoregion lacks a protective cap of thick sandstone, and the weak sandstones, siltstones, and shales exposed have been worn down into a hilly region. These rocks resist groundwater infiltration and yield acidic soils, so the forests tend to be dry and lack a karst influence. With groundwater scarce, streams contract in late summer and fall more than in most ecoregions.

The Shale Hills are known as the Warrior Coal Field. The Black Warrior River drains the region, and vast deposits of Pennsylvanian-aged coal lie beneath the surface rock. Mining here last century made Alabama one of the nation's top coal producers. Area residents enjoyed a strong economy and the greater region enjoyed an inexpensive source of energy. However, the boom closed soon after World War II and most mines were shuttered. Today, Alabama ranks 14th among coal-producing states, but the impact of surface mining on the Shale Hills is still extensive. Abandoned mines are widespread, surface mining is expanding, and polluted runoff pours into streams and rivers.

Beyond the mined areas, the Shale Hills support agriculture and forestry, but even here the original ecosystems are long gone. Descriptions from early last century tell of dry oak and pine "woods" (probably Shortleaf Pine) in the uplands, Virginia Pine groves at bluff margins, and a mix of broadleaf species in the ravines where water is more available. These ravines would have prevented wildfires from spreading extensively through the region.

Perhaps due to its history of intensive resource extraction, the Shale Hills have very few protected natural areas and biologists have written little about its natural history. This is lamentable, for the Shale Hills is one of only three ecoregions exclusive to Alabama.

Then again, mining in the Shale Hills has led to some of the Southeast's most important fossil finds. At the Steven C. Minkin Paleozoic Trackway Site, tracks of giant amphibians, crustaceans, insects, fishes, and other creatures have been recovered in fine-grained shale. These impressions were left on mud flats within Pennsylvanian swamp forests and provide insights into the activities of Alabama's ancient species.

A GIANT OF CONSERVATION

Every year hundreds of people make a pilgrimage to the Sipsey Wilderness to stand at the base of one of Alabama's largest trees, a colossal Tulip-poplar 21 feet (6.4 m) in circumference and 153 feet (50 m) in height. Such majestic trees were common in the primeval southeastern forests. This giant's survival is attributable to geology—the tree lives within a narrow box canyon from which timber extraction was difficult. The Sipsey's deep canyons also provide sanctuary for unusual species and provide some of the most enchanting footpaths in Alabama.

The Sipsey Wilderness lies within the heart of the Dissected Plateau, another level IV ecoregion endemic to Alabama. Its sandstone caprock is thinner than the caprock of the Southern Table Plateaus. Small streams have etched through the cap, sliced through the soft rocks beneath, and created a dissected landscape of narrow, weaving canyons. Canyon trails wander beneath looming sandstone cliffs and tall forest. Massive sandstone blocks fallen onto the canyon floor force

Larry Brasher (Birming-ham-Southern College) and students at the giant Tulip-poplar, Sipsey Wilderness, Lawrence County. Courtesy of Lamar Marshall/Wild South.

detours of the streams and trails. Smaller canyon creeks have their own box canyons. Each reaches an abrupt dead end where a small stream spills into the canyon from the plateau. This stream and its splash-zone gradually wear away weak rocks beneath the caprock, producing an overhang and rockhouse. The heap of boulders at the canyon's head reveals that when a rockhouse becomes too deep, the cap above will tumble into the canyon. This process of head-cutting by the stream will continue as long as there is rainfall and a caprock to dissect.

The Dissected Plateau's canyons are ecosystems shielded from prevailing winds, storms, and the summer sun. Streams and groundwater drawn into them provide moisture all year. Orchids and other wildflowers needing moist conditions populate the canyon floor, while moist cliff faces cultivate vertical gardens of ferns, liverworts, and mosses. In a few locations, small fly larvae, known locally as dismalites, spin small webs and cling to the wall in dense colonies. By night, larvae emit blue lights to lure tiny flying insects into their webs to be eaten. The spectacle of the canyon wall at night dotted with thousands of small blue lights has attracted ecotourists to the privately owned Dismals Canyon Conservatory for many decades.

The canyon's moist, cool environment sustains a variety of the South-Central Interior Mesophytic Forest and populations of several unusual trees, including the Eastern Hemlock. Resembling the firs and spruces found farther north, hemlocks seem out of place in Alabama. Indeed, the nearest substantial hemlock population is several hundred miles away in the Blue Ridge level III ecoregion of the Southern Appalachian Mountains. The Black Birch is another rare canyon dweller. In the Sipsey, these trees grow on large boulders and ledges, where their roots wrap across the rock surface seeking soil- and moisture-filled crevices. Many who know this tree cannot resist scraping the bark of a twig to enjoy a refreshing whiff of wintergreen. These two species are relicts of the forests pushed south during the Pleistocene's ice ages. As happened elsewhere in Alabama during the Holocene warming, these populations lingered in cool, moist, and fire-free conditions.

Forests are still common on much of the Dissected Plateau's surface, partially because the William B. Bankhead National Forest covers a fifth of the ecoregion. Like many eastern national forests, when the Bankhead was established in 1919, much of it (40 percent) had been recently logged or cleared for agriculture. During the first decades of management, a chief priority was the cultivation and harvest of timber. By the 1960s, many of the Bankhead's degraded forests contained large volumes of mature merchantable trees. Pressured by the timber industry, the US Forest Service committed to scaling up timber extraction from the Bankhead and other national forests. What the service didn't anticipate was that while the trees had been growing, the public's love of the national forests had grown, too.

In Alabama, a grassroots effort emerged to protect an area surrounding the West Fork of the Sipsey River within the Bankhead. The movement was led by Mary Ivy Burks, a Birmingham native and graduate of Birmingham-Southern College. Burks helped create the Alabama Conservancy (now the Alabama Environmental Council) and was its first president. The Conservancy's aim was to have the Sipsey's forests designated as a US wilderness area. National Wilderness Preservation System lands are protected from commercial resource extraction and are managed only to preserve their wildness. In the language of the 1964 Wilderness Act establishing the system, wilderness areas are "untrammeled by man, where man himself is a visitor who does not remain." Burks' strategy was a long shot, since few believed there were areas worthy of wilderness designation in the eastern United States. Yet in 1975, after years of toil by Burks and her followers, Congress established the Sipsey Wilderness, the first such wilderness in the East. The victory for conservation was much bigger than anyone initially realized. Burks' success quickly inspired dozens of similarly successful efforts in the East, an episode now known in conservation history as the Eastern Wilderness Movement.

One benefactor of Mary Burks' success is the Flattened Musk Turtle, a shy creature of rocky pools within headwater streams of the Black Warrior River. Though famous as one of Alabama's few endemic vertebrates, it is small and unassuming. It measures less than 4.5 inches (11.4 cm) long and has an olive head with black freckles, two fleshy chin whiskers, and a brown, flattened shell. This latter feature may help it squeeze into submerged rock crevices where it seeks prey (snails are a favorite) and shelter. Despite its size, herpetologists believe the turtle lives 20–40 years. Few ever glimpse this turtle, due to its size and nocturnal habits. Your best chance of spotting it would be sitting quietly alongside a stream pool in the Sipsey Wilderness on a summer day and watching carefully for its nose breaking the surface to steal a gulp of air.

The turtle's timidity keeps it safe from predators, but this has not protected it from sediment and nutrient pollution caused by deforestation, agriculture, poultry factories, and mining outside the national forest. Moreover, its rarity and diminutive size have encouraged harvesting for the illegal pet trade. Despite its Sipsey stronghold, the turtle's population has declined and it is now a federally listed threatened species.

Other famous inhabitants of the Dissected Plateau's streams include a collage of darters found almost exclusively here, including the Sipsey, Locust Fork, Warrior, Warrior Bridled, Rush, and Tuscaloosa darters. Several mussels and snails are also endemic. And, there is the Black Warrior Waterdog, a rare and poorly understood aquatic salamander reaching nearly 10 inches (25 cm) in length and surviving in only 14 streams in the Black Warrior watershed, about half of which are in the Dissected Plateau.

As is the case everywhere, the care with which we manage the lands surrounding our streams will determine whether vulnerable aquatic species endure. Fortunately, protection of the Sipsey Wilderness and improving management of the Bankhead's forests are giving the turtles, waterdogs, darters, hemlocks, birches, and others a shot at survival. With enough time and protection, the Sipsey's giant Tulip-poplar will have many peers standing just as tall and stately.

HARD TIMES AHEAD

On our tour of the Southwestern Appalachians ecoregion, we've found plateaus still protected by a caprock and others in their last stages of disintegration. The resulting variation in topography and geology sustains ecological and species diversity unrivaled within Alabama's highlands.

We're headed next to the last mountainous level III ecoregion left to explore, one that has fallen on hard times. The Piedmont has lost most of its mountains, original ecosystems, and many of its people. Nevertheless, it is Alabama's most geologically complex ecoregion and offers many surprises for students of biodiversity, including the state's most majestic mountains.

12

Piedmont

The Lonely Monadnocks

It's a November dusk on Cheaha Mountain, Alabama's highest peak. I am hiking a footpath to a precipice known as Pulpit Rock to watch the sunset. A cold front passed through two days ago, and while it's been sunny all afternoon, the chilling air forewarns of another cold night. Typical of such late afternoons in fall and winter, the air is still and the forest is hushed. I hear only my footfalls and the lonely songs of a few winter crickets. Earlier, squirrels and chipmunks rustled in the understory harvesting acorns, but they are now in their nests and burrows awaiting the chilly night. The canopy above is also empty. Summer songbirds left weeks ago for warmer latitudes, and most broadleaf trees have shed their last leaves. From a cluster of Virginia Pines I hear the thin calls of a Golden-crowned Kinglet hunting insects in hasty preparation for the night. This tiny bird is one of the few that can survive in Alabama's high-elevation forests in winter. Though the forest seems otherwise empty, it won't remain so for long. Foxes in their dens are stirring, raccoons in hollowed trees are waking, and owls are counting down the moments until it's time to hunt.

At the rocky precipice I find a comfortable sitting rock and admire the view. The falling sun has shifted from pale white to weak gold, and the eastern sky is darkening. Below is an expanse of forest stretching for miles without interruption. Mountains connected to Cheaha rise sharply from the lowlands and bound the southern and western horizons. Their eastern slopes are in deep blue shadow. The golden sunlight illuminates the short, gnarled trees around me. They are stunted from the thin soils and strong winds endured on the rock outcrop. Many are easily over a century old, but though they've seen a lot of history, the mountain onto which they cling is incomprehensibly ancient in comparison and has its own stories to tell.

Peaking at 2,407 feet (734 m) above sea level, Cheaha belongs to the

Sunset and Talladega Mountains from McDill Point, Talladega National Forest, Cleburne County. Courtesy of Bob Farley/F8PHOTO.

Talladega Mountains, a series of tall ridges rising 500–1,500 feet (152–457 m) above the surrounding terrain. Called monadnocks by geographers for their abrupt appearance in an otherwise level landscape, the mountains stretch 35 miles (56 km) as the crow flies across east-central Alabama. Cheaha leads the chain on its northeastern bearing, followed by Cedar, Talladega, Horn, and Rebecca mountains. Collectively, they form the northern edge of the Piedmont level III ecoregion in Alabama.

Alabama's corner of the Piedmont is a triangular wedge entering from central Georgia and coming to a point near the state's center. From Alabama, the Piedmont stretches northeast through Georgia, the Carolinas, and Virginia. Throughout, it is a transitional, hilly landscape stuck between the Coastal Plain and the mountains of the Ridge and Valley and Blue Ridge level III ecoregions. Alabama's monadnocks and other isolated mountains are anomalies. In this context, the literal translation of the term *Piedmont*, from the French for "foot of the mountain," seems sensible.

Geologists have determined that the Piedmont has survived a violent geologic past that distinguishes it from Alabama's other ecoregions. The Piedmont's rocks tell this story. Igneous and metamorphic rocks prevail here and nowhere else in Alabama. They were created during the early phases of the Allegheny Orogeny, the arising of the Southern Appalachian Mountains. As the Gondwana and Laurasia supercontinents collided, massive slabs of oceanic crust and continent were thrust upon one another and became mountains. Magma from deep below infiltrated and cooled in the faults through the mountains' foundation. Sedimentary and igneous rocks caught up in the collision were transformed by intense heat and pressure into metamorphic rocks with different properties.

Some geologists estimate that the Piedmont's young mountains were as tall as today's Rocky Mountains. However, this mountain range was not to last. Several hundred million years of erosion wore most of the Piedmont into a hilly region whose chaotic past is betrayed only by its metamorphic and igneous rock. Thus, the Piedmont has a long and violent history during which the epithet "foot of the mountain" would be inappropriate.

But what of the Talladega Mountains? Why are they the only survivors of the Piedmont's former grandeur? The monadnocks are composed of the Cheaha Quartzite, a metamorphosed sandstone. During the transformation, silicon-dioxide from the sandstone recrystallized into an interlocking framework extremely resistant to erosion. Other sedimentary rocks caught up in the Allegheny Orogeny met similar fates. Shales became slate, phyllite, or schist, depending on the intensity of the pressures applied. Limestones and dolostones became marbles of various colors. Igneous rocks were also transformed, in one case producing a green rock known as the Hillabee Greenstone. Of them all, the Cheaha Quartzite is the most durable. Thus, the monadnocks persist because they are wearing down more slowly than the surrounding region.

The sun has fallen below the southwestern horizon now, silhouetting the monadnocks to the south against a wash of fading orange. The air's chill is sharper, the sky's blue is deeper, and stars and planets are twinkling. Several notes from a distant Great Horned Owl's call drift in from the ridge behind me. The light is fading and it's time to head back, so I take in one last view of the lonely monadnocks and hit the trail.

LIONS, FIRES, AND BEARS, OH MY!

The monadnocks and their foothills form the core of the Talladega Upland, the smallest of the three Piedmont level IV ecoregions in Alabama. Though its extent is small, the Talladega Upland is at the center of the largest complex of public natural lands in Alabama. The largest piece is the Talladega National Forest (TNF), whose Shoal Creek and Talladega Ranger Districts protect 233,585 acres (945 km^2) of forest across five counties.[1] Within the national forest, Cheaha Wilderness Area is a 7,245-acre (29 km^2) roadless area encompassing more than 1,200 feet (366 m) of elevational change. Crossing much of the national forest is the Pinhoti National Recreational Trail. This still-expanding trail system will extend the Appalachian Trail from northern Georgia into central Alabama, where the Southern Appalachian Mountains begin. The Choccolocco and Hollins Wildlife Management Areas (WMA) are also within the national forest. Pulpit Rock and the peak of Cheaha Mountain lie within Cheaha State Park, which is small in size but provides facilities to help the public explore these mountains.

The Talladega Upland supports several forest types, many of which can be seen along the Talladega Scenic Drive through the national forest. The drive is a 26-mile (42 km) paved mountain road beginning near

Heflin and following the ridge of Horseblock Mountain. It climbs up and over Cheaha Mountain and ends atop Talladega Mountain. Scenic overlooks and a few trail crossings along the way allow one to study how forests respond to changing elevation, aspect, and slope position.

The monadnocks' ridgelines and flanking slopes are a challenging environment for trees. The thinner atmosphere of these high elevations forces average temperatures to be several degrees lower than those of the surrounding lowlands. Though precipitation is high due to a minor degree of orographic lifting (see chapter 3), the shallow rocky slopes dry out quickly due to rapid draining and near-continuous wind exposure. Also limiting biological productivity is the Cheaha Quartzite—it weathers slowly and provides few nutrients as it does.

The Southern Piedmont Dry Oak-Pine Forest ecosystem prevails in these tough conditions. Chestnut Oaks occupy the thinnest soils at the highest elevations, but where conditions are slightly less austere, other oaks share the canopy, including Black, Post, Scarlet, and White. Clusters of Virginia and Shortleaf Pine are present. Young of both species need full sunlight, so the pines mark where fires and tree-toppling storms have swept across the ridge.

The Southern Piedmont Mesic Forest occupies moist slopes at lower elevations, especially north-facing slopes sheltered from the sun. Wildfires are naturally infrequent since the moist understory does not easily burn. American Beech, Northern Red Oak, Tulip-poplar, and other trees more typical of the northern Appalachians are common. Tangled heath thickets of Mountain Laurel and Catawba Rhododendron occupy steep wet soils.

Below 1,900 feet (579 m), the foothills attending the monadnocks host the mountain Longleaf Pine woodland, an ecosystem also found in the Ridge and Valley ecoregion. Originally, wildfires sweeping through these woodlands every 5–10 years kept them open. Without fire, more competitive and fire-intolerant trees invade and usurp the sunlight young longleaf need to grow. In hilly and mountainous zones of Alabama's uplands, longleaf woodlands tend to occupy the south-facing slopes where sunlight exposure keeps soils and fuels dry.

Herbaceous plants and small shrubs enjoy the sunlight reaching the longleaf woodland's understory. In healthy mountain longleaf wood-

Dawn within a Southern Piedmont Dry Oak-Pine Forest, Cheaha Wilderness, Clay County. Photo by R. Scot Duncan.

land, two dozen plant species can grow per square yard, which is not as many as in the longleaf woodlands near the coast but more than in most other highland ecosystems. Many understory plants of the mountain longleaf woodlands are species not found in the lowlands. Among them is Eastern Turkeybeard, an unusual plant that is one of only two species in its genus. Most of the time, turkeybeard can be mistaken for a species of bunchgrass. Its long slender leaves are an extreme adaptation to reduce moisture loss in dry environments. However, when it

The rare Eastern Turkeybeard, Shoal Creek Ranger District, Talladega National Forest, Cleburne County. Courtesy of Michelle Reynolds/F8PHOTO.

reproduces, turkeybeard erects a stalk that can top 4 feet (1.2 m) in height and flaunts a crown of small six-petaled white flowers. Turkeybeard was once more common, but with canopy closure in the eastern forests due to decades of fire suppression, this sun-loving plant is now critically imperiled in the East.

You'll need to explore the roads crisscrossing the foothills of the national forest to find healthy longleaf woodlands. The US Forest Service

(USFS) subdivides the forest into compartments in which it matches management goals with environmental conditions. When driving through the national forest, you can see in quick succession even-aged Loblolly Pine plantations, open longleaf woodlands, mixed forests of oak and pine, and clearcut plantings of young Longleaf Pines.

The latter compartments are a sign of the changing priorities of the USFS. When established in 1936, around 30 percent of the TNF lacked forest cover and most of the rest was second-growth forest. For decades, the priority was timber extraction from naturally regenerated stands and timber production on plantations of Loblolly Pine. Since the 1990s, the maintenance and expansion of longleaf woodland has become a priority. The reasons for this turnaround include pressure from environmentalists; the infusion of young biologists and foresters into the USFS who have cross-trained in disciplines including forestry, ecology, and wildlife science; and a mandate to protect endangered species inhabiting the national forest, especially the Red-cockaded Woodpecker, or RCW (see chapter 9 for details about the RCW's unusual habits).

Thanks to the efforts of USFS biologists and others, the number of RCW clusters (a nesting pair and its family of helpers) in the Shoal Creek Ranger District has nearly quadrupled since the turn of this century. To provide for continued expansion of RCW populations and other longleaf woodland species, the USFS is converting many loblolly plantations and oak forests back to the longleaf woodlands they replaced. As of 2009, there were several thousand acres of high-quality mountain longleaf woodland in the Shoal Creek Ranger District alone; thousands more will be added as restored woodlands mature.

The TNF provides refuge for other unusual species, some of which inhabit stream valleys between foothills. The valleys host rocky streams and the Southern Piedmont Small Floodplain and Riparian Forest ecosystem. These shady oases shelter plants intolerant of the fire and dry conditions of adjacent uplands, including uncommon species such as Grass-of-Parnassus, False Hellebore, and White Fringeless Orchid. The larger streams harbor endangered aquatic species like the Finelined Pocketbook, Southern Pigtoe, and Holiday Darter.

One of the oddest species in the TNF and Talladega Upland is the

Red Crossbill, a small bird sporting one of the strangest of all bird beaks. Like all finches, crossbills have stout bills for crushing seeds. However, the crossbill's upper and lower mandibles curve sharply near the end to cross each other. This enables crossbills to pry out seeds lodged deeply within green, closed cones of pines and other conifers. Red Crossbills are common in the conifer forests of the western and northern United States but are extremely rare in the Southeast. The TNF hosts the southernmost of the few resident populations in the Southern Appalachian Mountains. Finding them is difficult even for the most dedicated birders. The Talladega's population is small and its flocks roam continuously, seeking locations where maturing pinecones are plentiful.

After an absence of many decades, American Black Bears are returning to the Talladega Upland. While bears were once common throughout the Southeast, hunting and habitat loss forced them into the remotest regions such as the Mobile River Delta. Today, bears are appearing in new areas throughout Alabama. Those in the Talladega Upland likely arrived from nearby populations in Georgia. Some are young males searching for unoccupied territories, but sightings of fe-

Black Bears have returned to the mountains of northeastern Alabama. Courtesy of Steve Hillebrand/US Fish and Wildlife Service.

males with cubs suggest a population has established. The mountains and hills of the Talladega Upland offer enough habitat to sustain a large bear population.

If there are bears, could there be mountain lions? Florida Panthers—known also as cougars or mountain lions—once occupied the entire Southeast. Habitat loss and hunting eradicated panthers from most of the Southeast by the mid-20th century. The last confirmed panther in Alabama was shot in St. Clair County in the 1940s. Currently, only 100–120 individuals survive in the swamps and thickets of southern Florida. Years of conservation efforts are paying off and the population is growing, but with little habitat in southern Florida to absorb new panthers, young panthers are dispersing widely in search of their own territories.

In November 2008, a hunter in Troup County, Georgia, shot, killed, and reported what he assumed to be a released or escaped cougar. Wildlife officers in Georgia and Alabama occasionally receive reports of panther sightings, and while some are probably hoaxes or misidentifications, a few are thought to be escaped or illegally released panthers of the western subspecies. Just to be sure, Georgia wildlife officers sent a tissue sample of the Troup County panther off for genetic analysis. To everyone's surprise, the slain cat was a Florida Panther. What's more, it was a young male in excellent physical condition that had not been raised in captivity. This was the first solid evidence that Florida Panthers could disperse this far north from their Florida stronghold.

Troup County lies within the Piedmont ecoregion and borders Alabama. It is premature to assume the slain panther is part of a surge of Florida Panthers recolonizing Georgia and Alabama. However, the incident reminds us that large endangered species could repopulate the Southeast, if allowed. With its large size, careful stewardship, and ample deer and hog populations, the Talladega Upland could be ideal for the reintroduction of panthers to Alabama. With enough time, they might return to Alabama on their own.

PHOENIX FORESTS
South of the Talladega Upland is the Southern Inner Piedmont level IV ecoregion, a terrain of moderately tall hills sometimes called the

Ashland Plateau. To its southeast lies the Southern Outer Piedmont. Known also as the Opelika Plateau, it is a level IV ecoregion of low, worn hills and slight topographic relief. Both ecoregions extend to North Carolina, but in Alabama, the ecological differences between them are subtle. The bedrock geology of both is mainly schist and gneiss, but the Southern Inner Piedmont contains a few granite outcrops. A large fault lies at the boundary between the two ecoregions, but there is no significant surface hint of that in Alabama. Given their ecological similarities, I'll consider these two ecoregions together, referring to them as the "Lower Piedmont."

Unlike the Talladega Upland, where steep slopes protected its forests and woodlands, the comparatively level terrain of the Lower Piedmont was one of the first regions settled by agriculturalists in Alabama. By the time biologists first studied the ecoregion, most of its forest had been felled for over a century. After decades of study, discussion, and debate, an emerging model of the pre-agricultural landscape suggests wildfires were frequent and woodlands of fire-tolerant oaks and pines (longleaf and shortleaf) were abundant. Pine woodlands prevailed where fires or wind damage from hurricanes was common. Fire-intolerant broadleaf trees occupied stream valleys and northern slopes of steep hills. Interspersed throughout the uplands were small-scale ecosystems including Piedmont Seepage Wetlands and Piedmont Upland Depression Swamps.

In addition to clearing the forests, the Piedmont's first agriculturalists also depleted the topsoil. Soil conservation was unknown to most in the 18th and 19th centuries. Farmers plowed up and down slopes instead of following contours, creating channels for soil to erode with every heavy rain. The land's fertility declined as its topsoil disappeared. The heavy clays exposed at the surface had few nutrients and would contract and damage fine roots upon drying. As the soils gave out, many farmers gave up. Land abandonment became widespread.

The situation for agriculturalists in the Lower Piedmont worsened for other reasons. After the Civil War, many young men didn't return to the family farm—many had died and some were disfigured and unable to farm, while others were drawn to industrial jobs in cities. The Boll Weevil arrived in the 1920s and devastated cotton production. The mid-century emergence of industrial agriculture in the Midwest made family farming unprofitable. Between 1900 and 2000, the amount of Piedmont agriculture in Alabama dropped from 75 percent to 10 percent. Thus, it is an understatement to say the Lower Piedmont and its people have endured hard times. However, this ecological story isn't a complete tragedy.

Idle land does not stay empty for long, no matter how heavily degraded. "Weeds" surviving in woodlots and fencerows invaded abandoned farms. The first were grasses, wildflowers, and vines that could eke out an existence on tired soils. As topsoil was rebuilt, pines and other pioneer trees established. Within a few years, abandoned fields supported young forests brimming with fast-growing trees. Decades later, oaks and hickories overtook the pioneer trees. Perhaps taking their cue from nature, many landowners established monocultures of Loblolly Pine on degraded soils to sell to paper mills. As for their biodiversity, the new forests, especially the plantations, fall far short of what the original fire-prone woodlands of the Lower Piedmont would have provided. Nevertheless, these postmodern ecosystems are rebuilding the topsoil and providing habitat for some of the Lower Piedmont's remaining wildlife. A new and distinctive ecology for the ecoregion is emerging.

A Lower Piedmont Sampler

Hints of the Lower Piedmont's original ecology can be found throughout Alabama. Wind Creek State Park is a 1,445-acre (5.8 km²) preserve on Martin Lake, a reservoir on the Tallapoosa River. While most people visit for the lake, there is a fine stand of second-growth Longleaf Pine woodland and Southern Piedmont Dry Oak-Pine Forest on the hills above the reservoir. Farther up the Tallapoosa River is Horseshoe Bend National Military Park. Maintained by the National Park Service, this 2,040-acre (8.3 km²) park commemorates the 1814 battle between the Creek Red Sticks led by Chief Menawa and federal troops led by General Andrew Jackson. While its mission is largely cultural, the park protects 3.5 miles (5.6 km) of the Tallapoosa River and many acres of second-growth Southern Piedmont Large Floodplain Forest. Hilly uplands harbor remnants of longleaf woodland that are being restored after decades of fire suppression. West of Horseshoe Bend, the 37,291-acre (151 km²) Coosa WMA protects game species and provides public hunting land on the eastern margin of Mitchell Lake. Within the WMA, one may visit the Coosa River Longleaf Hills, a 9,746-acre (39.4 km²) Forever Wild tract of dry oak-pine forest, mesic forest, and longleaf woodland, the latter harboring one of the few RCW populations surviving outside of Alabama's national forests. The tract also protects portions of Weogufka and Hatchet creeks, two of the Piedmont's most pristine streams.

In the Lower Piedmont's south are the 320-acre (1.3 km²) Coon Creek and 3,518-acre (14.2 km²) Yates Lake West tracts on the margins of the Yates Reservoir of the Tallapoosa River. These Forever Wild preserves protect mesic, riparian, and dry oak-pine forest for research, education, and recreation. The only other significant preserve in the Southern Outer Piedmont is Chewacla State Park, near Auburn. Located on the Fall Line, the park is on one of the southernmost extensions of the Piedmont in the United States. The park packs considerable topographical variation into its 696 acres (2.8 km²) and shelters longleaf woodland plus mesic, riparian, and dry oak-pine forests. Evidence of the Piedmont's tumultuous geologic past can be seen in the distorted schist formations exposed in the stream valleys and in the cliffs of a quarry visible from a ridge overlook.

These locations preserve a sampling of the original biodiversity of the Lower Piedmont, but all have endured considerable disturbance, including logging, agriculture, fire suppression, warfare, and flooding from dammed rivers. As a rule, none of the Lower Piedmont's original ecosystems survived unscathed. Or have they?

DEFIANT WILDFLOWERS
A community of Piedmont wildflowers breaks that rule and two others. Rule one: none of the original communities of the Lower Piedmont survived intact. Rule two: trees dominate southeastern terrestrial ecosystems. Rule three: wildflowers cannot grow on bare rock.

Despite the Lower Piedmont's transformation, one of its original terrestrial ecosystems survived nearly intact. It is a community of herbaceous plants living in shallow depressions on an otherwise sheer exposure of granite. These Southern Piedmont Granite Flatrock and Outcrop ecosystems (hereafter, "granite outcrop") harbor some of the Piedmont's rarest terrestrial organisms.

The origins of these ecosystems date back to the Allegheny Orogeny. Granite outcrop ecosystems occupy the tops of plutons, rock formed from magma injected into the fractured crust beneath building mountains. The cooled magma becomes granite that is exposed at the surface as the mountains erode. Because granite is very resistant to weathering, some plutons rise above the surrounding plains. Plutons also emerge throughout the Piedmont and adjacent plains of Georgia, the Carolinas, and Virginia. Those in Alabama formed mostly during the Allegheny Orogeny, but some farther north date back to the Taconic Orogeny (see chapter 5).

In Alabama, granite outcrop ecosystems are hidden within the Piedmont's second-growth forest and pasture, mainly in Clay, Randolph, Chambers, and Lee counties. Most are privately owned, but near the small town of Almond there is a 400-acre (1.6 km²) granite outcrop ecosystem known as the Almond Outcrop. It is the largest granite outcrop ecosystems in Alabama and one of the southernmost plutons surfacing in the Southeast.

The granite outcrop wildflower community breaks rule one, above, because a big slab of granite cannot support crops, forestry, or ranching. Thus, several granite outcrop communities have survived relatively unscathed. It breaks rule two because—as seen in similar ecosystems in the state—punishing extremes can prevent a forest from developing. Summer temperatures at the rock surface can top 130° F (54° C), the granite quickly sheds rainwater, and soils are mostly absent because granite weathers into fine particles very slowly and rain and winds wash most of these particles away.

The rebellious wildflowers break rule three because they colonize divots and fissures on the outcrop and build their own soils. Across the outcrop there are rounded, shallow depressions—usually not more than a couple yards across—where soil and water accumulate and sustain plants. The shallowest are sparsely vegetated while the deepest are packed with species. Surprisingly, the plants have made these depressions. It is a process of primary succession beginning where there is just enough of a depression for a scant layer of mineral particles to accumulate and a thin sheet of water to persist for a few weeks in spring. These are all the resources needed for Elf Orpine to start a wildflower party.

Elf Orpine is a small scarlet to pink succulent with tiny white flowers. Though a single plant may go unnoticed, the radiance of a large patch will add a festive flair to even the dullest granite slab. It is an annual species whose seedlings germinate in fall, produce a few small leaves, then suspend growth for the winter. Seedlings tolerate drought in fall; frost, snow, and ice in winter; and weeks of submersion in spring. It is a lot for a tiny seedling to endure, but this way the seedlings are ready to mature quickly when temperatures rise in early spring. At that point, the little plants quickly produce a few leaves, a flowering stalk, and several petite flowers. In this harsh environment they cannot produce enough nectar to attract large mobile pollinators, but they can make just enough sweet stuff to attract ants that pollinate them capably. Seeds mature just as the exhausted plants wither in the summer's heat. From midsummer to fall, the entire species is dormant in seed form.

The Elf Orpine of a granite depression complete this cycle year after year, century after century. Over time, the population grows and the soil layer thickens with the decay of their tissues and the trapping of windblown and waterborne sediments. Eventually, perhaps after millennia, soils deepen enough for small, shrublike fruticose lichens to establish and initiate a new phase of community development. Full-grown Elf Orpine are present for just a few weeks each year, but fruticose lichens provide a year-round presence that hastens sediment capture. With time, enough soil accumulates that fast-growing annual plants establish. Unable to compete with them, Elf Orpine is relegated to the depression's margins or dies out completely.

Thus far, the rock surface has remained relatively unchanged. But now, decay of the roots and stems left by the annuals yields weak organic acids that dissolve a minuscule granite layer each year. Over decades and centuries, depressions deepen and hold more water and soil. This milder environment now sustains a diverse assemblage of perennial plants. Some depressions even hold enough water during spring to support semiaquatic plants. In exceptionally rare cases, a depression

becomes deep enough to support a colonizing tree whose stunted canopy shades out most of the perennial wildflowers.

As with glade plants elsewhere, the attributes of granite outcrop species evolved to help them endure a harsh environment but rendered them incapable of surviving in ecosystems where more competitive species dominate. Some, including the Blackfoot Quillwort, Wooly Ragwort, and Horned Bladderwort, inhabit various outcrop ecosystems in eastern North America. Roughly a dozen, however, only inhabit the Piedmont's granite outcrop ecosystems. Those outcrop specialists ranging into Alabama include Spotted Phacelia and the federally listed Little Amphianthus (Pool Sprite). Several other species, including Elf Orpine and Oneflower Stitchwort, only reside in the granite outcrops and the sandstone glades of the Southwestern Appalachians ecoregion.

The biogeography of granite outcrop species prompts a series of profound questions. Why are some species endemic to granite outcrops? Why are others found in several types of outcrop ecosystems? How did widespread granite outcrop species spread across the Southeast when the intervening terrain is inhospitable and their seeds do not seem to

A blush of Elf Orpine seedlings in a granite outcrop depression. Photo by R. Scot Duncan.

disperse long distances? Though we may never have the answers, the posing and pursuit of such questions deepen our appreciation for these species and their ecosystems.

ENDEMIC LESSONS

Like the lands they drain, the Piedmont's rivers have also lost much of their biodiversity. Over millions of years, the ecoregion's rivers and their tributaries etched into the landscape several hundred miles of deep valleys and furnished them with cascades, shoals, and shallow sunlit runs. The Tallapoosa River is center stage, draining about half of Alabama's Piedmont. To its west, the Coosa River makes a short dash across the Piedmont and drains about a third of the ecoregion. To the east, the Chattahoochee River drains a fifth of Alabama's Piedmont and forms part of the border with Georgia. Each river's run through the Piedmont is its last as a highland river before dropping over the Fall Line.

A handful of aquatic species are endemic, or nearly so, to the Piedmont's rivers. Aquatic endemism here is lower than in the headwaters of these rivers, but there is enough to suggest the ecoregion has been an important zone of speciation. However, the narrow river valleys promoting speciation became a liability for the river's biodiversity. These valleys made it easy to dam the rivers and fill valleys with reservoirs. By the end of the past century, 10 large dams had been built within Alabama's Piedmont.

The harnessing of the rivers brought economic prosperity to a region with few other natural resources left to exploit. Dam construction and hydropower production brought jobs, and large reservoirs with ample shoreline attracted developers and new residents. Nevertheless, damming is catastrophic for river ecosystems. Dynamic, shallow, sunlit streams are transformed into still, deep, dark reservoirs (see chapter 7). The story of what happened to the ecoregion's endemic fishes reveals how damming affected species in a variety of ways.

Several of the ecoregion's endemics seem to have adjusted to the new landscape. The Tallapoosa Darter, Tallapoosa Sculpin, and Tallapoosa Shiner are surviving well because of their widespread distribution prior to damming and their preference for small streams beyond the reser-

voirs' reach. The Apalachicola Redhorse and the Greater Jumprock are two species of sucker endemic to the Chattahoochee and large rivers to which it connects (the Flint and Apalachicola). They prefer shallow sections of midsized and large streams, where they slurp up invertebrates from the bottom with their downturned mouths. Both were widespread before damming, and while they are gone from the main river, their populations in adjacent tributaries seem stable.

Other endemics are facing extinction. The Tallapoosa's Lipstick Darter, the Chattahoochee's Halloween Darter, and the Coosa's Holiday Darter inhabit shallow-water habitats in free-flowing streams and rivers. With the loss of such habitats in the main rivers, each now survives in a few isolated, small populations in river tributaries. The situation is more desperate for the Shoal Bass, an endemic to the Piedmont and Southeastern Plains sections of the Chattahoochee and Flint rivers. Unlike basses popular with anglers, the Shoal Bass cannot survive in reservoirs and silted rivers. Instead, it needs rapids and shoals in the main channel of rivers and large streams, habitats very nearly gone in the Piedmont. In Alabama, the bass survives only in a few of the Chattahoochee's larger tributaries.

So, there is bad and good news for the Piedmont's aquatic species. Hundreds of miles of river habitat were drowned, several mussel and snail species went extinct, and populations of many aquatic species were diminished and fragmented—some to the point of endangerment. Additional extinctions may be inevitable. More people are mov-

Tallapoosa Shiner, an endemic to the Tallapoosa watershed. Courtesy of Bernard R. Kuhajda/ Tennessee Aquarium Conservation Institute.

ing to the South, the demands for water are increasing, and less water flows through the Piedmont's rivers each year. In this century, major tributaries in the region—including those sheltering rare or endemic species—will be next in line for the building of reservoirs to supply water to growing cities. At least for now, none of the known endemic fishes of the Piedmont have gone extinct and the populations of many are stable.

Awkward Age

It's an awkward time for the Piedmont's biodiversity. During the past two centuries, the ecoregion has lost its original forests, soils, settlers, and rivers. Then again, much biodiversity has survived, and there are reasons to hope that better times are ahead: much of the Talladega Upland is wild and protected, longleaf woodlands are being restored, pockets of the Lower Piedmont's biodiversity are being saved, forests have expanded, and rare species are being championed. Hopefully, the needs of human culture and biodiversity can both be met.

The next level III ecoregion to explore is our last. In several ways, the Interior Plateau is similar to the Piedmont. Both have a long history of deforestation, agriculture, and river transformation, yet still retain uncommon species and ecosystems in their rugged terrain and undammed streams. In contrast, while the Piedmont is the remnants of ancient eroded mountains, the Interior Plateau is what's left of an ancient seafloor.

13
Interior Plateau

GHOSTLY FISHES

One of the strangest, rarest, and most highly endangered fishes on the planet lives below a swatch of northern Alabama farmland. That's right, *below* the farmland. The Alabama Cavefish has lived in the absolute darkness of the karst underworld for millions of years. During this time, it has fully adapted to an ecosystem alien and inhospitable to most other earth inhabitants. It is ghostly white and lacks eyes. Both are adjustments to the scarcity of nutrients and energy deep underground. Resources formerly allocated to growing and maintaining eyes and pigmentation were diverted to more essential systems including the sensory organs used to detect prey in the dark.

Biologists have only seen the Alabama Cavefish in Key Cave, a grotto nestled within a small national wildlife refuge on the banks of the Tennessee River in Lauderdale County. From the gravel roads above, there's no indication of an underground world with strange fishes. Instead, much of the refuge and surrounding terrain is a level expanse of fields and farm roads typical of rural northern Alabama—hardly the unsullied landscape one might expect for a spot of such importance to the state's biodiversity. This paradox typifies Alabama's Interior Plateau. Though it's the most thoroughly plowed and grazed level III ecoregion in the state, there are rare and sometimes strange species living throughout. How is this possible?

The Interior Plateau ecoregion extends through central Tennessee, western Kentucky, and parts of southern Indiana and Ohio. In general, it is a level terrain of limestone and a few other sedimentary rocks formed mainly during the Mississippian Subperiod when this portion of the continent was a shallow tropical ocean with abundant reefs. The bedrock remained relatively level during the Allegheny Orogeny, but tectonic pressures forced uplifting centered on central Tennessee. The

The Alabama Cavefish, one of the rarest fishes on the planet. Courtesy of Danté Fenolio.

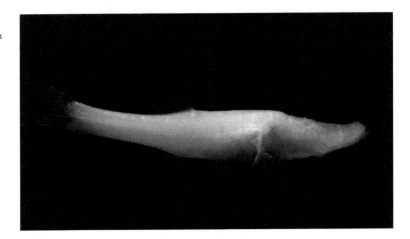

ecoregion largely corresponds to the area geologists refer to as the Interior Low Plateaus Province, the Alabama portion of which is known as the Highland Rim Section.

Because it is a flat to rolling terrain with fertile soil, agriculture dominates the ecoregion. However, wherever limestone is naturally at the surface, its dissolution by rain and groundwater creates a karst terrain of caverns, sinkholes, springs, and other features harboring uncommon ecosystems and promoting the evolution of unusual species. Most of these karst features cannot support agriculture, so they provide refuge for many species.

Alabama's portion of the Interior Plateau has two other major anomalies contributing to the survival of unique species. One is Little Mountain, a series of steep, sandstone-capped hills rising above the plateau and providing dense forests and steep canyons. The other is the Tennessee River, the principal geographic feature of Alabama's Interior Plateau. The river and its tributaries sustain dozens of aquatic species native to the Mississippi River Basin and found nowhere else in Alabama. In addition, there are many snails, mussels, crayfishes, and fishes endemic to Alabama's stretch of the Tennessee River. With its vast size and rich biodiversity, the Tennessee River is the ecoregion's biodiversity centerpiece. We'll begin our exploration of the Interior Plateau by studying the river's ecosystems, a few of its most unique species, and how settlement of the region impacted both.

The view from the crest of the I-65 bridge over the Tennessee River is impressive. The river is over a half mile wide at the crossing and bounded by the forests and fields of Wheeler National Wildlife Refuge. From the span's peak you can see the river stretching northwest toward Decatur and east toward the distant Jackson County Mountains. For many travelers, the crossing is a welcome change of scenery from the flat agricultural expanses. However, very few of the thousands crossing the river each day know the Tennessee is one of the most biodiverse North American rivers.

Prior to the 20th century, the Tennessee River roamed freely over the northern Alabama landscape. For millions of years, it sluiced and dissolved away the plateau's weak bedrock and carved a moderately deep river valley. Resistant rock sometimes forced the river to widen into broad shoals. In one section between Decatur and Florence, the river shoaled continuously for nearly 50 miles (80 km), tumbling across layers of the Fort Payne Chert. New species evolved in the river and in the springs, streams, and small rivers joining it along its Alabama route.

Though most of the region's species still survive, the Tennessee River Valley's aquatic habitats began deteriorating as agriculturalists settled the region. At first, the biggest impact was stream siltation caused by deforestation and poor soil conservation strategies on farmland. Bottom-dwelling organisms declined then disappeared as rocky habitats were smothered. Among them was the Harelip Sucker, a fish of rock and gravel habitats in large, clear streams. The fish was once common throughout the east-central United States and was frequently caught by fishers. From what ichthyologists have deduced, the fish went extinct as its food base of small mollusks and crustaceans dwindled due to siltation.

The other fish species not surviving the 19th century was the Whiteline Topminnow. This small fish had the misfortune of living solely in Big Spring. Downtown Huntsville grew around the spring, harnessed it to supply water, and transformed the location into a city park. In 1889, a biologist collected and preserved 25 topminnows, and they are all that survives of the species. Ichthyologists know very little about the topminnow's life history, but it is no surprise it couldn't endure the lining of the spring with concrete, competition with introduced exotic fishes, and several complete dewaterings. As far as scientists know, the Whiteline Topminnow and Harelip Sucker are the only two Alabamian fishes to have gone globally extinct. They were just the beginning of an extinction wave that swept through the Tennessee River Valley in the next century.

As the agrarian economy grew after the Civil War, so did the clamor for river navigation, hydropower, and a river that wouldn't flood. Damming of the Tennessee River began after World War I and by the end of the 1930s, the Tennessee Valley Authority (TVA) had impounded the entire length of the river in Alabama. Taming of the river spurred industry, economic prosperity, and the growth of towns into cities. However, the damming and impacts caused by subsequent development triggered the extirpation or extinction of nine fishes and over a dozen mussels. Ichthyologists know most about the impact on fishes. Their survival stories reveal to us how 20th-century development affected the region's aquatic biodiversity and highlight the challenges to come.

The Boulder Darter's story is typical for many survivors. It is a small, mottled-brown fish whose breeding males have a blue wash on their fins. The darter needs rocky habitats with a strong current, the habitats most quickly lost when the Tennessee was dammed. The drowning of the shoals and similar habitats pushed the darter and many other species high into tributary watersheds. Today, the darter survives only in the Elk River, where it struggles with sediment pollution from agriculture and, increasingly, urban sprawl. A decade-long captive breeding and release program has bolstered populations in the Elk River watershed and reestablished darters in Shoal Creek, near Florence, where biologists first discovered the species. Though it still is highly endangered, its situation has improved.

Another survivor has gone extinct *twice*. The small Spring Pygmy Sunfish is a karst-dependent species inhabiting dense aquatic vegetation in springs. Adults are dappled-brown, but breeding males display iridescent blue-green bars across their flanks and shiny flecking across the face. Biologists discovered the fish at Cave Spring in 1937, only months before the completion of Pickwick Dam. The impoundment flooded the spring, eradicating the only known sunfish population. In

1941, biologists rediscovered the fish many miles upstream at Pryor Spring. However, this population was lost when the spring's vegetation mysteriously died. After 32 additional years of extinction, the fish was rediscovered in 1973 in Beaverdam Spring. Biologists have also found it in a second spring. Clearly the species is vulnerable to extinction due to its habitat specificity and limited distribution. To make matters worse, individuals live only a year. This hinders population growth and makes the species highly vulnerable to disturbances during critical times, especially spawning. Scientists studying the species have proposed that the Spring Pygmy Sunfish be placed on the US Endangered Species List to help it avoid a third extinction.

The Slackwater Darter has the most epic survival story. It is a handsome fish—breeding males sport a black mask, reddish-orange flanks, and a dorsal fin banded in blue and red. It survived damming because it prefers small, slow-moving streams beyond the reservoir's reach. However, the darter is threatened by agricultural practices affecting the seepage springs where it spawns. Each winter, darters embark on a perilous journey to these springs. The fish ascend shallow streams that only flow after it rains. Between storms, darters are stranded in stream puddles that can dry out if rains are infrequent. Ledges blocking its route must be surmounted when the stream swells with swift floodwater.

Spring Pygmy Sunfish has been declared extinct twice, but still survives. Courtesy of Emily Horton Reilly.

Slackwater Darters make heroic migrations to their breeding habitats. Courtesy of Bernard R. Kuhajda/Tennessee Aquarium Conservation Institute.

After several weeks, the darters arrive in a spawning habitat with no competition and few predators. This strategy worked well in the past, but most breeding springs are now in pastures. Others were drained, converted to fishponds, or polluted with agricultural chemicals and manure. Populations have declined steeply, and the Slackwater Darter is listed as threatened by the US Fish and Wildlife Service (USFWS).

It is a positive sign these and other endangered species survived a century of development in the Tennessee River Valley. Perhaps we already lost the most sensitive species and today's endangered species have a good chance of surviving the 21st century and beyond. One thing is for sure, however. They won't survive without our help.

SILVER LININGS

Despite the great losses to biodiversity, damming of the Tennessee River created new opportunities for native species in the region. Exhibit A is Wheeler National Wildlife Refuge (NWR). Established in 1938 to provide habitat for overwintering waterfowl (ducks and geese) and migrating birds, Wheeler was the first wildlife refuge on lands primarily devoted to reservoir management. The 35,000-acre (142 km²) refuge encompasses river margins and embayments and adjacent lands. Around 650,000 residents and tourists visit each year for educational programs, hiking, hunting, fishing, and observing wildlife.[1] Nearly 50 species of mammals, 75 species of "herps" (reptiles and amphibians), and 285 species of birds use the refuge. Prominent among the latter are large wintering populations of waterfowl, including the southernmost wintering population of the Southern James Bay population of the Canada Goose. Each autumn, dozens of waterfowl species arrive by the thousands after breeding in the tundra and Midwest. On the refuge, they find habitats carefully managed to provide food throughout the winter and a safe haven from hunting.

Refuge managers maintain shallow impoundments along the river's margin specifically for waterfowl. In summer, impoundments are drained and planted with seed crops, especially corn. In autumn, the impoundments are flooded and returning ducks dabble or dive for grains and other foods.[2] These habitats and adjacent fields also attract thousands of overwintering Sandhill Cranes. In recent years, a handful of Whooping Cranes, one of the rarest birds in North America, have joined them. The refuge's abundant wildlife attracts many birds of prey, including Bald Eagles, Red-tailed Hawks, and Northern Harriers. Even larger predators prowl the waters below.

American Alligators have a complicated and somewhat comical history in the refuge. Locals had occasionally seen alligators in the Tennessee River Valley before refuge establishment, but no one knows whether these were natural occurrences. By the 1960s, alligator sightings were regular on the refuge. Refuge managers suspected these were releases, since during this era it was common for tourists in Florida to buy young alligators as pets and release them when the alligators became too dangerous or costly to maintain. Because the American Alligator was an endangered species at the time, the USFWS established several populations using Louisiana alligators as stock. Since Wheeler already had a small population, the service released 50 alligators in 1979. When the service notified the public, well after the fact, many panicked residents complained loudly to their US congressman, Ronnie Flippo. Defending his constituents, Flippo demanded the alligators be removed from the refuge. An earnest recapture program in 1980 using baited traps was a lesson in futility. With Wheeler being huge and overflowing with prey, only two alligators were captured. Trapping was discontinued, and now Wheeler sustains a large alligator population, though they are rarely seen.

Sandhill Cranes over-wintering at Wheeler National Wildlife Refuge, Morgan County. Courtesy of George W. Ponder III/Beaux Point Photography.

There is much more to Wheeler NWR than flooded fields and impoundments. The refuge protects extensive tracts of bottomland forest, or South-Central Interior Large Floodplain, where large creeks join the Tennessee River. Two of these, Beaverdam Swamp and Dancy Bottoms, have some of the most pristine tracts of swamp and floodplain forest in northern Alabama. Though river impoundment altered their hydrology, these ecosystems provide important refuge for many species, including at-risk aquatic fauna.

The chronicle of what happened to Blackwell Swamp illustrates how large-scale ecological tinkering rarely turns out as planned. The swamp is a contrived ecosystem created to establish additional swamp habitat by damming Blackwell Creek and flooding several hundred acres of bottomland forest. Everything was fine until North American Beavers found the new wetland. Beavers eat the stems, leaves, and inner bark of woody plants. They build dams across small streams using vegetation and mud, then use the flooded area to reach more trees and shrubs and float their harvest to their dams and lodge. Until recently, beavers were still recovering from decades of fur trapping. Today, without trappers or predators controlling their numbers, they are abundant. The flooding of Blackwell Creek created perfect conditions for beavers, and they have converted the swamp into an open lake. At the lake's northern end is the latest tupelo stand the beavers have killed. Decaying trunks tower above the lily-carpeted lake. Birds nest in cavities within the standing dead trees, while turtles bask on fallen trunks. Though the Blackwell Swamp project didn't work out as planned, at least some species have benefitted from these changes.

THE PRESETTLEMENT PLATEAU

Beyond the Tennessee River, the levelness of the Interior Plateau is evident. Small country roads immerse one in a sea of cotton, hay, and food crops. The soil is fertile, fields are large, and trees are few. Tall roadside grasses swirl in the wake of passing pickups. The landscape seems characteristic of Texas, Oklahoma, or Kansas where tallgrass prairies once prevailed. The biogeographic disorientation can be acute for birders. Perching on the fences and power lines stringing the roadways are midwestern prairie birds, including Scissor-tailed Flycatchers, Dickcissels, Lark Sparrows, and Horned Larks. The landscape calls into question whether the region was forested in presettlement times.

Despite appearances, forests blanketed the region until the 19th century. Broadleaf forests prevailed in the Eastern Highland Rim and the Western Highland Rim level IV ecoregions. Both ecoregions extend northward across Tennessee and into Kentucky. Together, these ecoregions cover 80 percent of Alabama's Interior Plateau. Both have alkaline soils, level or slightly rolling topography, and similar ecosystems. Still, there are noteworthy ecological differences tied to their geology.

The Eastern Highland Rim covers the entire eastern half and the lower western half of Alabama's Interior Plateau. The area south and immediately north of the Tennessee River is underlain by limestone, but the far eastern portion is underlain by the Fort Payne Chert, a formation of limestone and chert. Both areas sustained the southernmost extent of the Southern Interior Low Plateau Dry-Mesic Oak Forest. None of the original forest remains, but documentation from the early

Prairie-like landscape of the Tennessee River Valley, Key Cave National Wildlife Refuge, Lauderdale County. Photo by R. Scot Duncan.

20th century describes a forest transitional in character—drier than the mountain forests to the east and moister than prairie woodlands to the west. The abundance of oaks and hickories increased westward across this gradient. Except on chert outcrops, trees were species tolerant of alkaline soils, including White and Post Oak, and Mockernut and Pignut hickory. Hilltops and outcrops supported species adapted to drier conditions, including Chinkapin Oak and Eastern Red Cedar. Forests were kept open where wildfire was occasional. Woodlands, including stands of Shortleaf Pine, prevailed where wildfire was frequent.

The southern portion of the Western Highland Rim in Alabama is similar to the Eastern Highland Rim, but the northern portion is a dissected, hilly region on the Fort Payne Chert. The Western Highland Rim's original forests were much like those of the Eastern Highland Rim. However, its dissected hills provided sheltered, moist conditions sustaining a variation of the South-Central Interior Mesophytic Forest. These forests harbored a great diversity of broadleaf trees including American Beech, Shagbark Hickory, Black Walnut, and the rare Smoketree.[3] Some species survive on slopes too steep for agriculture, including the forests at Shoal Creek Preserve, a Forever Wild tract near Florence.

Unlike the karst ecoregions of the Southern Appalachian Mountains, the level plains of the Interior Plateau harbor limesinks. Limesinks, more formally known as Central Interior Highlands and Appalachian Sinkhole and Depression Ponds, are shallow surface depressions created when the limestone below has dissolved and collapsed. They are more frequent in the Eastern than in the Western Highland Rim. The variable ecology of these isolated wetlands bolsters the ecoregion's biodiversity. Some are many acres in size, while others are so small that farmers plowed them into their fields. Some permanently hold water, while others are wetted seasonally. Some are open, while others support wetland trees and shrubs. Small limesinks lack fish and are ideal breeding habitats for amphibians, while others are large ponds with open, shallow margins attracting wetland birds. Though limesinks are havens for rare or endemic species, there is a karst surface feature of the Interior Plateau even more famous for its botanical diversity—the limestone glade.

LIMESTONE GLADES AND BARRENS

If you travel through the Moulton Valley on State Highway 24, you will pass mile after mile of the cropland and rolling pasture typical of the Eastern Highland Rim. Turn north on Lawrence County Road 42, a narrow gravel farm road, and you'll encounter the rural landscape more intimately: fields thick with planted grain, fencerows crowded with small trees, rutted entrances into gated pastures, and small herds of cattle mildly curious about your presence. After a mile or so there is a sign marking the entrance to the Nature Conservancy's (TNC) Prairie Grove Glades Preserve.

From the road, the tract doesn't look like anything special. Just behind the rusted barbed wire there is second-growth forest choked with young Eastern Red Cedars and other pioneer trees. But hike into the preserve on the path from the parking area and the forest opens to reveal a landscape of multicolored wildflowers and botanical oddities.

In spring and early summer, the ground is a patchwork of color and texture. There are clusters of grasses and blooming wildflowers, exposures of blue-gray limestone, beds of sun-bleached gravel, and mounds of deep green moss.

These are the limestone glades and barrens of the Interior Plateau, also known as the Central Interior Highlands Calcareous Glade and Barrens ecosystem.[4] They are distributed throughout the east-central United States, mainly in the Interior Plateau and secondarily in the Ridge and Valley. Though widespread, they are rare—at least in the modern landscape. Bedrock near or breaching the surface helps maintain these ecosystems. The thin, alkaline soils don't offer enough space, nutrients, or water for most trees to survive. However, dozens of unusual, hardy plants tolerate these droughty soils and enjoy unlimited sunlight. Several species at Prairie Grove are Alabama endemics, including Alabama Gladecress and Lyreleaf Bladderpod. In barrens environments, the soil is deeper and grasses cover more than half the ground. Barrens plant species are usually more common and geographically widespread than glade species.

Because glades and barrens are now scarce, many of Prairie Grove's species are imperiled. Moreover, though the preserve is protected, its rare plants are not safe. Surrounding these open habitats is a wall of Eastern Red Cedar, the archenemy of this ecosystem. They loom at the margins, threatening to shade out small, sun-loving wildflowers and grasses with their thick, evergreen canopies. Cedars can tolerate shallow, rocky soils, and many younger cedars have advanced into the barrens. A few have colonized glade habitats where faults through the bedrock provide greater soil depth. The decay of needles beneath older cedars has created an island of topsoil. Shrubs and small broadleaf trees cluster at their base and help the "tree island" expand. If unchecked, cedars will eventually overtake the glade. So, what has historically prevented glades from disappearing?

Most ecologists studying glades and barrens agree that periodic, severe summer droughts kill many invading cedars. However, since glades and barrens have survived for millennia, cedars should have had ample opportunity to snuff out these ecosystems. Many ecologists, myself included, believe recurring wildfire also helped maintain these and similar southeastern ecosystems. Eastern Red Cedar might be tough, but it is not fire-tolerant. Glades and barrens could only burn regularly if wildfires ranged across the greater landscape. Thus, much of the post-Pleistocene landscape of the Interior Plateau must have been woodland kept open by periodic wildfire.

Though ecologists have studied how southeastern glades and barrens ecosystems are sustained, very few have offered ideas on their genesis. Sometimes their origin is obvious. Those on rocky slopes exist because of the geologic processes (uplift and erosion) creating the hosting topography. Herbaceous plant communities persist there because water and soil are shed quickly and only a few hardy trees like cedars can colonize. Other glades and barrens, including those at Prairie Grove, have a weak slope or none whatsoever. It has been suggested that the limestone beneath such glades is more durable and weathers more slowly than adjacent limestone, resulting in areas of slightly higher elevation with thin soils. Research is needed to test this and other hypotheses on the origins of these unique ecosystems.

A Hidden Ecoregion

There are places in northwestern Limestone County where the farm-land roads depart from their monotonous straightaways and plunge down steep hillsides into narrow, hidden valleys. This realm is the Outer Nashville Basin level IV ecoregion, and it is one of Alabama's underappreciated treasures. Here, tectonic forces have cracked open the plateau, and a dramatic landscape of uplifted plateau, steep escarpment, and low stream valleys has emerged.

The Outer Nashville Basin extends into Tennessee to surround the Inner Nashville Basin level IV ecoregion; together, the two ecoregions are often termed the Central Basin. The Central Basin originated when the Southern Appalachians were forming and tectonic pressure on the Interior Plateau created a broad area of uplift known as the Nashville Dome. The rock formations fractured during this upheaval were eroded away, leaving a craterlike basin, 50 miles wide and 100 miles long (80 by 161 km), with the relatively flat Inner Nashville Basin at its center. The Outer Nashville Basin is in an intermediate stage of erosion, as evidenced by its tortuous landscape. The Eastern and Western Highland Rim ecoregions, which were the Nashville Dome's lower slopes, form a plateau perimeter around the basin.

In Alabama, the Outer Nashville Basin is a gorgeous landscape. The Elk River and its tributaries have cut through the upper plateau surface, creating steep escarpments dropping as much as 300 feet (91 m) to the valley below. South-Central Interior Mesophytic Forests blanket many slopes, but the forest cedes to cropland and pasture on the fertile valley floors. The smaller valleys wind in unexpected directions, ending in narrow coves surrounded by steep forested slopes. Though this landscape begs for exploration, there are no preserves open to the public, which is a major reason the ecoregion is so poorly known in Alabama.

The Paint Rock

Another stunning landscape materializes where the easternmost margin of Alabama's Eastern Highland Rim meets the Jackson County Mountains of the Southwestern Appalachians level III ecoregion. The boundary between these ecoregions is beautifully complicated. Clus-

ters of mountains, even a few single mountains, arise like islands from the plateau. Larger island clusters send peninsula-like ridges into the surrounding flatlands, transforming them into a network of narrow branching valleys. The mountains are cloaked with thick forests contrasting sharply with the fields and pastures of the valleys. The juxtaposition of pastoral valleys and wild mountain slopes is exquisite.

Flowing through one of these valleys is what many aquatic biologists regard as the most important and pristine tributary of the Tennessee River—the Paint Rock River. The river and the streams it gathers bring the Eastern Highland Rim's valleys deep into the Jackson County Mountains. They also shelter about 100 fish species and nearly 50 mussels. About two dozen aquatic species, including the Palezone Shiner and Alabama Lampmussel, are critically imperiled because they are endemic to the Paint Rock watershed or they lost most of their habitat when the Tennessee River was dammed.

The Paint Rock River does have its problems. The same agriculture it supported for nearly two centuries is polluting the river with silt and

Pasture and forest-cloaked mountains, Paint Rock River Valley, Jackson County. Photo by R. Scot Duncan.

Late evening on the Paint Rock River, Jackson County. Courtesy of Alan Cressler.

livestock feces. In places, the river was straightened to speed drainage during floods. However, due to the Paint Rock's substantial biodiversity, several organizations and agencies are protecting and restoring the river and its watershed. At its Roy B. Whittacker Paint Rock River Preserve, TNC has planted pasture with riparian forest trees and restored the meander of a major tributary. Farther up the valley, TNC has partnered with ranchers to keep thirsty cattle out of the Paint Rock by providing fencing and water tanks and has teamed with other landowners to stabilize eroding stream banks. The uplands surrounding the Paint Rock, part of the Southwestern Appalachians ecoregion, are also being protected. TNC has purchased tracts on Sharp and Bingham mountains to safeguard forests, subterranean ecosystems, and slopes feeding groundwater to the river. Fern Cave National Wildlife Refuge near the mouth of the valley protects other escarpment slopes and a winter hibernaculum for endangered Gray Bats. In the river's headwaters, the Alabama Department of Conservation and Natural Resources manages Skyline WMA and the Walls of Jericho, a Forever Wild tract. The beauty, the biodiversity, and the conservation practiced in the Paint Rock watershed are inspiring.

BIG STARS OF LITTLE MOUNTAIN

Within the deep canyons of Little Mountain is found French's Shootingstar, a lovely little plant whose presence in northwestern Alabama is linked to an ancient coastline. Early each spring, the shootingstar emerges from dormancy and grows low to the ground. By late spring, older plants will develop a thin stalk several inches long on which hang several delicate flowers. At the base of each petal is a strip of deep maroon bordered in yellow, but the rest of the petal is an immaculate white turning pinkish with age. The flower strongly recurves its petals skyward while aiming its yellow pistil and stamens downward. As the name suggests, the overall display could resemble a meteorite streaking earthward. However, these shootingstars will never see the sky, for they live exclusively beneath the lip of sandstone overhangs provided by the Cumberland Acidic Cliff and Rockhouse ecosystem. Why the shootingstar lives here is clear: light is so scarce and the soils can be so dry that virtually no other vascular plants vie for this habitat. What's more, the rockhouse protects the plants from erosion and being smothered by the thick mats of leaf litter accumulating on the forest floor just a few feet away. How rockhouses came to be in the Interior Plateau is less evident, and the explanation takes us back to 340 million years ago.

When the Southern Appalachian Mountains arose, the area that is now the Southwestern Appalachians and Interior Plateau ecoregions was a shallow tropical sea created as the crust sagged under the weight of the growing mountains. Vast reefs developed and became the limestones so plentiful in these ecoregions. As the young mountains grew, they shed great volumes of sediment into the nearby oceans. Sands, clays, and muds became the sedimentary rocks of the Pride Mountain Formation. Just offshore, sands accumulated as a chain of barrier islands like those along the Gulf Coast. This island chain became a thick rock formation dubbed the Hartselle Sandstone.

The Pride Mountain and Hartselle Sandstone formations breach the surface of northern Alabama in several locales, but nowhere as

French's Shootingstar, Little Mountain, Colbert County. Courtesy of Frederick R. Spicer Jr.

prominently as in the Little Mountain level IV ecoregion of the Interior Plateau. Little Mountain isn't really a mountain, it's a series of tall, dissected hills spanning 75 miles (121 km) east to west across northern Alabama. The northern boundary of the ecoregion is steep, while the southern boundary slopes gently into the Moulton Valley. The Hartselle Sandstone serves as a protective caprock to some of the softer rocks of the Pride Mountain Formation.

The dry, sandy soils on the caprock sustain the Southern Interior Low Plateau Dry-Mesic Oak Forest that was once widespread in the Interior Plateau. These upland forests would have developed into woodlands where and when fires were frequent. The most intriguing ecosystems of Little Mountain occupy the valleys and canyons weaving through the hills. These sheltered slopes harbor another variation of the South-Central Interior Mesophytic Forest and hundreds of animals and plants rare elsewhere in the Southeast. Among the latter are orchids, lilies, azaleas, American Ginseng, Goldenseal, American Columbo, Bosch's Filmy Fern, Eastern Leatherwood, and Kentucky Yellowwood, to name a few. There's so much plant diversity that plant enthusiasts return to Little Mountain year after year.

The rockhouse homes of French's Shootingstar occupy the steepest slopes of Little Mountain's canyons. They form because canyon streams continuously transport sediments out of these canyons. As a result, stream valleys widen, deepen, and migrate upstream through a process called head-cutting. Softer and lower rocks of the Pride Mountain Formation erode first, while the upper and harder Hartselle Sandstone yields more slowly. These differential rates of erosion create the overhangs. Eventually, overhangs collapse and send a jumble of boulders into the valley. However, while they persist, rockhouses provide a unique environment for plants and animals, including the shooting-star.[5]

French's Shootingstar is known from Arkansas and four midwestern states, but its presence in Alabama is a biogeographical puzzle. None of Alabama's neighboring states have populations. What's more, Alabama's only population is on Little Mountain, despite the proximity of rockhouses a dozen miles away in the Dissected Plateau level IV ecoregion. For whatever reason, this little plant survives in Alabama, thanks to those barrier islands from long, long ago.

You can search for French's Shootingstar at Cane Creek Canyon Nature Preserve. This 700-acre (2.8 km²) private preserve is the only nature sanctuary open to the public on Little Mountain. Miles of trails thread through the dry oak forests of the caprock and lush mesophytic forests in the canyons. Jim and Faye Lacefield, two conservation superstars, own the preserve. TNC protects it in perpetuity with a conservation easement. Jim has also dedicated much of his life to educating the public about Alabama's geology and its relationship to ecology and human culture. He authored *Lost Worlds in Alabama Rocks: A Guide to the State's Ancient Life and Landscapes*. Written for the layperson, it is one of the most important natural history guides ever written about Alabama and is an essential reference for many students of biodiversity, myself included.

LEGACIES

As Jim Lacefield would point out, Little Mountain and all of Alabama's ecoregions developed from the geological legacies of ancient environments. The rivers, rocks, and rains shaping Alabama's Interior Plateau will endure long into the future, but what will happen to the ecoregion's biodiversity legacy? The Interior Plateau is already Alabama's most heavily transformed level III ecoregion. Its most unique ecosystems and species are confined to the rugged margins where agriculture, urban sprawl, and river impoundments haven't reached—yet. Whether future generations will enjoy the Alabama Cavefish, French's Shootingstar, and other treasures depends on whether the ecoregion's current tenants are good stewards of the remaining forests, glades, and free-flowing streams.

Our journey through the state's rivers and its six level III ecoregions is complete. We've covered a lot of ground, from tall mountain peaks to low coastal marshes, but our last adventure will be in Alabama's greatest wilderness—the Gulf of Mexico.

14
Gulf of Mexico

The Greatest Wilderness
By most measures, the open water of the Gulf of Mexico is Alabama's greatest wilderness. The expanse between its sandy beaches and the deep waters beyond the Continental Shelf has the lowest human population density and the most intact suite of ecosystems. Explorations beyond the sight of land or below the surface require complex technology. And if wilderness can be rated by the presence of large animals undaunted by man, then the Gulf is unrivaled.

For many, the Gulf is a realm whose mystique is reinforced by periodic newspaper headlines announcing sightings of large and strange species: "US Scientists Net Giant Squid in Gulf of Mexico," "Biologist Discusses Big Fish Story About Killer Whales off Alabama's Gulf Coast," "Whale Sharks Approach Alabama Coast in Unprecedented Numbers," and "Giant Deep Sea Jellyfish Filmed in Gulf of Mexico." The thought of such creatures so close to our shores can excite a potent blend of intellectual curiosity and primal fear.

Are these creatures lost wanderers from faraway exotic locations? Or are they offshore neighbors with whom we're not well acquainted? For many years the northern Gulf of Mexico was relatively unexplored, and even scientists couldn't answer these questions. For that matter, biologists hardly understood the life cycles of the Gulf's most common species, even those supporting the seafood industry. Our understanding of the northern Gulf of Mexico's ecosystems has greatly advanced in recent decades. In this chapter, we'll explore Alabama's marine wilderness, from ancient reefs to floating forests.

The Physical Gulf
Unlike terrestrial ecosystems where most of the biology dwells within a few meters of the surface, oceanic ecosystems extend from the surface

The Gulf of Mexico, Alabama's greatest wilderness. Courtesy of Alan Cressler.

to the seafloor. Depths off the coast of Alabama range to more than 6,500 feet (1,981 m) and offer a considerable volume of habitat for biodiversity. Marine species do not live in this space randomly; instead, they occupy a particular range of depths.

The seafloor of the Continental Shelf is a submerged, shallow plain extending around the perimeter of the Gulf of Mexico. South of Alabama, it is mainly an expanse of sand and mud (this was dry land during the last ice age) extending 65 miles (105 km) offshore. Across most of the shelf, the slope is slight, and at 50 miles (80 km) due south of Alabama, depths range from only 160 to 200 feet (49–61 m).[1] Beyond this point, the seafloor begins descending rapidly across a rugged terrain dubbed the Continental Slope. At about 65 miles (105 km) offshore, depths have dropped to 3,300 feet (1,006 m). And at 85 miles (137 km) offshore, the approximate distance between Birmingham and Montgomery, depths have nearly doubled to 6,500 feet (1,981 m). This is a dramatic transition. The change in depth across the Continental Slope south of Alabama is nearly 2.5 times the total range in elevation on land in Alabama and exceeds the average depth of the Grand Canyon by more than 600 feet (183 m). Beyond the slope, the floor of the Gulf continues to drop and reaches depths exceeding 10,000 feet (3,048 m) at 300 miles (483 km) south of Alabama. Off Alabama and northwestern Florida, the Continental Slope forms the DeSoto Canyon, a cleft in the shelf that brings a portion of the Gulf's deep Central Basin—and its wildlife—closer to shore than elsewhere along the northern Gulf Coast.

The Continental Slope is a transitional zone in other ways. Exquisitely clear, tropical waters (known as "blue water") here meet the shelf's nutrient-rich waters (known as "green water," referring to its abundant plankton). The line where the Gulf's blue and green water ecosystems converge varies seasonally in response to the strength of the Gulf's currents. The mixing of these waters in combination with the variable depths stimulates vigorous growth of small pelagic (open ocean) plants and animals, and attracts large oceanic animals.

The primary current governing much of the Gulf's biology is the Loop Current. It enters the Gulf from the Caribbean Sea via the Yucatan Channel, a 122-mile (196 km) gap between the Yucatan Peninsula and the western tip of Cuba. The Loop Current brings tropical waters into the Gulf and pushes them north toward the Louisiana coast. At mid-Gulf, the current begins a clockwise turn and eventually exits the Gulf via the Florida Straits, a 90-mile (145 km) wide channel between the Florida Keys and northern Cuba. The current is swiftest near the Gulf's center, but it slows when approaching and flowing over the Continental Shelf due to resistance from the seafloor. While the general path of the Loop Current is simple, the current isn't a single ribbon of water with distinct boundaries. Instead, it is a collection of water bodies moving in a general direction, but with variable speeds, sizes, temperatures, and headings.

The strength and position of the Loop Current vary seasonally and govern the distributions of oceanic species. In late fall, winter, and early spring, the Loop Current remains in the central Gulf for two reasons. First, the cool atmosphere over the Continental Shelf chills the surface waters. These waters resist the current because cold water moves more slowly than warm water. Second, during this time the Mississippi and

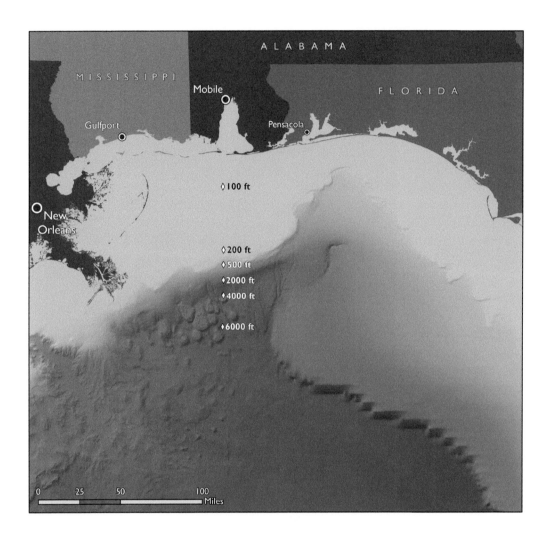

ALABAMA

MISSISSIPPI

Mobile

FLORIDA

Gulfport

Pensacola

◊100 ft

New
Orleans

◊200 ft
◊500 ft
◊2000 ft
◊4000 ft

◊6000 ft

0   25   50   100
                                      Miles

Seafloor topographic variation in the Gulf of Mexico south of Alabama. Courtesy of Ed Brands/University of Minnesota Morris.

Mobile rivers spill huge volumes of water into the northern Gulf due to heavy winter rains and melting snows over the eastern and midwestern United States. This influx of freshwater further chills the northern Gulf and pushes the Loop Current southward. During the late spring, summer, and early fall, the situation reverses and the Loop Current flows over more of the Continental Shelf than at any other time of the year. During this time, many oceanic species ride the current into Alabama's offshore waters. Among them is the largest shark species in the world.

BIG SHARK, SMALL PREY

Sharks ranging greater than three dozen feet in length (11 m) invade Alabama's offshore waters every summer. They are Whale Sharks, and they arrive ravenous, not for fishes or beachgoers, but for some of the tiniest creatures alive—plankton. Whale Sharks are gentle giants that spend their days in warm waters around the world leisurely swimming near the surface with their huge mouths yawning wide to strain their food from the water: microscopic plants (phytoplankton), small drifting animals (zooplankton), and the occasional larger animal that gets in the way. How can an animal weighing several dozen tons survive by eating organisms so small? The answer to this question reveals how oceanic ecosystems work and why the northern Gulf of Mexico is so productive.

Biologists often call phytoplankton the "grass of the sea." There are hundreds of species, but all absorb nutrients from the water and capture energy through photosynthesis. Phytoplankton are the starting point

Whale Sharks are regular visitors to the Alabama coast. Courtesy of Andy Murch/Elasmodiver Co.

for the oceanic food chain. Zooplankton are their primary predators. Some zooplankton species remain minute their entire lives, while others, like fishes and mollusks, are zooplankton only during a larval stage. Zooplankton are food for slightly larger animals that become food for larger animals, and so on. A few large creatures skip links in the food chain and simply graze on the plankton. Whale Sharks do this by filtering great volumes of water each day and migrating seasonally to seek out phytoplankton blooms. Their immense body size enables them to live this way.

Most oceanic waters are generally lacking in nutrients, but the waters in the northern Gulf are nutrient-rich, especially in the spring and early summer near the discharge of large continental rivers. The combination of warming water, plentiful sunshine, and abundant nutrients triggers a phytoplankton population explosion each summer. These plankton blooms are predictable events that transform the northern Gulf into the watery equivalent of a fertile, tropical savanna attracting herds of migrating oceanic animals, including Whale Sharks.

Quicksilver Ecology

Migrating into the northern Gulf with the Whale Sharks are species of predatory fishes. They come for the vast schools of anchovies, herrings, sardines, and other baitfish growing in the northern Gulf on a generous diet of zooplankton. Anglers pursue them for their meat and the fight given when caught on hook-and-line. However, there can be problems with eating these predators—problems of toxin exposure and species conservation. The ecological stories of three taxa in the Gulf—mackerel, tuna, and billfishes—illustrate how the northern Gulf is not immune to problems of pollution and overfishing.

One of the most commonly captured fishes of the northern Gulf is Spanish Mackerel. These are long, sleek fishes growing to about 2.5 feet (0.8 m) with silver sides and short fins. Those along the Alabama coast arrive from their wintering waters off southern Florida just as biological productivity increases and the Loop Current reaches its northern extent. They patrol the surface over the Continental Shelf, usually where the Gulf is less than 328 feet (100 m) deep. When the northern Gulf cools and baitfish become scarce, Spanish Mackerel migrate south with the Loop Current to complete a 500-mile (805 km) annual migration.

Spanish Mackerel form immense schools where baitfish are common. Where they have surrounded a shoal, the surface churns with the silver flashes of striking mackerel. Enthusiastic mackerel will even leap through the air during their attacks. The commotion attracts the attention of terns, birds that dive into the water from the air to catch small fishes the mackerel have herded together. Experienced anglers look for circling terns to find mackerel schools.

The Kingfish, or King Mackerel, is another sought-after fish. It is the Gulf's largest mackerel, reaching lengths just over 5 feet (1.5 m). Its migration resembles that of the Spanish Mackerel, but Kingfish prefer deeper waters of the Continental Shelf. Kingfish feed in small groups or large schools, sometimes targeting smaller surface fishes or diving deep to hunt squid. Anglers value Kingfish for the strong fight they give on hook-and-line.

Though there was commercial overharvest of the Gulf's mackerel in the past, both mackerel now have strong populations and fishing is sustainable. However, Spanish Mackerel are commonly eaten, while Kingfish are not. The reason stems from a major difference in life expectancy between the two species. Most Spanish Mackerel only live a few years, but Kingfish live much longer (maximum recorded age for each species is 12 and 20 years, respectively[2]). The difference is enough for Kingfish to accumulate high levels of methylmercury, one of the most toxic of human pollutants.

Mercury is naturally rare in most ecosystems but is a pollutant released through waste incineration and the burning of coal. Mercury molecules are heavy and do not linger in the atmosphere. Much of the mercury settling on the ocean or washed there by rivers is converted by bacteria into methylmercury, a toxic compound that easily enters the food chain when the bacteria are eaten. Once absorbed by the body, methylmercury is not easily excreted and is stored in animal tissues. Thus, older or longer-lived animals will have accumulated more methylmercury than younger animals. Furthermore, methylmercury moves up the food chain in a process known as bioaccumulation. At each higher step in the chain, there are greater concentrations of methylmercury.

Within vertebrates, methylmercury damages the brain and nervous system, causing an array of problems. Fetuses and children are especially vulnerable because their nervous systems are still developing. The US Environmental Protection Agency (EPA) and Food and Drug Administration (FDA) advise fish and shellfish consumers on how to avoid or reduce consumption of methylmercury. They recommend that children, pregnant women, women planning on reproducing, and nursing mothers completely avoid King Mackerel.

The story of the Atlantic Bluefin Tuna and its fishery tells a different story about the perils of overconsumption. Tuna are in the same family as the mackerel, Scombridae, but are more rotund, larger, and swifter. The largest species in the Gulf, the Atlantic Bluefin Tuna, grows to 10 feet (3.1 m) and weighs over 1,400 pounds (635 kg).

The Atlantic Bluefin Tuna is one of the most athletic of the ocean's fishes. They can swim faster than 40 miles per hour (64 kph) and react quickly when chasing prey or evading predators. Bluefin achieve this by maintaining a body temperature much higher than that of the surrounding water. Most fishes have body temperatures matching their environment, but all tuna have an elaborate counter-current mechanism in their circulatory system helping them retain heat generated by their swimming muscles. However, these adaptations come at a cost. To maintain the supply of heat, tuna must swim constantly, maintain a voluminous circulatory system, and eat voraciously.

Like mackerel, tuna need one another to locate and herd prey. Packs of bluefin will advance through near-surface waters in a semicircle, with the left and right flanks in the lead. Once tuna overtake enough baitfish, they orbit the school and herd the fish into an increasingly dense shoal—often called a baitball. Tuna then take turns darting through the baitball to snatch prey until none remain. Feeding tuna schools are so big and noisy they often attract billfishes, sharks, and whales that take their own swipes at the doomed baitfish. Unfortunately for the tuna, the commotion can attract commercial fishing vessels that encircle tuna schools in nets or yank them from the sea with various forms of handgear.

For a long time, the Atlantic Bluefin Tuna was not highly prized. In the early 1970s, bluefin caught for canning yielded a mere $0.05

per pound. Prices for the fish then soared in the mid-1990s, peaking at $9 per pound with the popularization of sashimi, a pricey Japanese delicacy of thinly sliced raw fish flesh. Unfortunately, such high prices can drive even common species to ecological extinction, where a species survives but cannot fulfill its ecological role. Following a pattern repeated for species after species, fishing pressure on Atlantic Bluefin Tuna increased as prices soared. Prices declined as more tuna were captured, but this only encouraged fishers to land more tuna to maintain revenue. Today, the size of the bluefin's populations in the western Atlantic has dropped to about a quarter of what it was in 1975.

The National Marine Fisheries Service (NMFS) now classifies the Atlantic Bluefin Tuna as a species in jeopardy and one still harvested unsustainably.[3] Conservation is difficult due to its high market value and the need for international cooperation. International management of the Atlantic Bluefin Tuna is predominantly through the International Commission for the Conservation of Atlantic Tunas, or

Tuna hunt in fast-moving schools. Courtesy of Danilo Cedrone/United Nations Food and Agriculture Organization.

ICCAT, but the commission has been unsuccessful in reversing the species' decline.

Fortunately, the Atlantic Bluefin Tuna's primary spawning grounds are in the northern Gulf of Mexico within US territorial waters. Each summer, bluefin arrive and spawn over the Continental Slope and adjacent deeper waters south of Alabama. After spawning, adults return to the Atlantic, but young will feed and grow in the shelter of the Gulf. Because of the importance of the northern Gulf to the bluefin fishery, the United States prohibits fishing for the species in its waters. However, at the rate Atlantic commercial fishing fleets are capturing bluefin elsewhere, it is widely believed this protection will be insufficient to prevent the fishery from crashing.

A different story of ecology, fishing patterns, and contamination involves Swordfish and marlin. Together these are known as billfishes, in reference to their long, lancelike upper jaws. Billfishes are long, sleek apex predators of blue water habitats. They feed alone or in small groups, targeting squid and fishes including Dolphinfish (also known as Mahimahi), mackerel, and baitfish. They slash and stab prey with their sharp beaks, but swallow many smaller prey whole. Though they often feed near the surface, billfishes have a heat-retaining, counter-current circulation system for their brain and eyes, and this helps them make deepwater hunting forays where temperatures are cold. That said, billfishes generally prefer warmer waters, which is why they migrate into the northern Gulf each summer.

Two species of marlin inhabit the Gulf of Mexico, and sportfishers pursue both for the thrilling fights and spectacular leaps they make when caught on hook-and-line. The cosmopolitan Blue Marlin is the larger species, reaching lengths of 14 feet (4.3 m). The White Marlin inhabits the Atlantic's warmer waters and reaches lengths just over 9 feet (2.7 m). The international marlin sportfishing industry earns billions of dollars annually. Because sportfishers release most marlin alive and the United States bans commercial fishing in its waters, you might expect marlin populations to be stable. The NMFS concludes otherwise, however, and considers both species to be struggling from unsustainable harvest.

The culprit is longline fishing, a method used by commercial fishing fleets to capture large pelagic species. In longlining, a buoyed line is stretched great distances, sometimes for miles, across the ocean's surface. Attached to the mainline are shorter lines, each ending with a baited hook. It is an indiscriminate technique that captures many large creatures, including toothed whales, sharks, sea turtles, pelagic birds, and marlin. The non-target species are called bycatch.

Longline fishing has come under scrutiny due to its role in depleting populations of bycatch species. Some progress has been made to reform longline fishing through regulations and the use of bait and gear that is less attractive or injurious to bycatch species. However, it is not yet clear whether these strategies will be widely adopted or sufficiently effective to reverse the decline of marlin populations.

Swordfish are a similar cosmopolitan apex predator of the open ocean. They can grow to lengths of nearly 15 feet (4.6 m), but those in the western Atlantic and Gulf of Mexico tend to be smaller. Swordfish can tolerate colder waters than marlin, so they can hunt at deeper depths and higher latitudes.

Longline fishing for Swordfish in the North Atlantic last century cut populations down to critically low levels. Concerned over the species' plight, several US conservation organizations educated the public and rallied pressure on the federal government and international community to protect the species. The campaign was successful. The United States adopted stiff protective regulations and ICCAT and its participating countries agreed on regulatory measures aimed at rebuilding the Atlantic populations. As a rare example of success in international fisheries management, regulators declared the Atlantic population of Swordfish rebuilt in 2006. Ready to order Swordfish next time you are dining out? You may want to reconsider. As apex predators, swordfish (and marlin) accumulate high levels of methylmercury, and the FDA and EPA have issued the same warnings about consuming swordfish as they have for Kingfish.

Alabama's Sharks

Human fear of sharks is sometimes warranted, for sharks are efficient carnivores and several larger species occasionally attack humans. However, much of the public's fear of sharks is founded on misconception.

Not all sharks, for example, are dangerous. Of the several dozen species in the northern Gulf of Mexico and its estuaries, very few have ever attacked people. Even large sharks are not "out to get" people. In most shark attacks, people put themselves in risky situations. The fact is, sharks are so abundant in Alabama's waters that if they were as dangerous as many assume them to be, few swimmers would emerge unscathed. There are over a dozen common shark species off the Alabama coast, and each has more to fear from us than we have to fear from them.

In the blue water zone of the Gulf of Mexico, there are several common pelagic sharks, including the Oceanic White Tip, Blue, and Silky sharks. All have cosmopolitan distributions, attain lengths of greater than 10 feet (3.1 m), and are solitary hunters aggregating only when drawn to concentrations of prey. Large pelagic sharks have few predators, but among them is another pelagic shark, the Short-finned Mako. Makos are powerfully built and can exceed 13 feet (4.0 m) in length. They have an effective heat-retention system similar to the tuna's, which helps them reach attack speeds greater than 20 miles per hour (32 kph). At full maturity, they hunt some of the swiftest pelagic animals, including billfishes, tunas, and other sharks.

Different sharks inhabit the Continental Shelf. Among the most common are the Atlantic Sharpnose and Blacknose sharks, and *Mustelus* sp., an unidentified species of smooth dogfish. All are less than 5 feet (1.5 m) at maximum length and feed on small fishes and invertebrates. They migrate, sometimes in large schools, to deeper waters for the winter. The Atlantic Sharpnose Shark is the most common and occupies a variety of depths, including the surf zone, which is why it is the most frequently captured shark in Alabama.

Two larger and common sharks of the northern Gulf are the Blacktip and Sandbar sharks. Blacktips reach lengths of 8.4 feet (2.6 m) and populate nearshore environments at depths less than 100 feet (30.5 m). Blacktips attack baitfish shoals by darting through them while spinning rapidly along their length. These sorties can lead to spectacular leaps above the surface. The Sandbar Shark is a bottom dweller preferring smooth seafloor habitats at depths greater than 98 feet (30 m). It grows to a maximum size of 7.5 feet (2.3 m) and consumes invertebrates and small fishes.

None of these aforementioned species pose much of a threat to humans, but two other common species, the Bull and Tiger sharks, are exceptions. Named for their shortened snout and belligerent disposition, Bull Sharks are thought to be responsible for the majority of attacks on humans in the Southeast.[4] Bulls can reach a formidable 11.5 feet and can survive in freshwater for weeks. In marine waters, the only animal competing with a full-grown bull is the Tiger Shark. Tigers are the largest carnivorous shark of nearshore and offshore waters in the northern Gulf, with maximum recorded lengths of 18 feet (5.5 m). With stripes and spots along their flanks and dorsal surface, they are one of the easiest sharks to identify. They prefer murky waters near estuaries where they can easily ambush sharks, rays, sea turtles, dolphins, birds, and large invertebrates.

If large carnivorous sharks are ubiquitous in the northern Gulf and its estuaries, what are the odds of being attacked? Dr. George Burgess has the answer. He is director of the International Shark Attack File (ISAF), a program hosted by the Florida Museum of Natural History and the American Elasmobranch Society to compile information about shark attacks from around the world. Between 1999 and 2009, there were 455 shark attacks in the United States, of which nine (2 percent) were fatal. These included four attacks in Alabama, none of which were fatal. Given the millions of people who spent time in and

Blacktip Shark captured by Dauphin Island Sea Lab researchers. Courtesy of John J. Dindo/Dauphin Island Sea Lab.

on Alabama's coastal waters over this period, the odds of being attacked are minuscule. According to Burgess, compared with the odds of being killed by a shark, residents of coastal US states are 325 times more likely to die in a vehicular collision with a deer, 77 times more likely to be fatally struck by lightning, and 45 times more likely to be fatally attacked by a dog.

That said, ISAF offers research-based precautions to reduce the risk of a dangerous encounter with a shark. Swim in groups, because the activity will frighten most sharks away. Avoid swimming at night, dusk, and dawn when sharks are most active but cannot see as well and may mistake you for prey. Do not swim when bleeding (including menstruation) because slight amounts of blood can attract a shark from afar. Finally, avoid areas where baitfish are abundant and sharks may be feeding.

In reality, sharks have far more reason to fear us than we have reason to fear them. Despite their strength, agility, and abundant habitat, many species are in steep decline. In 2006, when the International Union for Conservation of Nature (IUCN) first assessed the status of the world's sharks and their close relatives the rays, they estimated 20 percent were already threatened with extinction. IUCN classified more species as threatened the following year when more data became available. A separate review of 64 pelagic sharks and rays determined a third of species are threatened.

Overfishing is causing the declines. Many sharks are killed as bycatch in the longline fishery for tuna and Swordfish, while others are targeted by longlining for their liver oil, meat, skin, and fins. Growing demand in Asia for shark steaks and shark fin soup has forced much of this decline. Some finning operations capture sharks on longlines, cut off their fins, and dump the dying animal overboard. Unable to swim, the maimed sharks are attacked by other sharks or drift helplessly to the ocean floor. Most targeted sharks have international distributions and undergo long migrations. This makes regulation of their harvest dependent on international agreements to which compliance can be spotty and difficult to enforce. Many sharks are killed during recreational fishing competitions, known as rodeos or derbies. Sportfishing for some species is popular and lucrative for fishing guides. And finally, day-to-day subsistence and recreational fishing depletes shark populations in estuarine and nearshore waters. It is a rare and noble angler who releases a shark alive.

Should we let sharks vanish? Wouldn't we all relax more when swimming in the Gulf if we knew large sharks were extinct? Many biologists would answer by pointing to unsettling ecological trends emerging where shark populations have declined. Along the US mid-Atlantic Coast, populations of larger sharks have declined 10–50 percent since the 1970s, and some species are approaching ecological extinction. The prey populations of these large sharks have increased, especially rays, skates, and smaller sharks. One of these prey species, the Cownose Ray, is a conspicuous summer visitor to the shallows of the mid-Atlantic and Gulf Coast. Cownose Rays are voracious predators of bivalves, including commercially and recreationally important species such as Bay Scallops, Soft-Shelled Clams, and Eastern Oysters. Due to increased predation by Cownose Rays, fisheries for these mollusks along the Atlantic Coast are beginning to collapse. Thus, decimation of shark populations is disrupting our ecosystems and negatively affecting recreation and the seafood industry. So, yes, we should save sharks.

Cetacean Surprises

I spent most of my life unaware that whales were living within 85 miles (137 km) of my hometown—Pensacola, Florida. In truth, everyone was unaware of the Gulf's cetacean (whale and dolphin) diversity until the end of the past century. Surveys in the 1990s revealed that 28 species, more than a third of the world's cetaceans, inhabit or stray into the northern Gulf. Today, few outside the scientific community are aware that dozens of the planet's most intelligent species live in the northern Gulf.

Cetaceans in the Gulf are drawn to particular oceanic habitats. Most are attracted to the productive waters above the Continental Slope, especially cool-water gyres developing when the Loop Current collides with the slope and causes upwellings of nutrient-laden water to rise from below. The phytoplankton blooms at the gyre's surface become the foundation for a bountiful, but localized and temporary, oceanic habitat. Cetaceans also aggregate off the southeastern Louisiana coast where the nutrient-laden waters of the Mississippi River reach the Continental Slope.

Dolphins are the most numerous of the Gulf's cetaceans, and with 14 species they are also the most diverse cetacean group. The Bottlenose Dolphin is the common species in nearshore waters, while schools of the Atlantic Spotted Dolphin are found over deeper sections of the shelf. The most abundant species in the northern Gulf is the Pantropical Spinner Dolphin, whose population size throughout the entire Gulf numbers 50,000–100,000. Schools of thousands churn the surface waters as they hunt, often accompanied by other dolphins and tuna schools.

The Gulf's offshore dolphins feed on the abundant schools of small fishes inhabiting the surface above the Continental Slope and adjacent waters. Squid are also an important food of dolphins and other toothed whales. By day, squid dwell in the dark depths to avoid detection by predators. By night, squid migrate closer to the surface to feed. Though squid stick to the shadows, dolphins find them with sonar, a form of echolocation. Dolphins even use sonar to momentarily stun prey.

One of the biggest surprises to biologists was that 100–300 Killer Whales, or Orcas, inhabit the Gulf of Mexico. Popularized by movies, documentaries, and amusement parks, Orcas may be the most well-known species in the dolphin group. Perhaps it shouldn't have come as a surprise that Orcas reside in the Gulf; some biologists consider them to be the most widely distributed mammal species on the planet. Orcas are big dolphins—males attain lengths of 32 feet (9.8 m)—with a reputation for being intelligent and fearless. They are also highly social, maintaining complex relationships organized around the matriline (a female and her descendants). Several matrilines, known as a pod, will travel and hunt cooperatively for other marine mammals, sharks and other fishes, sea turtles, and squid. Their elaborate communication through clicks, whistles, and pulses helps them coordinate hunts and maintain their intricate relationships.

The Sperm Whale is the most common large whale of the Gulf, with an estimated population size of 1,350. Sperm Whales—easily recognized by their huge head—occupy the world's tropical and temperate oceans. They are particularly common along the Continental Slope near the mouth of the Mississippi River. Males attain lengths greater than 50 feet (15.2 m) and weights of 45 tons (40,823 kg) and thereby are the largest of the toothed predators on the modern planet. Sperm Whales specialize in hunting squid near the ocean floor, though they also eat sharks, skates, and other fishes. They are famous for hunting giant squid of the genus *Architeuthis*, whose largest individuals exceed 32 feet (9.8 m) in length. Sperm Whales can stay below water for more than an hour while hunting in the abyss, though dives averaging 35 minutes are more typical. They often emerge from hunts scarred with large circular sucker marks from battles with squid. Female Sperm Whales travel with related females in groups of a dozen or so. Young males leave their natal group to join pods of other young males of similar age. As individuals of these so-called bachelor groups mature, their pods split into increasingly smaller groups until eventually fully grown males hunt and travel in solitude.

The northern Gulf also has a resident population of the Bryde's (pronounced "BROOD-us" or "BROO-days") Whale. Bryde's Whales attain lengths of 55 feet (16.5 m) and weights of 45 tons (40,823 kg). They prefer warm waters and feed on small pelagic fishes. The population along the northern Gulf Coast numbers around 40 individuals; most sightings are south of the western Florida Panhandle. Bryde's Whales are baleen whales, the cetacean group including the largest animals on the planet. Baleen whales lack teeth, but their upper jaws grow numerous thin, hanging sheets of keratin (the material making up our fingernails and hair) called baleen. Each sheet unravels into fine strands along one margin to function as a sieve. Feeding strategies vary among species, but all involve taking into the mouth large volumes of water and prey, expelling the water through the baleen, then swallowing the trapped prey.

Individuals of other baleen whales, including Blue and Humpback, occasionally enter the Gulf, but biologists think these whales took a wrong turn during migration. Why there are not larger populations of baleen whales in the Gulf is a matter open to speculation. Some researchers suspect the Gulf produces insufficient quantities of their preferred food or the Gulf is too far afield from their Atlantic migration routes.

Marine mammalogists are just beginning to understand the ecological roles of cetaceans in the northern Gulf. Most knowledge about their ecology and behavior is from studies in other regions. Much more research is needed to understand cetaceans in the context of the northern Gulf of Mexico.

FLOATING FORESTS

Whales might be the largest animals living in the northern Gulf, but they are not the largest organisms, at least by one definition. On vast expanses of the western Gulf are large, floating mats of brown algae of the genus *Sargassum*. There are two nearly identical algae species forming this *Sargassum* community, *S. natans* and *S. fruitans*. Neither has a common name, but collectively they are called gulfweed. Gulfweed reproduces vegetatively—a piece of one plant breaks off and grows into a new individual. Because offspring are genetic clones of the parent, and because there are several million tons of each species in the northern Gulf, *S. natans* and *S. fruitans* are the largest clonal organisms in the Gulf and possibly the entire planet.

Gulfweed mats are among the most action-packed habitats on the surface of the Gulf, and they are essential to maintaining some of the Gulf's pelagic biodiversity. *Sargassum* is a multi-branched, golden-brown plant with leaflike appendages and gas-filled spheres keeping the plant afloat. The plants grow into entangled mats akin to floating forests. Because the open ocean provides no reliable place for small creatures to hide, dense concentrations of baitfish and crustaceans aggregate beneath and within gulfweed mats. They, in turn, attract predatory fishes, especially Dolphinfish, King Mackerel, and tuna species. These predators will burst from the blue haze below to snatch small animals straying too far from the sheltering weed. Ironically, many of them matured in the gulfweed jungle below which they now hunt. The predatory fishes attract larger hunters, including sharks and billfishes. As an important nursery habitat for young pelagic fishes and a hunting ground for larger predators, *Sargassum* is a keystone species (keystone genus, really) for the pelagic ecosystem of the northern Gulf.

The importance of the Gulf's *Sargassum* ecosystem reaches into the Atlantic Ocean. Each spring and early summer, gulfweed in the northwestern Gulf that survived the winter grows into giant rafts. At peak, this gulfweed expanse is larger than any other aggregation of the algae in the world. By midsummer, winds and currents send patches of the floating forests eastward, where they are pulled into the Loop Current. The mats often become reconfigured into long weedlines forming along the boundaries of waters differing in current speed or temperature. Within a month, the Loop Current carries drifting mats and weedlines through the Florida Straits and into the Atlantic. Much of this gulfweed accumulates in the Sargasso Sea, a large gyre of calm water off the Southeast's Atlantic Coast named after the gulfweed. The Loop Current exports about 1 million tons of *Sargassum* to the Atlantic annually, a massive infusion critical to sustaining the gulfweed ecosystem of the Sargasso Sea.

Sargassum contributes to the Gulf's biodiversity in another way. Within the gulfweed mats are animals completely dependent on this ecosystem. These specialists are champions of camouflage, able to remain hidden while in plain sight. They are so thoroughly adapted to gulfweed that their average survival time outside the floating jungle would be measured in seconds.

I learned about *Sargassum* specialists as a child on the shore of Pensacola Bay. Every few years in late summer or early fall, southerly winds and strong incoming tides would bring gulfweed into the bay. As *Sargassum* clumps passed by our dock, we'd scoop them up with dipnets and dump them into buckets and tanks to look for small animals hiding in the algae. Among them were tiny brown Sargassum Crabs, whose shells and legs are dappled with yellow to resemble sun flecks, and small shrimp, whose colors and shape perfectly matched the algal stems. These creatures were so well hidden that we'd usually resort to picking up the gulfweed and shaking them out of hiding.

Of all the *Sargassum* specialists, the one we prized most was the Sargassumfish. This is a stout, toadlike fish decorated with flared fins, fleshy projections resembling *Sargassum* leaves, and an intricate pattern of browns to match the gulfweed's colors and patterns. Taken together, the animal's profile says "plant" instead of "fish." The Sargassumfish is crafty in even more elaborate ways. It has a modified fin that it dangles

Right: Though well camouflaged, Sargassum Crabs are quick to defend themselves. Courtesy of Danté Fenolio.

Below: The Sargassum-fish's elaborate shape helps it blend perfectly into floating gulfweed. Courtesy of Danté Fenolio.

as a lure in front of its mouth. The fish will gently wave this lure while otherwise remaining perfectly still. Unsuspecting prey attracted to the lure never have a chance.

We'd also catch opportunists who had hitched a ride with the gulfweed, including slender pipefish that would entwine themselves around gulfweed stems and penny-thin filefish with emerald green eyes. The parade of fantastic gulfweed critters would slow as the tide peaked, whereupon we'd release our captives so they could catch the last mats of *Sargassum* leaving the shallows.

Most regular beach visitors to the shores of the northern Gulf Coast are only familiar with gulfweed when winds and tides have cast weedlines onto the beaches in summer and early fall. Many consider the beached gulfweed to be a nuisance, but those schooled in coastal ecology savor the sight and smell of rotting gulfweed. They know small invertebrates eating the decaying algae will feed hungry migrant shore-birds, and nutrients released through decay will aid the growth of Sea Oats and other dune-building plants. Even in death *Sargassum* sustains biodiversity.

CONTINENTAL SHELF LIFE
Below the Gulf's surface are a diverse suite of seafloor ecosystems. They begin in the shallow waters of the outer beach. If you snorkel in these waters on a typical midsummer day, you'll find an expanse of pale-yellow sand rippled by wave action. It's beautiful, but seemingly devoid of animals. Yet life is here, hidden just beyond the visible. Remaining undetected is a necessary survival strategy for many seafloor species, whether they are the hunters, the hunted, or both.

Holes on the seafloor mark the entrances to worm burrows. Irregu-lar marks in the sand betray the locations of buried sand dollars, Speck-led Crabs, or snails such as Lettered Olives and Atlantic Augers. If you are persistent, you might spy one of the fishes hiding under the sand surface waiting for unsuspecting prey. Perhaps it is a Gulf Flounder, whose contorted head places both eyes on its flattened upper surface. Maybe it is the chubby Southern Stargazer, which delivers an electrical shock to any predator venturing too close. Or, it could be a Lesser Elec-tric Ray or Atlantic Stingray, species remaining buried by day but pa-trolling the seafloor for burrowing invertebrates by night. You can even *hear* the biodiversity if you listen carefully—the waters vibrate with snaps and crackles made by fishes and crustaceans feeding in deeper waters. And, if you are lucky, you may even hear the squeals and clicks of Bottlenose Dolphins feeding farther offshore.

Farther out in the Gulf the seafloor ecosystem is more complex. The bottom can be sand, mud, or a mix of both; sand and gravel; or sand

and shell. Off eastern Alabama, the shelf's seafloor is the westernmost extent of a massive sand sheet centered south of the Florida Panhandle. Off western Alabama, the seafloor is muddier due to sediment deposits from the Mobile and Mississippi rivers. On the firmer substrates throughout, sessile creatures like sea pansies, gorgonians, and sponges lightly populate the seafloor. They filter plankton from the current, while crabs, sea urchins, and sea stars creep among them searching for prey or carcasses. In muddier substrates, clams and worms stay hidden in burrows and filter plankton from seawater.

Waters over Alabama's Continental Shelf, but beyond the reach of the Loop Current, experience seasonal changes just as pronounced as any on land. During winter and early spring, rivers deliver to the shelf large amounts of cool rainwater laden with silt and the final products of plant decay. Though phytoplankton surge at the surface, photosynthesis on the seafloor is impossible due to the murky waters above. Consequently, corals and other creatures needing light are rare on the Continental Shelf. Instead, the seafloor is populated by scavengers and filter-feeders, such as worms and clams, harvesting organic compounds and detritus. During midsummer, surface waters become so hot that phytoplankton populations decline. Dead plankton drift to the bottom, where they sustain the filter-feeders. Autumn brings a gradual transition into winter, but cyclones in late summer or early fall can trigger a warm-season infusion of organic compounds and detritus from the continent. The seasons are also marked by the migrations of fishes, shrimps, and crabs keeping up with their preferred combination of salinity, temperature, and food, all the while dodging attacks by dolphins, sharks, and other predators.

Just recently, marine scientists discovered a series of elongated low ridges rising above the expanse of muddy sand at depths ranging from 66 to 105 feet (20–32 m). The ridges are a few feet tall, about 328 feet (100 m) wide, and can stretch over a mile in length. They are the rubble of oyster reefs established 9,000–10,000 years ago as sea levels rose after the last ice age. While oysters cannot survive at these depths today, their ancient shells provide a hard substrate where sessile creatures grow profusely. The ridges and their tenants attract schools of bottom-dwelling fishes seeking shelter and food.

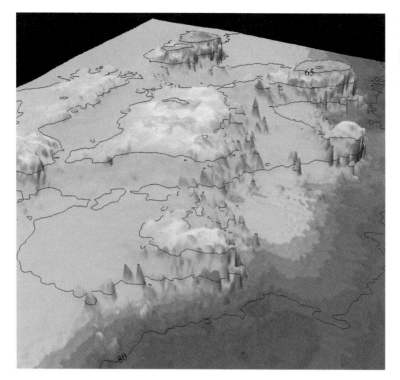

Scanning imagery of the complex topography of the Pinnacles, a "drowned" reef. Courtesy of US Geological Survey.

At the margins of the Continental Shelf, at depths of 200–330 feet (60–100 m), a seafloor ecosystem like no other in the northern Gulf exists. Known generally as the Pinnacles, it is a reef ecosystem built on the skeleton of a reef that drowned 11,000 years ago near the end of the last ice age. The former ecosystem was dominated by corals and other reef-builders no longer surviving at these dark depths. Their remains provide a hard substrate for sessile filter-feeders including soft corals, sea fans, and sponges. Some are short, squat structures that are former patch reefs. Others are long, tall, and flat-topped. Still others are narrow pinnacles rising 65 feet (20 m) above the seafloor. The height of these reefs reveals how reef-building organisms desperately tried, and failed, to keep up with rising sea levels.

While anglers have known about the Pinnacles for decades, biologists only began exploring them recently. They have discovered that the uppermost horizontal surfaces support the most dense and diverse communities of sessile invertebrates. Steep, descending slopes sustain fewer organisms because it is difficult for animals to anchor. Reef bases maintain hardy species that can endure periodic envelopment in seafloor currents conveying clouds of sediments to the deep Gulf. The margins of many pinnacle structures have overhangs, arches, and small caverns sheltering small fishes. Biologists have tallied over 146 fish species at the Pinnacles, including such oddities as Red Barbier, Short Bigeye, and Reticulate Moray. Over half of these species are typically associated with Caribbean coral reefs. Biologists believe the Loop Current brings larval fishes from the Caribbean into the Gulf each year, and the Pinnacles are the first suitable habitat they encounter. The larvae settle, grow, and then migrate shoreward to other reef-like habitats.

Seafood Sagas

Red Snapper and shrimp species are two of the Gulf Coast's most highly prized seafoods. Red Snapper are valued for their taste and are eagerly pursued by recreational and commercial fishers. Among fishes commercially harvested in the Gulf from 2000 to 2009, Red Snapper was second only to tuna (all tuna species combined) for the revenue generated per pound when sold at the dock (landing revenue). In Alabama over this interval, Red Snapper yielded higher landing revenues per pound than any other seafood. However, the big money is in the shrimp fishery. From 2000 to 2009, the average annual total revenue generated by the landing of Gulf Coast shrimps was over $400 million. The shrimp fishery produced over half of all seafood landing revenue along the Gulf Coast and was more lucrative than any other fishery. In Alabama during this time, average annual landing revenue for all shrimp species combined was $37 million. Clearly, the two species are important players in the seafood industry of the northern Gulf Coast.

An interesting ecological story has unfolded about the relationship between Red Snapper and shrimps. Prior to the mid-20th century, the interaction between shrimp species and snapper was minor—Red Snapper have a broad diet and are one of many fishes consuming shrimps. But last century, the ecological connection between shrimps and Red Snapper changed, becoming stronger than ever before. This new dynamic arose as our fisheries for the species developed.

Adult Red Snapper are bottom dwellers inhabiting waters 33–620 feet (10–190 m) deep. They aggregate around hard structures

Red Snapper reeled in from the floor of the Continental Shelf. Courtesy of Jimmy Stiles.

on the seafloor and feed on invertebrates and smaller fishes. In summer, adults gather near the Continental Shelf edge to spawn. After a planktonic stage, larval snapper settle over mud or sand bottoms of estuaries and shallow waters of the Continental Shelf.

After World War II, commercial and recreational snapper fishing became a lucrative component of the Gulf Coast seafood industry. From the 1960s to the mid-1970s, anglers along the Gulf Coast landed about 7 million pounds annually. Photographs from that period show proud anglers posing dockside with row after row of huge snapper. Unfortunately, the harvest was more than the population could bear.

By the late 1980s, landings had declined to just over 3 million pounds. Trophy photographs from the 1980s show proud anglers with fewer and much smaller snapper. In 1988, federal scientists completed the first analysis on the Gulf's Red Snapper population and concluded the fishery was in steep decline and being fished unsustainably. Immediate and sweeping regulations were imposed to protect the long-term commercial viability of the fishery. Despite a long series of increasingly restrictive fishing limits, many years passed with little clear evidence of recovery. Fisheries biologists and regulators were increasingly worried the Red Snapper fishery would fail.

Meanwhile, the Gulf Coast shrimp fishery was booming. Along the northern Gulf Coast, the harvest of Brown, Pink, and White shrimp was one of the most lucrative fisheries in the United States, a status the fishery still enjoys today. From the western Florida Panhandle to eastern Texas, Brown Shrimp comprise the majority of landings (80 percent), followed by White Shrimp (18 percent) and Pink Shrimp (2 percent). Because of their economic importance, shrimps became the most thoroughly studied group of marine invertebrates in the Gulf, if not the entire world. One of the most surprising discoveries was how much shrimps depend not only on healthy estuaries and offshore environments but also the region's rivers.

Brown Shrimp, for example, live about one year and reside as adults on the Continental Shelf in waters more than 50 feet (15 m) deep and beyond 10 miles (16 km) offshore. Adults hide in burrows by day but emerge at night to scavenge organic debris, hunt small invertebrates, and mate. Each female can release up to a million fertilized eggs into the water. The eggs hatch into minute zooplankton bearing no resemblance to the adult. Like most crustaceans, maturing shrimp periodically molt and metamorphose to adjust their size and body form, respectively. After several weeks, the young finally morph into a shape resembling an adult, though they are merely 0.15 inch (4 mm) long.

In early spring, postlarval Brown Shrimp begin a long migration to the coastal estuaries. Each day they ride an incoming tide toward the shore and into the estuaries. Beginning in March, they arrive and take

White Shrimp, one of Alabama's most prized seafood species. Courtesy of National Oceanic and Atmospheric Administration.

shelter within seagrass meadows and salt marshes. At first, they feast on the surplus of decaying plant material that accumulated over the winter, but when they are about 1 inch (2.5 cm) long, they begin hunting tiny invertebrates. Growth is rapid, and in May–July when the young browns are 3–4 inches (7.6–10.2 cm) long, they leave the estuaries and head back into the Gulf to begin their adult phase. The important ecological lesson is that the size of the emigrating shrimp population depends on the health of the seagrass meadows and salt marshes nurturing young shrimp, and the health of these nurseries depends on the amount and quality of water delivered to the coast from rivers.

During their migrations to deeper waters, shrimps hide in burrows by day, then head along the bottom to deeper, offshore environments during the early evening. These tactics make it difficult for sight-oriented predators to catch them, but there is no escape if they are in the path of an outrigger trawler.

Most shrimps in the United States are harvested from large boats called outrigger trawlers that slowly drag a cone-shaped net (trawl) along the seafloor. The usual configuration is an otter trawl, where the trawl is weighted by two heavy plates called otter boards or doors. Migrating shrimps are swept into the trawl, while those still in their burrows are rousted out by the dragging of the doors and chains connecting them.

A Gulf Coast outrigger trawler working at sunrise. Photo by R. Scot Duncan.

Shrimp trawling is highly destructive to marine ecosystems because it does not discriminate between target and non-target animals.[5] The amount of bycatch in shrimp trawls varies, but in the northern Gulf it averages a whopping 84 percent of the biomass harvested. Biologists have identified 450 species of bycatch in Gulf Coast shrimp trawls. Most of the bycatch biomass is fish, nearly all of which die while workers pick shrimps from the haul. A few marketable fish might be kept, but most are dumped overboard. And here is where the ecological relationship between the Red Snapper and shrimp species has dramatically changed. During the 1980s, studies of the bycatch problem revealed that shrimp trawling was decimating a large proportion of the juvenile Red Snapper population. Had researchers found the reason that Red Snapper populations were not rebounding?

While some scientists researched the effects of bycatch on snappers, others studied a related problem: shrimp trawls were capturing and drowning endangered sea turtles—lots of them. Fisheries biologists developed the turtle excluder device, or TED, a trapdoor installed on the trawl to allow large animals to escape without much loss of shrimps. The shrimp industry complained TEDs released too many shrimps, but federal regulators made TEDs mandatory on all commercial trawls in federal waters.

After several years of study, snapper biologists concluded that shrimp trawling was one of several important barriers to restoration of the fishery. Inspired by the success of TEDs, scientists developed bycatch reduction devices, or BRDs, to help fishes escape from trawls, while not greatly lowering the number of shrimps captured. Again, the shrimp industry resisted, voicing concerns over the cost of BRDs and the loss of shrimps through them. While shrimpers, regulators, and politicians argued, Red Snapper populations languished.

In the late 1990s, federal legislation passed strengthening the regulation of US fisheries. Soon, BRDs became required on most commercial shrimp trawls in federal waters, and the amount of Red Snapper captured by commercial and recreational fishers was more tightly regulated than ever. At the same time, the importation of inexpensive farmed shrimps from other countries caused a steep drop in shrimp trawling in US waters.

These events conspired to benefit the Red Snapper. In 2009, a federal assessment of the northern Gulf Coast's Red Snapper population provided the first good news for the fishery in over 20 years. While the fishery was still depleted, overfishing was no longer occurring, and the population was on the path to recovery. The species is still in a precarious position, however. With the current management plan, it is estimated the fishery will not be fully rebuilt until 2032.

ARTIFICIAL REEF DEBATE

Because Red Snapper congregate and feed around seafloor structures, adding structures to the seafloor of the Continental Shelf should help bolster their populations, right? Coastal residents have created such artificial reefs to enhance fishing success for thousands of years. Sunken objects are quickly colonized by barnacles, bryozoans, and sponges, which later give way to slower-growing, longer-lived species. As the reef ecosystem matures, small fishes are drawn to it for food and shelter, and they attract the larger fishes sought by anglers.

Above: Fishes clustering near one of Alabama's artificial reefs. Courtesy of Kevan Gregalis/Dauphin Island Sea Lab.

Right: Artificial reef installation offshore from Gulf Shores, Baldwin County. Courtesy of David Walter.

Alabama's artificial reef program may be the largest in the world. Off the Alabama coast, there are 1,197 square miles (3,100 km²) of shelf habitat open for reef creation by citizens, businesses, and local and state governments. Since 1953, over 15,000 artificial reefs have been built out of materials as diverse as rubble from demolished bridges, ships, planes, and combat tanks. All reef-building materials are cleaned of pollutants and inspected to ensure compliance with state and federal regulations. Increasingly, the structures used are custom designed to provide fish habitat. Alabama's program has fostered a lucrative recreational fishing and scuba diving industry. Though unnatural, these reefs have enhanced the Continental Shelf's habitat diversity. Catch rates of Red Snapper and other fishes are much higher at artificial reefs than in adjacent areas lacking reef. Is this a win-win for Red Snapper and the coastal economy?

Maybe not. In the late 1980s, a handful of scientists began questioning whether artificial reefs were helping fisheries, and the Red Snapper fishery took center stage in the discussion. The idea that adding reef habitat for snapper (or other reef species) will aid snapper recovery is the "production hypothesis." At its core is the assumption that the snapper population is limited in size due to the scarcity of reef-like habitat. However, a rival idea, the "attraction hypothesis," proposes that artificial reefs attract snapper, but do not bolster their survival or reproduction. What's more, artificial reefs may cause the decline of snapper and other species because they concentrate fishes where anglers can readily catch them. This problem is compounded if reef creation stimulates more fishing than there had been previously.

Given the difficulty of studying populations of Red Snapper, this debate will not be resolved soon. In the meantime, many wisely advocate that to safeguard the recovery of the Red Snapper fishery, the fishery should be managed with the assumption that artificial reefs make snapper more vulnerable.

Over the Edge

Beyond the edge of the Continental Shelf is the domain about which scientists know the least. The seafloor of the Gulf's Continental Slope and Central Basin is perpetually cloaked in darkness, subjected to intense pressures from the waters above, and sparsely occupied by strange animals bearing only faint resemblance to their cousins in the shallows. Scientists use sonar for mapping the seafloor and unmanned robots to photograph animals and to sample the water and sediments. Biologists drag otter trawls across the seafloor and retrieve unfamiliar animals whose ecological roles are poorly understood. New species are discovered frequently. It was recently calculated that, fishes aside, over half of the midsized species captured are new to science or had been captured only a few times previously.

Despite the darkness, these abyssal ecosystems still depend on sunlight for energy. The energy arrives as detritus drifting down from the lighted waters above—from fish scales to dead whales, all is consumed. Scientists also know the Continental Slope has many canyons and domes, and the biodiversity on their surfaces declines with increasing depth. At greater depths, pressure increases, temperatures decrease, food becomes scarcer, and biomass and species diversity decline. There is, however, a remarkable exception in the form of yet another diverse ecosystem built from the remains of ancient life.

Scattered across the lower Continental Slope are many locations where crude oil and natural gas naturally seep from reservoirs deep below. As made clear by our dependency on them for fuel, these hydrocarbons store great amounts of energy. Chemoautotrophic bacteria at seeps survive by harnessing usable energy from captured hydrocarbons. Some have evolved mutualisms with tube worms, clams, and mussels crowding around the seeps. The bacteria survive in these animals' guts, where they enjoy shelter and a steady supply of hydrocarbon-rich water. In return, the bacteria give their hosts energy left over from digesting the hydrocarbons. The bacteria and their hosts are the foundation for an ecosystem independent of today's sun and providing an ecological oasis relative to the surrounding seafloor.

The events leading to the formation of hydrocarbon seeps began with the Gulf of Mexico's formation during the Jurassic Period. At first, the Gulf was shallow and frequently cut off from the Atlantic Ocean. During periods of isolation, trapped seawater evaporated and left behind salt and other minerals. Millions of years of accumulation created a massive layer known as the Louann Salt. Once the Gulf permanently filled with seawater, thick layers of sediment and plant and animal remains washing off the continent buried the Louann Salt. Older sediments at the bottom were compressed into rock, while newer sediments accumulated above. The intense pressure also transformed plant and animal remains into oil and natural gas.

As sediments piled higher, pressure on the Louann Salt intensified. However, salt resists compression and doesn't become rigid like true

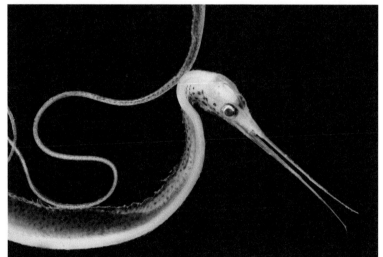

Clockwise from top left:

The Fangtooth inhabits some of the Gulf's deepest waters.

The glowing bulb on the Ghostly Seadevil's face attracts prey in the darkness of the deep Gulf.

Snipe Eel's jaws curve outward for catching small prey in near darkness.

All courtesy of Danté Fenolio.

rock. Once the weight of the sediments above exceeded the weight of the salt, enormous bodies of salt began rising up through the sediments above to form large seafloor domes, or diapirs. As the salt rose, it fractured the surrounding sedimentary rock, freeing oil and gas molecules. The natural buoyancy of the hydrocarbons forced them upward through the fractures. Where fractures reach the seafloor, hydrocarbons escape and support the seep ecosystems.

ISLANDS OF LIFE OR DEATH?

The hydrocarbon reservoirs in the Gulf of Mexico are indirectly responsible for another distinct offshore environment, the production platform. The extraction of oil and gas from hydrocarbon reservoirs in the northern Gulf is important to the local and national economies. To access these reservoirs, large production platforms (often called rigs) are built atop an elaborate system of support legs and trusses to provide a base for drilling and extraction infrastructure. In 2010, there were about 4,000 production platforms in the US waters of the Gulf. Throughout much of the northern Gulf, several platforms are within sight from land at any one time. Even at night, they light up the horizon with gas flares and bright lights.

Like artificial reefs, the support system below new rigs becomes encrusted with organisms building a new ecosystem. Unlike natural and artificial reefs, the rigs provide structure all the way from the seafloor into the clear near-surface waters. Thus, the rigs provide the only habitat in the northern Gulf where corals and colorful tropical fishes are abundant.

When a rig is fully colonized, the waters below and around it teem with life. Biologists have tallied nearly 50 fish species below production platforms and have estimated that 28,000 individual fishes can be found around a typical rig, a number far higher than would be found in the open ocean. Anglers flock to production platforms seeking bottom-dwelling fishes, like snapper and grouper, or pelagic fishes, including mackerel, Cobia, and Amberjack, circling rigs at shallower depths. These larger fishes attract sharks and Bottlenose Dolphins, while Loggerhead Sea Turtles hunt slow-moving creatures hiding within the reefs along the support beams. The popularity of recreational rig fishing and scuba diving has fostered a sizeable tourism industry for states with offshore drilling. Overall, many have come to view the production platforms as beneficial to fisheries, biodiversity, and the economy.

Then there was the explosion of the Deepwater Horizon/Macondo

252 production platform on April 20, 2010. In the desperate days and months that followed, we mourned the death of the 11 rig workers and watched in horror as crude oil gushed into the Gulf. The spill caused severe social, economic, and ecological harm to the northern Gulf Coast. It is safe to say that the reputation of oil rigs as "Islands of Life," as a 2006 US Minerals Management Service poster promoted them, is forever stained. Though Deepwater Horizon was a floating platform without a reef-like ecosystem beneath, the damage inflicted on the Gulf by the spill illustrates that any biodiversity and economic benefits from production platforms must be measured against the damage already incurred and the knowledge that catastrophic spills are a real and constant possibility.

15

The Future of Alabama's Biodiversity

THORNY ISSUE
It's time to tackle the thorny issue of what will happen to Alabama's biodiversity this century. Conservation is a difficult subject because it is about conflict—conflict between different views on how to value other organisms, how natural resources should be managed, and who should make sacrifices when biodiversity is to be saved.

Humans are hard on other species. We've been pushing southeastern plants and animals toward extinction for millennia. Today, vulnerable species and natural ecosystems largely survive at the margins, where agriculture, suburbs, impounded rivers, and tree plantations haven't yet reached. All species, even common ones, face a range of novel threats, from competition with exotic species to rapid climate change. Considering the pace of human population growth and the accompanying demand for natural resources, the future for Alabama's remaining biodiversity can seem bleak.

However, it's too early to give up on Alabama's biodiversity. The state's ecosystems and native species have survived some rough periods, and the outlook for conservation has been significantly worse than it is now. Consider the situation in 1900. The 19th century had been cruel to Alabama's ecosystems and there was little reason to expect anything different from the next century. Settlers had eradicated several species and pushed many others to the edge of extinction. Railroads crisscrossing the landscape had aided the expansion of cities and industry. The damming of rivers had begun, most primeval forests had been logged, and wildfire as a seasonal phenomenon had been largely extinguished. There were no laws of significance protecting species, ecosystems, or natural resources, and advocates for conservation were few.

In today's Alabama, agriculture is no longer expanding, ecological knowledge has advanced, terrestrial and riverine ecosystems are being restored, forests have grown back on clearcuts, federal laws protect species from extinction, and a community of impassioned conservationists labors for the protection of biodiversity. These are encouraging trends for Alabama's biodiversity.

This chapter is a prediction of what could become of Alabama's biodiversity this century. I'll review the most daunting threats to the state's flora and fauna, and why the loss of species and ecosystems should distress everyone, not just conservationists. I'll also share reasons for my optimism about Alabama's biodiversity, and finish by reviewing the exciting species discoveries made this century and why more discoveries are on tap.

Loss

Highway border signs welcome visitors to "Alabama the Beautiful." To many, this boast is a nod to the state's wealth of gorgeous natural landscapes. Alabama's natural beauty, from its coastal beaches to its lush mountain forests, is inextricably tied to its biodiversity. Begin depopulating its landscapes of the flora and fauna, and the beautiful becomes the banal. Can Alabama remain a stronghold for biodiversity this century? Will the state's greatest boast be legitimate in the year 2100?

Maybe, maybe not. With its wealth of native species, one might assume Alabama's biodiversity is secure, but the numbers tell a different story. In historical times, the state has lost more species to full extinction than any other state except Hawaii. Gone forever are about 90 species, including such treasures as the Carolina Parakeet, Southeastern Elk, Bachman's Warbler, and Whiteline Topminnow.

Many of Alabama's surviving species are nearing extinction. Alabama is ranked fourth among states for the number of its imperiled species. Nearly 15 percent of its species have narrow distributions, small populations, or other characteristics making them vulnerable to extinction. As of October 2012, the US Fish and Wildlife Service (USFWS) listed 120 of the state's species as threatened or endangered and, therefore, protected under the Endangered Species Act (ESA). There are many candidate species whose populations have declined sharply but must wait for a position on the Endangered Species List. Hundreds of other species are officially recognized as needing review for possible protection. Alabama's Department of Conservation and Natural Resources (ADCNR) maintains its own list of imperiled species not federally protected by the ESA. These state and federal lists of endangered and threatened species are growing.

Why are so many of Alabama's species facing extinction? Conservation biologists around the world have documented five factors at the root of species decline, and each is well manifest in Alabama. These are (1) habitat loss and degradation, (2) invasion of ecosystems by non-native species, (3) overharvesting of species, (4) pollution, and (5) rapid climate change.

As is true elsewhere, the greatest threat to the state's biodiversity has been habitat loss and degradation. Humans have modified southeastern ecosystems for thousands of years, and the impact has been severe:

- Paleoindians caused or contributed to the extinction of dozens of species.
- Damming of Alabama's rivers caused the majority of Alabama's modern extinctions.
- Alabama's forests have been logged repeatedly and are substantially less biodiverse than they were originally.
- Settlers converted nearly all of Alabama's prairies to agriculture and pasture.
- Most unprotected land along the immediate coast has been developed with little regard for ecosystem preservation.
- Over 97 percent of the Coastal Plain's seepage bogs, prime habitat for pitcher plants and other rarities, has been lost.
- Alabama has lost 50 percent of its original wetlands, and surviving wetlands bear scars of drainage canals.
- Coastal erosion has been accelerated by boating channels dredged through marshes and a deep shipping channel connecting Mobile to the Gulf.
- Alabama's cities are sprawling outward with little attempt to conserve natural resources, ecosystems, and green space.
- Urban sprawl, highways, railways, gas pipelines, and transmission lines are slicing undeveloped lands into ever-smaller fragments.

Most of these trends will continue throughout this century in order to meet the needs of an ever-expanding population.

Non-native invasive species (NNIS) are a significant threat to Alabama's biodiversity. NNIS are fast-growing species humans have introduced—intentionally or accidentally—from other regions of the world. While most non-native species are benign and noninvasive, many have escaped human control. Unchecked by the predators and diseases of their homeland, NNIS spread quickly. Their complete eradication is usually impractical, if not impossible. Many native species cannot compete with them, while others become prey or parasitic hosts. The list of established NNIS in the Southeast is long. Those remaining in disturbed habitats, including Mimosa, Brown Anoles, and European Starlings, are less worrisome than those invading Alabama's intact ecosystems. There are far too many examples of the latter:

- Cogon Grass is taking over the forests of southern Alabama, including longleaf woodlands.
- Kudzu vines creep into forests from open habitats and smother trees.
- Red Imported Fire Ants devour any animals they can subdue, including endangered turtle hatchlings.
- Chinese Tallow trees are transforming wetlands into dry forests.

Dams, like this one near Guntersville, caused most of Alabama's historic extinctions. Courtesy of Alan Cressler.

- Feral hogs plow through lowland forest soils, eating the roots of many native plants, including some of the most imperiled.
- Amazonian Apple Snails, voracious herbivores, are poised to invade the Mobile River Delta from a pond in Mobile.
- Chinese Privet shrubs overtake the understory of moist forests.
- Hydrilla, Eurasian Milfoil, and Alligator Weed are choking waterways.
- Lionfish, voracious reef predators, have arrived on the Alabama coast and threaten reef ecosystems.
- Free-ranging Domestic Cats, probably the most abundant carnivores in the world, kill one billion birds in the United States annually.

The overharvest of species has also taken a toll on Alabama's biodiversity. Some species had or have the misfortune of being tasty and abundant or easy to capture. Among those hunted to extinction is the Passenger Pigeon, a species that some biologists believe was once the most abundant North American bird. Others, including the Gulf Sturgeon, American Black Bear, and Mississippi Diamondback Terrapin, were severely overharvested but still barely survive. A few still face overharvesting for food, including Atlantic Bluefin Tuna, Red Snapper, and many oceanic sharks. Populations of Alabama's freshwater turtles have been decimated in recent years by commercial harvesting for sale to Asian countries.

Kudzu, shown here invading a pine woodland, is one of many non-native invasive species transforming Alabama's ecosystems. Courtesy of B. William Garland.

Other species were overharvested for different reasons. In the 19th and early 20th centuries, hunters shot hundreds of thousands of Carolina Parakeets, herons, and egrets to collect showy feathers for decorations, especially on women's hats. Fur trappers nearly drove North American Beavers to extinction. Some species declined when overharvested unintentionally, including juvenile Red Snappers and Kemp's Ridley Sea Turtles swept up by shrimp trawlers. Hunters eradicated the Red Wolf and Florida Panther from Alabama because they were considered a threat to livestock and humans.

Pollution also contributed to the decline of Alabama's biodiversity. Pollutants come in many forms, from eroded sediments, to toxic organic compounds, to heavy metals. Aquatic animals have been especially hard hit because most pollutants end up in creeks and rivers. Since the 19th century, rivers have been receiving sediments, herbicides, pesticides, sewage, and industrial waste. Many municipal or industrial pollutants are poured directly into streams and rivers, while other pollutants arrive in stormwater washing off adjacent lands. Conditions became so severe that the US Congress passed the Clean Water Act in 1972. In subsequent decades, water quality in Alabama's streams and rivers improved markedly and many streams are well into a recovery.

Though conditions have improved, pollution is still a serious threat to public health and biodiversity in Alabama. In 2010, Alabama ranked eighth in the United States for toxic releases into waterways, and third for releases of carcinogens. Furthermore, the state ranks 49th in the nation for money spent on enforcing pollution laws. The result is that Alabama's citizens and biodiversity are exposed to inordinate levels of toxins every day. Mercury, one of the most dangerous contaminants, isn't just a problem in the Gulf of Mexico (see previous chapter), it is also a pollutant of inland waters, where it accumulates in carnivores. The Alabama Department of Public Health (ADPH) annually updates the list of waterways where people should avoid or limit their consumption of fishes due to concentrations of mercury and other toxins in fish tissues. In 2011, ADPH posted mercury advisories for 86 waterways, including many of Alabama's most heavily fished waters. Another industrial pollutant of great concern is PCB (polychlorinated biphenyl), a compound listed by the US Environmental Protection Agency as a probable human carcinogen. Though its manufacture was banned in 1979, these stable molecules still contaminate many waterways. In 2011, ADPH renewed a PCB and fish-consumption advisory for much of the Coosa River and several other waterways. The potential impact of these industrial contaminants on public health is disturbing, as is their impact on Alabama's biodiversity.

Nutrients are among the most troublesome contaminants in Alabama's waterways. Most aquatic ecosystems naturally have a limited supply of key nutrients, especially phosphate and nitrate. However, a heavy load of nutrients enters Alabama's streams from fertilizer misuse on agricultural areas and lawns, plus discharge from wastewater treatment plants and livestock and poultry production facilities. The nutrient overload triggers populations of phytoplankton, cyanobacteria, and green algae to surge. Streams become tinted bluish-green with the overgrowth of phytoplankton. By midsummer, stretches of some streams and rivers become carpeted with thick mats of green algae and cyanobacteria that smother habitats and render streams unappealing. Eventually, the nutrient stew flows into coastal estuaries and the Gulf of Mexico and triggers plankton blooms. When these short-lived plants die, their decomposition by microbes exhausts the dissolved oxygen in coastal waters (like us, the microbes need oxygen to survive). Mobile species leave, while the rest, including many fishes and shellfishes, become stressed or die. These processes contribute to the low-oxygen events plaguing Mobile Bay (see chapter 8).

While industrial and nutrient pollution is disturbing, the most hazardous pollutant for the Southeast's aquatic biodiversity today is sediment. Sediment pollution in Alabama began hundreds of years ago. Early agriculturalists, including Native Americans of the Mississippian Cultural Period and pioneers, did little to conserve soil, and sediment runoff clogged streams and rivers and smothered habitat. Today, the superabundance of sediments washing into streams from construction sites, agricultural lands, and developed areas is a burden too heavy for many stream ecosystems to bear. Gravel and sand buries stream-bottom habitats, while fine sediments suspended in the current abrade the gills of animals and block light from reaching bottom-dwelling plants and animals. Those organisms most acutely sensitive to sediment pollution,

Cahaba River, Jefferson County. Sediment pollution in Alabama's rivers reduces their biological, economic, and cultural value. Courtesy of Hunter Nichols/Hunter Nichols Productions.

including several endangered species, live or hide their eggs in crevices on the stream bottom that quickly fill with excess sediments.

The Deepwater Horizon/Macondo 252 well explosion of April 20, 2010, was the beginning of the largest oil spill in US history. About 4.9 million barrels of oil spewed from the well during 150 anxious days that summer. Various forms of floating petroleum pollution washed onto beaches and into salt marshes. Long plumes of submerged oil and chemical dispersants enveloped seafloor ecosystems far offshore. As this environmental tragedy unfolded, many feared the spill would be utterly catastrophic for the northern Gulf's ecosystems and the people depending on them for their livelihoods. At the time of this writing, nearly two years later, it is still unclear how extensive the damage to the Gulf has been. What happened to the Whale Sharks spending summers in the northern Gulf? How has the Atlantic Bluefin Tuna spawning been affected? What happened to floating forests of gulfweed? How were the deepwater reefs along the Continental Shelf's margin affected? Will oiled salt marshes recover?

It will take many years of study before scientists can offer conclusions.[1] However, there is good news. There are more scientists studying the Gulf now than ever before and the Gulf's advocates have never been as numerous, vocal, or heeded. What's more, for reasons not yet fully understood, the Gulf has been far more resilient than many scientists anticipated. Shrimp and Blue Crab populations are rebounding and populations of heavily fished species including Red Snapper have increased as a result of the fishing ban imposed after the spill. These are reasons to be hopeful that the northern Gulf of Mexico will recover.

Last on our list of threats to Alabama's biodiversity is climate change. The scientific evidence is overwhelming and compelling that the carbon dioxide, methane, and other greenhouse gas pollutants released into the atmosphere are changing climate patterns across the planet. The most advanced models of how the planet's climate will change provide a range of outcomes, from best- to worst-case scenarios. The moderate scenarios, those usually considered most likely to be true, offer several grim predictions for the Southeast over the next 50–100 years:

- Summer maximum temperatures will increase 3°–7° F (1.7°–3.9° C).
- Winter minimum temperatures will increase 5°–10° F (2.8°–5.6° C).
- July heat index will increase 10°–25° F (5.6°–13.9° C).
- Relative sea level will rise 15–44 inches (38.1–111.8 cm).
- Tropical cyclones will be more powerful.
- Rainfall events will be more intense.
- More droughts and longer dry spells between rains will occur.
- Coastal regions will receive much less rainfall.
- Marine waters will acidify and weaken corals, shellfish, and other organisms that use calcium to construct their bodies and homes.

Such changes are commonplace through the arc of geologic time. Some are arguably trivial in light of the climate shift of the past 12,000 years as the ice ages subsided. Most species alive today have survived many episodes of planetary cooling and heating. So, why are biologists so worried about this new era of climate change? What is most troublesome isn't the changes themselves, nor even their magnitude. It is the *rate* of change that is alarming. Periods of climate change in the past 50 million years have been gradual, spanning thousands to millions of years. Slow transitions give species opportunities to shift their geographic range and adapt to changes in temperature and rainfall. In contrast, today's climate shift is occurring over decades and centuries.

Changes may manifest too rapidly for many species to adjust, especially if their populations have already been undermined by the other stresses we've imposed on them.

How will climate change affect Alabama's biodiversity? Many species will survive climate change just fine. They are adaptable species already thriving in many ecosystems. Some of the more sensitive and mobile species will migrate north or to higher elevations to keep up with optimal climate conditions. However, many of these migrants will be forced to navigate across highways, cities, and reservoirs in their path. Sensitive species unable to migrate will be among the most vulnerable to extinction, especially habitat specialists adapted to rare ecosystems like glades—these species have nowhere to go. Aquatic animals of shallow streams will experience higher summer water temperatures and habitat loss if there is more drought. Estuarine species will receive less freshwater from the continent if droughts increase. Rising sea levels and increased tropical cyclone strength may decimate barrier island species. As if this weren't enough, warmer temperatures may enable tropical and subtropical NNIS from southern Florida and Texas to invade Alabama and other southeastern states.

These five threats will challenge Alabama's most vulnerable species and the conservation biologists protecting them this century. Unless current trends are reversed, Alabama will likely lose many species to extinction and a handful of ecosystems may disappear completely.

CONSEQUENCES

It is often asked whether extinction really matters. Many ecologists and more than a few economists would answer that our civilization depends on the free resources—known as ecological services—provided by functioning, healthy ecosystems. Dozens of these services improve our daily lives:

- Plants provide oxygen and clean air.
- Filter-feeders clean river water that is extracted for drinking.
- Predators keep pest species in check.
- Insects pollinate many of our most important crops.
- Trees provide shade in summer.

- Plants prevent soil erosion.
- Natural landscapes allow groundwater reservoirs to recharge and keep rivers flowing during droughts.
- Wetlands absorb and clean floodwater.
- Healthy soils support agriculture.

Scientists have illuminated the connection between ecological services and species conservation. The formula is simple: ecosystems provide better services when they have a full complement of native species. Since every species in an ecosystem depends on many others, when one is lost the ecosystem and the services provided are weakened. Thus, our economy, environmental security, and way of life depend on biodiversity preservation.

Other arguments for protecting biodiversity point to the many native species people harvest for their livelihoods or for recreation. In Alabama, native pines and broadleaf trees are important timber species. Native oysters, shrimps, crabs, and fishes are harvested for the state's seafood industry. Offshore reefs support a lucrative sportfishing and diving industry. Mussels from the Tennessee River are used in the cultured pearl industry. Recreational and subsistence hunting and fishing are exceptionally popular in Alabama. Though its practice has almost disappeared, a few rural peoples regularly harvest wild plants to eat or use as medicine. None of this will endure without healthy ecosystems. Thus, much of the Alabamian way of life and economy depends on biodiversity preservation.

The potential use of species for new medicines, foods, and other products yet undiscovered is a common refrain in defense of biodiversity. Our pharmacies and pantries are already brimming with products discovered in nature. Though controversial, food scientists splice genes from wild species into crop or livestock species to confer advantages such as drought resistance or increased nutritive value. While it is impossible to know if and to what extent Alabama's native species might provide new products, these options are diminished every time we extinguish another species.

Another economic incentive for conservation is to safeguard and enhance Alabama's ecotourism industry. Birders, hunters, anglers, and photographers crisscross Alabama, stalking its plant and animal life.

Other ecotourism activities popular in Alabama include swimming, beach sitting, hiking, fall foliage watching, horseback riding, diving, snorkeling, camping, biking, geocaching, caving, and rock climbing. Residents and tourists spend huge sums of money pursuing these interests and thereby support the state and many local economies. However, tourists will not come to Alabama if the beaches are soiled, too many forests are clear-cut, or the rivers and reservoirs are too polluted. Once again, protecting biodiversity protects Alabama's economy.

The above are some of the commonly offered reasons that extinction is "bad" and biodiversity preservation is "good." Though compelling, these arguments cannot justify the protection of all species from extinction. Consider the Boulder Darter, an endangered fish with a small range in northern Alabama and adjacent Tennessee. It is unlikely this small drab fish will ever be economically important. Some might argue it is a good indicator of ecosystem health, but there are more direct ways of measuring stream quality. Furthermore, while it is true that if enough species are lost an ecosystem will unravel, the loss of a single species rarely triggers ecosystem failure. Sadly, if the Boulder Darter disappeared, only a handful of ichthyologists would notice. Are there any other compelling reasons, federal laws notwithstanding, to save these species?

Many would answer with a resounding "yes!" Several powerful arguments for species conservation emphasize the less tangible ways people value species and ecosystems. These values arise from a wide range of cultural practices and beliefs:

- Many find spiritual wisdom and transcendence through the study and contemplation of nature. Some view species as divine handiwork and species conservation as sacred duty.
- Some generously believe that other species have just as much right to survive as we do.
- Some value species as part of our extended evolutionary family, or value them as our only known living companions in the universe.
- Nature-inspired art fills museums, art festivals, and artisan workshops everywhere.
- Extinction takes away opportunities for future generations to appreciate a species and alters the future of life on earth for millions of years to come.

While these reasons range from the practical to the sublime, they all argue that our world is better off without extinction and that seemingly "useless" creatures like the Boulder Darter should be saved.

REVIVAL

The future for biodiversity can seem bleak. Frankly, the abundance of bad news about habitat destruction, failing ecosystems, extinctions, and population declines can leave one hopeless. However, it is far too early to write the eulogy. Quite to the contrary, there are many reasons to be encouraged—even excited—about the outlook for Alabama's biodiversity.

Biologists and conservationists have been fighting to protect biodiversity and save species from extinction for several decades. Backed by some of the strongest environmental laws on the planet, they have won inspiring victories. Consider Brown Pelicans, American Alligators, Bald Eagles, and Peregrine Falcons. All are formerly endangered species found in Alabama whose populations have recovered. North American Beaver and White-tailed Deer were nearly extinct a century ago in Alabama due to overhunting. Now they are nuisance species in many areas. The successes also include more humble species like the Tulotoma Snail, whose imperilment status was recently downgraded from endangered to threatened. These are stories to be celebrated.

Similar stories are being formulated as scientists bolster the populations of many struggling species in Alabama. Auburn University biologists are rearing Eastern Indigo Snakes for release into the Conecuh National Forest. Artificial nesting cavities for Red-cockaded Woodpeckers are being installed throughout the longleaf woodlands. Wildlife biologists have reintroduced Ospreys, a fish-eating raptor, on the state's rivers and lakes. The West Indian Manatee is doing well and biologists from the Dauphin Island Sea Lab monitor those summering on Alabama's coast. Believed extinct for 80 years, Interrupted Rocksnails were rediscovered and are being bred in captivity for reintroduction into the Coosa River. Members of the Alabama Plant Conservation Alliance are safeguarding and restoring populations of the state's most critically endangered species.

Protected species do a lot to help themselves recover when given

Left: Indigo Snakes are being reintroduced into Alabama's Longleaf Pine woodlands. Courtesy of Kevin Messenger/Alabama A&M University.

Below: US Fish and Wildlife technician prepares a Longleaf Pine for an artificial nesting cavity for Red-cockaded Woodpeckers. Courtesy of Beth Maynor Young/Cahaba River Publishing.

the chance. Black Bears from Georgia have settled in the mountains of northeastern Alabama. Unable to find enough nesting beaches, endangered Least Terns nest on gravel rooftops of shopping centers around Mobile Bay. Rare at mid-century, Cooper's Hawks have adapted to urban life and are common in southeastern neighborhoods.

A monumental effort to save imperiled southeastern species is ongoing at the Alabama Aquatic Biodiversity Center near Marion. This $2.5-million facility, the largest of its kind in the United States, was officially dedicated in October 2010 to breed imperiled aquatic mussels and snails for recovery efforts. The center's staff has pioneered methods for raising these species in captivity. The mollusks are released into streams to bolster or reestablish populations, or are studied to determine how pollutants in their home streams affect their survival, growth, and reproduction. Already the center has released thousands of imperiled mollusks, including Plicate Rocksnails, Spotted Rocksnails, and Alabama Lampmussels. The list of species propagated at the center is growing and will soon include critically endangered fishes. These efforts may save many of Alabama's most imperiled aquatic animals, but if the state's streams are too degraded to support released animals, these efforts are in vain. So what's being done to restore damaged rivers?

River and stream restoration is a chief focus of conservation biologists in Alabama and throughout the Southeast. One of the most important and daunting challenges is restoring habitats below dams on major rivers. Ideally, dam operators would release water in ways mimicking seasonal flows. Instead, the daily and seasonal demand for electricity takes priority in scheduling releases. However, these dams can only operate with a license from the federal government, and when that license is up for renewal, USFWS can negotiate for a release of water patterned to aid the recovery of endangered fauna downstream. Such restoration efforts have been implemented on the Coosa River, Elk River, and Bear Creek, and more cooperative projects need to be initiated this century.

Dams also impede the migration of fishes to their upstream spawning habitats. It has been known for a long time that fishes congregate below dams, presumably following migratory instincts. In 2009, a partnership of state and federal agencies, Auburn University, and the Ala-

bama Chapter of the Nature Conservancy (TNC) tested a method to help fish through two dams. Dams on large rivers have navigation locks allowing the passage of vessels. At Millers Ferry and Claiborne dams on the Alabama River, the locks were used to shuttle migrating species like Paddlefishes through the dam. This project is still experimental but may partially restore an ecological phenomenon lost for over 40 years.

Other statewide initiatives have added momentum to river resto-

Paul Freeman (TNC) holds a Paddlefish captured and tagged during a fish passage study at Millers Ferry Dam, Wilcox County. The cause of the fish's injury is unknown. Courtesy of Beth Maynor Young/Cahaba River Publishing.

ration. A small dam functioning as a bridge across the Cahaba River was removed in 2004, which has reconnected portions of the river for the first time in decades. Subsequent monitoring has documented upstream migrations by dozens of fishes, mussels, and snails in the years since. On the Paint Rock River, TNC and Alabama's Forever Wild program have purchased land to protect the river's watershed. TNC and Alabama's Division of Wildlife and Freshwater Fisheries partner with landowners to reduce erosion and pollution in the Paint Rock and other rivers. Altogether, these restoration projects do more than help imperiled species recover, they also benefit anglers, paddlers, and anyone downstream who values clean water.

Alabama's ecological restoration efforts are not restricted to rivers. An alliance consisting of TNC, Mobile Baykeeper, The Ocean Foundation, and the Alabama Coastal Foundation has begun establishing 100 miles (161 km) of oyster reef and 1,000 acres (4 km²) of salt marsh and seagrass meadow. This effort builds on TNC's experience with coastal restoration in and near Mobile Bay and capitalizes on the attention and restoration funding the Gulf Coast has received since the Deepwater Horizon/Macondo 252 oil spill.

Many terrestrial restoration projects are in full swing across Alabama but are largely restricted to Longleaf Pine woodlands. Most are

Burning of collected oil near the site of Deepwater Horizon/Macondo 252 explosion. Courtesy of US Coast Guard.

occurring within national forests, and secondarily within Alabama State Parks, Forever Wild preserves, and TNC's holdings. The reintroduction of frequent fires into these ecosystems through controlled burns is essential to their restoration. These burns also help seepage bogs, depression ponds, and other ecosystems nested within the woodlands.

Ecological restoration is challenging. It is expensive and requires decades of dedication. Practitioners rarely have the time and resources to study and compare different restoration techniques. Projects are complicated because they often involve coordination among numerous agencies, organizations, and other stakeholders. Many times the public can be so accustomed to an ecosystem's current impaired state that restoration is feared, resented, and opposed. It doesn't help that the first phases of restoration are usually messy—thinned forests and recently burned habitats may appear destroyed to those unfamiliar with restoration.

Another class of conservation projects enhances biodiversity through habitat creation, where constructed ecosystems replace existing ecosystems. Hundreds of reef species benefit from thousands of artificial reefs created along Alabama's coast to enhance fishing opportunities. Gaillard Island in Mobile Bay was created to provide much-needed nesting habitat for marine birds. However, habitat creation projects also can be controversial. Fisheries scientists debate whether artificial reefs encourage overfishing, and Gaillard Island smothered shallow-water habitat and altered the shape of the bay. Do the benefits of these projects outweigh the costs to biodiversity? The answer is rarely clear.

Because there is still land with imperiled species and ecosystems throughout Alabama, land preservation will be an efficient means of protecting Alabama's biodiversity this century. Land preservation is inexpensive and fast compared to the costs and complications of habitat restoration. Private lands cannot be counted on as part of any long-term conservation strategy unless owners have a conservation easement or safe-harbor agreement with the government or a conservation organization to protect particular species or ecological features.

Land preservation for biodiversity conservation is accomplished in many ways in Alabama. Among the nonprofit land trusts, the most important is the Alabama Chapter of the Nature Conservancy. Like most land trusts, TNC uses donations and grants to purchase land or implement conservation projects. TNC has over a dozen preserves in Alabama, plus several areas where they concentrate land purchases and other conservation actions. Land trusts can act more quickly than governmental agencies when desirable land is on the market. In many instances, TNC has purchased lands they later donated or sold at cost to Alabama's Forever Wild program.

The federal government protects the largest share of public land for biodiversity conservation in Alabama, most of which is within the national forests managed by the US Forest Service (USFS). The USFS established Alabama's four national forests (Bankhead, Conecuh, Tuskegee, and Talladega) for soil and water conservation and the protection of timber reserves for future generations. Commercial timber extraction is still a priority in some areas, but Alabama's national forests have become important preserves for rare species and ecosystems and their protection is a top priority. Alabama's three wilderness areas (Sipsey, Dugger Mountain, and Cheaha) are public lands managed by the USFS where only foot traffic is allowed and management is limited to trail maintenance. USFWS manages 10 national wildlife refuges in the state to protect vulnerable or imperiled species, the largest being Bon Secour, Cahaba River, Eufaula, Mountain Longleaf, and Wheeler. Several smaller refuges protect wetlands or caves important to particular species. The US National Park Service (USNPS) manages only one nature preserve in Alabama, Little River Canyon National Preserve. USNPS protects several culturally significant sites, portions of which are managed for conservation. Though closed to the public, military bases in Alabama include large areas of undeveloped land for training purposes. In recent years, the armed forces have partnered with others to implement conservation strategies for sensitive ecosystems and imperiled species on their bases. Lastly, the US Army Corps of Engineers manages many of Alabama's reservoirs and portions of adjacent lands. Some of the corps' holdings, including Haines Island Park near Franklin, are noteworthy preserves of floodplain forests and adjacent habitats.

An expansion of federal holdings this century could bolster natural resource conservation in Alabama. Expansion of wildlife refuges or the establishment of a national park would boost ecotourism, protect biodiversity and the natural beauty of the Alabama landscape, and share more of the responsibility for conservation with the state's government. However, some of Alabama's citizens protest vigorously when such plans are proposed. They are concerned about the influence of the federal government in the state, forced acquisition of lands (a practice used rarely, if ever), loss of timber industry jobs or taxes derived from timber sales, and preserving the character of modern rural life. Opposition by locals and prominent politicians to a recent proposal to expand the Cahaba River National Wildlife Refuge was so strong that the plan was shelved.

The state government also manages public lands in Alabama for biodiversity conservation. Alabama State Parks (ASP) protects key habitats—from coastal dunes to high mountain forests.[2] ASP's primary mandate is to provide recreational opportunities for the public. Historically, biodiversity conservation has been a secondary and often neglected priority. ASP's capacity to prioritize conservation is constrained because its parent agency, ADCNR, must rely on unpredictable funding sources including revenue through the issuance of various licenses and fees, royalties from oil and gas leases on state lands, a fraction of state sales taxes, public endowments, and federal funding. Consequently, ASP's budget is lean, its staff is small, and long-term conservation planning has been lacking. That said, in the past few years ASP has initiated several major biodiversity conservation projects within its most prominent parks, a trend that will hopefully continue.

Alabama's most successful land preservation program has been Forever Wild. Since its inception in 1992, the program has purchased 86 tracts totaling over 220,000 acres (890 km²) throughout Alabama. The program was approved through referendum by Alabama's citizens, who voted 84 percent in favor of the proposal. Much of the program's popularity is that it is paid for without taxes. Forever Wild's funding comes from the Alabama Trust Fund, a pool of royalties from oil and natural gas extraction in state-owned waters. The State Lands Division of ADCNR purchases selected land at appraised values from willing sellers, then manages it for biodiversity conservation and outdoor recreation. Some purchases are stand-alone preserves, while others augment state parks or wildlife management areas for public hunting.

The Forever Wild program came before the state legislature for reauthorization in 2011. Despite energetic and widespread support for reauthorization, the program became caught in the crosshairs of political debate. Some legislators and their constituents worried about the size and influence of state government, while others objected to funding conservation programs during a time of economic hardship and high unemployment. After weeks of political wrangling, the legislature agreed to let Alabama's citizens decide Forever Wild's fate during the November 2012 general election. Fortunately, the citizenry had the wisdom to reauthorize the program. The vote affirmed for everyone the popularity and importance of this program. The future of Forever Wild and ecotourism in Alabama is brighter than ever before.

When will enough land be preserved for conservation? This is an important question, and answers could be reached in different ways. Perhaps the answer should involve whether the ecological services sustaining Alabama's core economy and the health of its people are sufficiently protected. Or, perhaps the answer should address whether Alabama has enough outdoor recreational lands for the next few centuries. In terms of biodiversity conservation, the best answer is that enough land will be preserved and ecosystems restored when Alabama's native species are no longer imperiled.

Alabama isn't close to meeting any of these criteria. As of 2007, the state was dead last among nine southeastern states for the amount of public lands, both conservation and recreational. The average was 12.5 percent of state area—Alabama's total was just over 4 percent. Given the state's wealth of biodiversity, its citizens' love of outdoor recreation, and the growing importance of ecotourism to its economy, Alabama should do much more to protect its natural heritage.

DISCOVERY

There is one last and exciting trend to review regarding Alabama's biodiversity for this new century. Discoveries of new species are rare in most US states. Not so for Alabama and other states topping the biodiversity charts. At the time of this writing, there have been at least 76 species new to science discovered in Alabama since this century began.[3] This includes an azalea from the Red Hills; two pancake batfishes from the floor of the northern Gulf; a blue, burrowing crayfish from Perry County; the giant 11-inch-long Tennessee Bottlebrush Crayfish (28 cm; max. total length); two cave crayfishes from northeastern Ala-

bama; three flesh flies that breed in pitcher plants; four eyeless beetles
living in caves or deep in the soil; a freshwater diatom from the Cahaba
River; three parasitic nematodes from fishes in the Chattahoochee
River; the Tallapoosa Sculpin; the Alabama Bass; the Cypress Floater
(a mussel); one marine and two freshwater snails; 32 plants, most being wildflowers; and a trapdoor spider named after rock-n-roller Neil
Young (*Myrmekiaphila neilyoungi*). Though it doesn't add to current
biodiversity, paleontologists uncovered a new species of tyrannosaurid
dinosaur from the Cretaceous.

One of the most stunning discoveries this century was that of the
Pink Meanie. Several of these massive jellyfish washed ashore on Dauphin Island in 2000. Some were over 3 feet across (0.9 m), weighed
over 50 pounds (23 kg), and had tentacles over 70 feet long (21 m).
The Pink Meanie eats other jellyfish; one specimen was digesting 34
other jellyfish it had captured. That a creature so large could go undetected for so many decades indicates how little scientists know about
Alabama's marine wilderness.

Other discoveries this century are of known species turning up in
unexpected places. The Mitchell's Satyr, an endangered butterfly previously unknown from Alabama was found thriving in the wetlands of

the Fall Line Hills. Willow Flycatchers have bred in Alabama for the first time in recorded history. Perhaps most thrilling has been the finding of four river snails thought to be extinct—the Cahaba Pebblesnail, Cobble Elimia, Teardrop Elimia, and Oblong Rocksnail.

Such discoveries rarely involve a daring field expedition. More often, taxonomists find new species through painstaking investigations involving extensive field collections, reviewing museum specimens, comparisons of genetic variation, and careful measurements of minuscule anatomical features. The process may take years, and conclusions must withstand the scrutiny of other experts. Though the progression is tedious, such work is critical to ecological research and biodiversity conservation.

Taxonomists may be describing new species from Alabama for centuries to come. Consider the recent discoveries in the Great Smoky Mountains National Park. In 1998, over 200 scientists embarked on cataloging all species within the park. Within nine years, biologists had discovered 829 species completely new to science. That's an average of over 90 species per year. Imagine how many species would be discovered from a similar effort in such biologically rich areas as the Jackson County Mountains, Mobile River Delta, or Red Hills!

Not all interesting discoveries this century will be of new species. Within this millennium biologists have uncovered dozens of intriguing ecological findings in Alabama:

- The globally rare Eared Coneflower can be infected with a fungus that sterilizes the plant and populates its flower with pollen-mimicking fungal spores spread by unsuspecting insect pollinators.
- The salt marshes of the Mississippi Sound are wintering grounds for large populations of the Black and Yellow Rails. These species are so secretive that most birders who regularly explore the marshes have never seen them.
- The highly endangered Alabama Beach Mouse has a complex family life. Parents raise young together and at night parents often forage together. Subadults often accompany parents, but like human teenagers, they sometimes prefer travelling with other subadults.
- Tracking of acoustically tagged Gulf Sturgeon revealed these threatened, primitive-looking fish have a Gulf wintering ground about a mile

(1.6 km) offshore between Pensacola, Florida, and Gulf Shores. The sediments there are thick with marine worms and mollusks, a sturgeon's favorite fare.

This is just the beginning for what will be a century of continued biological discovery. A new generation of ecologists and taxonomists have already fanned out across the Southeast, schooled by their predecessors and equipped with new technologies, perspectives, and ideas. Their research will be essential to conserving Alabama's biodiversity in the new century.

CONCLUSIONS

By mid-century it should be clear whether today's generations did enough to protect biodiversity and the natural resources humanity needs to prosper. For centuries to come, future generations will study this time—*right now*—as a defining moment in our species' history. Will we muster the resolve to save biodiversity? I think we might for three reasons.

First, we have solutions to most of the world's environmental problems. We have the scientific know-how to end extinctions, minimize habitat loss, curtail pollution, and mitigate climate change. The barriers to fixing these problems are not scientific, they are political and economic. This means that saving biodiversity is a choice, not an unattainable fantasy.

Second, thanks to the heroic efforts of educators, tomorrow's generation will know more about the environment and their connections to it than any previous generation. Most leaders of today's world need to be convinced that saving biodiversity is important. In tomorrow's world, most leaders will simply need to be reminded.

Third, ending poverty, poor health, and armed conflict are goals that most governments and citizens around the world are striving to attain. What few realize is that ending these social ills works to protect biodiversity. People need social stability before they can focus on their long-term future and build a prosperous and sustainable economy. Because sustained prosperity is only possible when ecological services and natural resources are thoughtfully managed, biodiversity conservation eventually becomes a societal goal. This is the great lesson conservation biologists learned and shared with us last century: the protection of people requires biodiversity conservation, and biodiversity conservation requires the protection of the people.

We'll need many decades to save biodiversity and ourselves. It will be a chaotic transition, and setbacks are guaranteed. Not all species and ecosystems will survive. But for humanity to continue the marvelous adventure of life, it must provide for the needs of nature. We must preserve the natural world while helping our own species find peace and prosperity. For Alabamians, these efforts can begin right at home in Alabama the Beautiful.

Notes

Chapter 1

1. Since Stein's report, surveys in Tennessee and Arkansas have placed these states on par with Alabama for fish diversity. With new fishes discovered in all three states nearly each year, even the experts are unsure which state will come out on top.

Chapter 2

1. A subspecies is a discrete population genetically distinct from other populations of the species, but not sufficiently dissimilar to warrant designation as a separate species.

Chapter 3

1. The rising of air and its subsequent cooling involve simultaneous interactions between temperature, volume, and pressure. Air molecules near the ground are densely packed due to gravity's pull on the atmosphere. This is measured as air pressure. Temperature is a measure of the molecular collision rate in a substance and is used to gauge a substance's heat. The heating of air, or any substance, excites its molecules by causing molecular collisions with increasing force and frequency. When air warms near the earth's surface, the collective force of these collisions causes the air to attempt expansion because its molecules can travel farther and faster between collisions than before. But neighboring air pushes back, either because it is also warming and expanding, or because it is cooler and, thus, dense, heavy, and slow to budge. Above the heated air, the atmospheric pressure is lower and less resistive, and the heated air rises like a bubble. While rising, the air expands because its molecules now have more elbow room. Because the number of molecular collisions then decreases and the molecules slow down, the air subsequently cools. The air also cools as it gets farther from its source of heat, the ground that had been heated by the sun.

2. Water molecules tend to stick together at cool temperatures. For water molecules to exist as vapor, they must be warm enough to bounce off one another instead of sticking. This is why colliding water molecules within cooling air will stick together and form droplets.

3. The air molecules in cooling air slow down, reduce collisions with one

another, and crowd together more than molecules in warm air. Thus, cold air is heavier than surrounding warm air and sinks.

4. As this air drops, its molecules are increasingly compressed, collide more frequently, and, thus, are warmed even without added heat from an outside source. The descending air will also warm as it drops because it absorbs heat radiated from the earth's surface. The closer it comes to the surface, the more heat it absorbs.

5. Where the northern and southern Hadley cells meet is the Intertropical Convergence Zone, or ITCZ. Here, the amount of convective storm formation is most intense. The ITCZ shifts north of the equator during the northern summer and south of the equator during the southern summer, but always stays within the tropics.

6. Though destructive, storm damage in forested ecosystems triggers ecological succession and augments habitat and species diversity.

7. Uncharacteristically strong high-altitude winds over these oceanic areas prevent tropical storms and hurricanes from developing.

8. Rain shadows are the dry areas on the lee side of mountains where orographic lifting on the windward side has pulled most moisture out of the air.

Chapter 4

1. Scientists divide earth's long and tortured history into hierarchical units of eons, eras, periods, and epochs. For example, the Ordovician is one of seven periods in the Paleozoic Era, which is one of three eras in the Phanerozoic Eon. To keep track of these dates and names, geologists use a timeline known as the geologic time scale.

2. Dolostone, or dolomite rock, is like limestone but contains high levels of magnesium in addition to calcium carbonate. It also produces karst, but is less common at the surface than limestone.

3. The Appalachian Plateaus and Interior Low Plateaus provinces are better known in Alabama by their section names, the Cumberland Plateau and Highland Rim, respectively.

Chapter 5

1. This plate became North America when it assumed its modern configuration.

2. Extensive coal deposits developed elsewhere in the Mississippian, which is why geologists fuse it and the Pennsylvanian Subperiod into the Carboniferous Period.

3. Intense erosion is associated with uplift and mountain building because

fractured and freshly exposed rock erodes quickly. Erosion slows once debris has washed away, soils develop, and vegetation covers exposed rock.

4. Biological productivity is the sum of all energy captured by organisms in an ecosystem over time, including energy inputs of sunlight and detritus. Primary productivity refers specifically to the capture of solar energy by plants.

5. Volcanism in western India at nearly the same time as the Chicxulub impact may have played a minor role in the K-Pg extinction. The Deccan Traps covered at least 193,000 square miles (500,000 km²) and for about one million years spewed ash and gasses into the atmosphere at volumes that could have cooled earth's climate by several degrees. Though this cooling would have stressed Cretaceous ecosystems, the conventional view among geologists is that the Chicxulub impact was the main cause of the K-Pg extinction.

6. This land bridge facilitated the migration of terrestrial species between the continents. The flow was mostly north to south, but North America received a few species, including the Virginia Opossum.

7. Paleobotanists have learned much about Pleistocene plant distribution by studying pollen preserved in sediments from wetlands surviving the ice ages.

8. Ancient trash heaps, or middens, from these sites contain mussel shells that provide aquatic biologists valuable clues about the prehistoric distribution and sizes of mussel species.

9. Scientists believe seasonal, lightning-sparked fires became common in the mid-Holocene with the stabilization of the modern warm climate pattern. It's to be expected such storms were also frequent in the Southeast during earlier warm climate phases (e.g., interglacial stages of the Pleistocene).

Chapter 6

1. The process of mate choice influencing species evolution is known as sexual selection and explains many traits with seemingly no survival advantage. If females consistently choose the gaudiest males in a population, males become more colorful with each generation. Some taxa, like darters, have elaborate colors but relatively minor ecological differences. Many ichthyologists conclude sexual selection is driving the speciation of these and other fish taxa.

Chapter 7

1. A watershed is all the land draining to a common point in the land-

scape. A large watershed contains smaller watersheds, each of which contains even smaller watersheds, and so on until no further subdivisions are possible. *Basin* is a term geographers often use for a large watershed draining to the coast.

2. Fully aquatic animals that don't breathe air still need oxygen to survive. Oxygen enters water at the water–air interface via diffusion. Some creatures need high oxygen levels to survive, while others tolerate variable conditions.

3. When two streams join, the combined waterway retains the name of the larger stream. If equally matched, the combined waterway gets a new name.

4. The last miles of the Tennessee in Alabama flow through the Coastal Plain.

5. Compared with Coastal Plain flooding during the Tertiary, this flooding was less extreme, occurred in the past three million years, and would have depressed freshwater diversity in small coastal rivers but left the rest of the Coastal Plain's aquatic fauna intact.

Chapter 8

1. NatureServe is a nonprofit international organization whose goal is to provide the scientific information and tools needed by conservation practitioners. Its online database provides a thorough description of each ecological system.

2. At present, a rock and sand wall has been built across the Katrina Cut, the breach on Dauphin Island's western end. Allegedly it was built as a temporary barrier to prevent oil from the Deepwater Horizon spill from entering the Mississippi Sound. However, many want the wall to remain to help restore the island and protect the oyster reefs and salt marshes in the sound. The wall's fate is in question as state and federal agencies deliberate.

3. Prior to the post–World War II development boom, there was an abundance of suitable breeding bird habitats along the US Atlantic and Gulf coasts. Today, the paucity of such habitats has contributed to the precipitous decline of many shore-nesting species.

4. Petit Bois Island was once part of Dauphin Island but broke away sometime in the 18th century, probably during a cyclone.

5. These sands become acidic because decaying plant debris contributes organic acids to the groundwater, and there are no clays or other fine particles to capture calcium and other alkaline minerals that would neutralize these acids. Some clay particles have been transported to the claypan while others and any alkaline minerals have been washed away by the abundant groundwater.

6. Published estimates on the size of the delta vary from 185,000 to 250,000 acres (749–1,012 km²), depending on the ecological boundaries used. Recent state and federal documents usually provide the latter statistic.

7. Scientists believe these bears belong to the Florida subspecies, while those in northeastern Alabama are the Southern Appalachian subspecies.

8. These predators have unnaturally large populations in the delta due to humans. The hogs and ants are introduced species, and crows and raccoons are abundant due, in part, to overhunting of their predators.

9. Perdido Bay's stronger tidal mixing is partly artificial. A 6-mile-long peninsula (10 km) once extended from its western shore and forced a circuitous connection with the Gulf through a channel known as Old River. In 1867 an industrious citizen dug through the peninsula to create a more direct connection. The peninsula became Ono Island.

Chapter 9

1. Geologists and many biogeographers divide the Southeastern Plains into upper and lower sections along the boundary between Cretaceous and Tertiary deposits.

2. The word *hardwood* traditionally refers to broadleaf tree species, while *softwood* refers to conifers. Still in widespread use, both are misnomers. Some conifers, like Longleaf Pine and Eastern Red Cedar, have harder wood than some broadleaf species.

3. Keystone species have an effect on community structure disproportionate to their number or size.

4. These habitats are not part of the Sand Hills level IV ecoregion of Georgia, South Carolina, and North Carolina, though they share a similar ecology.

5. The NatureServe characterization of the Southern Coastal Plain Sinkhole emphasizes steep slopes with exposed limestone. However, exposed bedrock is often not present in the sinkholes of Alabama.

6. The roster includes 7 pitcher plants, 5 sundews, 4 butterworts, and 11 bladderworts.

7. Biologists have used the name "Red Hills" for areas largely corresponding to the Buhrstone/Lime Hills ecoregion. However, geologists use the name for a distinct physiographic district to the north, the Southern Red Hills. The physiographic areas included in the Buhrstone/Lime Hills ecoregion correspond to the Lime Hills District and Hatchetigbee Dome Subdistrict. There is also complication in the use of the name "Buhrstone" (buhrstone is a limestone strengthened by high silica content). Geologists la-

beled the region immediately north of the Buhrstone/Lime Hills ecoregion the Buhrstone Hills Subdistrict. Meanwhile, soil scientists reserve the name "Buhrstone Hills" for the area on the northern tier of the Buhrstone/Lime Hills ecoregion containing buhrstone (the Tallahatta Formation). Use of "buhrstone" in the naming of the ecoregion was in deference to soil scientists.

8. There is also a small disjunct section of the Red Hills 10 miles (16 km) south of Jackson, within the Fred T. Stimpson WMA. While not within the Buhrstone/Lime Hills ecoregion, it shares the same geology, topography, and ecology as the Red Hills to the north.

9. This region is often called the Black Belt Prairie or, in older works, the Upper Prairie or Central Prairie Belt.

10. These oak forests belong to the East Gulf Coastal Plain Northern Dry Upland Hardwood Forest and the Southern Coastal Plain Dry Upland Hardwood Forest ecosystems. The former has proportionally more White Oak; the latter has proportionally more Laurel Oak.

Chapter 10

1. Such valleys are known as anticlinal valleys. The ridges bordering them expose multiple rock layers angled up and toward the valley center. Synclinal valleys are those where compression bowed the crust downward. The slopes bounding synclinal valleys are comprised of the same rock.

2. The districts include the Birmingham–Big Canoe, Coosa, and Cahaba valleys, and the Coosa, Weisner, Cahaba, and Armuchee ridges. The Ridge and Valley ecoregion includes two districts from the Appalachian Plateaus physiographic province, Wills Valley and Murphrees Valley.

3. At Alabama's latitude, the sun is always in the southern half of the sky and never reaches directly overhead.

4. There is an old record of a Boynton's Oak from Texas, but the museum specimen documenting it is lost, and no trees have been found in Texas since the alleged collection. Many conclude this record was an error.

5. A fifth population was established as a reserve population in a spring where the fish is not known to be native.

6. The Nature Conservancy protects a small patch of flatwoods and a colony of Green Pitcher Plants in the Coosa Bog Preserve near Centre. However, Alabama Leather Flower is not there.

7. Candidate species are those the USFWS believes warrant consideration for placement on the Endangered Species List.

8. New species included Cahaba Paintbrush, Cahaba Prairie Clover, Cahaba Torch, Sticky Rosinweed, and Deceptive Marbleseed. New subspecies

included Cahaba Daisy Fleabane, Ketona Tickseed, and Alabama Gentian Pinkroot—the latter has been featured on U-Haul trucks.

Chapter 11

1. See chapter 8, note 5, for an explanation of why sandy soils are acidic.

2. Elf Orpine also inhabits the granite outcrops of the Piedmont level III ecoregion; a description of its peculiar life cycle is in chapter 12.

3. Note there are several Turkey and Hurricane creeks in the state.

4. Tennessee and Alabama cooperatively manage the canyon since the border nearly bisects the Walls.

5. Researchers attempting to determine the origination of the fungus suspect cave-visiting tourists accidentally brought the fungus from Europe.

Chapter 12

1. The northern part of the Shoal Creek District lies within the Ridge and Valley ecoregion and includes the Dugger Mountain Wilderness Area.

Chapter 13

1. Hunting is limited to mammals with robust populations and is essential to controlling feral hog and White-tailed Deer populations.

2. The TVA inadvertently discovered this method for nurturing waterfowl when managing the river's margins for mosquito control.

3. Mesophytic forests also survive in the Eastern Highland Rim but are less common since the terrain is more level. The Interior Plateau's mesophytic forests are less diverse than those in the mountainous ecoregions to the east.

4. Some refer to these ecosystems as cedar glades, but this term is falling out of favor. It doesn't denote the associated rock, and abundant cedars are a symptom of an unhealthy calcareous glade (see below in the text). Finally, note that some apply the word *barrens* to grasslands created by Native Americans through frequent burning.

5. One of the larger rockhouses on Little Mountain is the Stanfield-Worley Bluff Shelter, one of the Southeast's most important Native American archaeological sites.

Chapter 14

1. This is a slope of about 3 feet per mile (0.57 m per km).

2. Fish are aged by counting annual growth rings on ear bones called otoliths.

3. NMFS is an agency of NOAA, the National Oceanic and Atmospheric

Administration. NMFS enforces the US Endangered Species Act for marine species.

4. On paper, Tiger Sharks are responsible for more unprovoked shark attacks than Bull Sharks, but it is believed the number of Bull Shark attacks is underrepresented since they are not as easily as those by Tiger Sharks.

5. Shrimp trawling also degrades marine ecosystems by smoothing the seafloor, removing or damaging sessile organisms, and thereby reducing habitat for fishes and other animals.

Chapter 15

1. Though many scientists are studying the spill's effects, much of the data is part of the government's legal case and has not been released.

2. Parks particularly important for habitat preservation are Cheaha, Chewacla, Buck's Pocket, DeSoto, Gulf, Guntersville, Monte Sano, Oak Mountain, and Wind Creek.

3. This tally comes from published scientific papers, articles in newspapers and magazines, and haphazard communications with scientists. Included are species described for the first time in the scientific literature published this century, plus species discovered since 2000 that are not yet described. I likely missed many species in my search.

Abbreviations

AABC	Alabama Aquatic Biodiversity Center
ACES	Alabama Cooperative Extension System
ACWCS	Alabama's Comprehensive Wildlife Conservation Strategy
ADCNR	Alabama Department of Conservation and Natural Resources
ADEM	Alabama Department of Environmental Management
ANHP	Alabama Natural Heritage Program
ASP	Alabama State Parks
CEC	Commission for Environmental Cooperation
DISL	Dauphin Island Sea Lab
FAO	Food and Agriculture Organization of the United Nations
FDNR	Florida Department of Natural Resources
FMNH	Florida Museum of Natural History
FNAI	Florida Natural Areas Inventory
GSA	Geological Survey of Alabama
ICCAT	International Commission for the Conservation of Atlantic Tunas
IUCN	International Union for the Conservation of Nature
MASGC	Mississippi-Alabama Sea Grant Consortium
MBNEP	Mobile Bay National Estuary Program
NMFS	National Marine Fisheries Service
NMNH	National Museum of Natural History
NOAA	National Oceanic and Atmospheric Administration
NRCS	National Resources Conservation Service
SEDAR	Southeast Data, Assessment, and Review
TNC	The Nature Conservancy
USACE	United States Army Corps of Engineers
USEPA	United States Environmental Protection Agency
USFS	United States Forest Service
USFWS	United States Fish and Wildlife Service
USGS	United States Geological Survey
USMMS	United States Minerals Management Service
USNPS	United States National Park Service

References

Chapter 1

National biodiversity rankings. Bruce A. Stein, "States of the Union: Ranking America's Biodiversity" (Arlington, VA: NatureServe, 2002).

Ecological systems in Alabama. Alabama Department of Conservation and Natural Resources (ADCNR), "Alabama's Comprehensive Wildlife Conservation Strategy (ACWCS)" (Montgomery, AL: ADCNR, 2005). Several ecosystems found in Alabama do not appear in the above report, but are listed by NatureServe Explorer, accessed June 2010, http://www.natureserve.org/index.jsp.

State size. United States Census Bureau, "State and Metropolitan Area Data Book: 2010," accessed June 2010, http://www.census.gov/prod/www/abs/smadb.htm.

Factors affecting species diversity among states. Bruce A. Stein, Lynn S. Kutner, and Jonathan S. Adams, *Precious Heritage: The Status of Biodiversity in the United States* (New York: Oxford University Press, 2000).

Mountainous areas in Alabama. Glenn E. Griffith, James M. Omernik, Jeffrey A. Comstock et al., *Ecoregions of Alabama and Georgia* [color poster with map, descriptive text, summary tables, and photographs] (Reston, VA: United States Geological Survey [USGS], 2001). The statistic I provide does not include the Interior Plateau ecoregion.

Civilization's dependence on nature. Of the many appropriate references, one of the most captivating is Jared Diamond's *Collapse: How Societies Choose to Fail or Succeed* (New York: Penguin Books, 2005).

Chapter 2

Ecological principles. Ecological details about the barrier island system come from many years of personal study and exploration. General principles can be found in most introductory textbooks on general ecology such as Thomas M. Smith and Robert L. Smith, *Elements of Ecology*, 7th edition (New York: Benjamin Cummings, 2009).

Beach mice. Alabama Agricultural Experiment Station (AAES), "Hurricane Opal Takes Its Toll on Beach Mouse Populations: Illustrates Importance of Deep Dune Systems on Gulf Coast Beaches," accessed June 2010, http://www.ag.auburn.edu/adm/comm/news/1996/hurricane.php; N. R. Holier, D. W. Mason, R. M. Dawson, T. Simons, and M. C. Wooten, "Re-establishment of the Perdido Key Beach Mouse (*Peromyscus polionotus trissyllepsis*) on Gulf Islands National Seashore," *Conservation Biology* 3 (2005): 397–404; Daniel C. Holliman, "Status and Habitat of Alabama Gulf Coast Beach Mice *Peromyscus Polionotus ammobates* and *P. P. trissyllepsis*," *Northeast Gulf Science* 6 (1983): 121–129; Smithsonian National Museum of Natural History, "*Peromyscus polionotus*: Oldfield Mouse," accessed June 2010, http://www.mnh.si.edu/mna/image_info.cfm?species_id=262; United States Fish and Wildlife Service (USFWS), "Perdido Key Beach Mouse (*Peromyscus polionotus trissyllepsis*), 5-Year Review: Summary and Evaluation" (Panama City, FL: USFWS, 2007); USFWS, "Alabama Beach Mouse (*Peromyscus polionotus ammobates*, Bowen 1968), 5-Year Review: Summary and Evaluation" (Daphne, AL: USFWS, 2009); USFWS, "Perdido Key Beach Mouse Recovery Action Plan" (Panama City, FL: USFWS, 2009); Michael C. Wooten, "Alabama Beach Mouse," in *Alabama Wildlife, Volume 3: Imperiled Amphibians, Reptiles, Birds, and Mammals*, edited by Ralph E. Mirarchi, Mark A. Bailey, Thomas M. Haggerty, and Troy L. Best (Tuscaloosa: University of Alabama Press, 2004), 183–185; and Michael C. Wooten, "Perdido Key Beach Mouse," in *Alabama Wildlife, Volume 3*, ed. Mirarchi et al., 185–186.

Economic impacts of hurricanes. National Oceanic and Atmospheric Administration (NOAA), "Extreme Events: Hurricanes and Tropical Storms," accessed June 2010, http://www.economics.noaa.gov/?goal=weather&file=events/hurricane. Values presented were standardized to 2007 dollar values.

Chapter 3

Weather and climate. My understanding of these topics has been profoundly shaped through conversations with my father, Robert (Bob) A. Duncan. He has maintained a lifelong interest in these topics and has kept daily field notes most of his life. His book on how weather and climate affect bird migration along the Gulf Coast is well respected among birders and ornithologists (Robert A. Duncan, *Bird Migration, Weather, and Fallout* [Gulf Breeze, FL: Self-published, 1994]). I also relied on Edward J. Tarbuck and Frederick K. Lutgens, *Earth Science*, 12th edition (Upper Saddle River, NJ: Pearson Prentice-Hall, 2009); and John M. Wallace and Peter V. Hobbs, *Atmospheric Science: An Introductory Survey*, 2nd edition (Burlington, VT: Academic Press, 2006).

Southeastern climate. Philip L. Chaney, "Climate," Encyclopedia of Alabama, accessed July 2010, http://encyclopediaofalabama.org/face/Article.jsp?id=h-1283; Ellen Chen and John F. Gerber, "Climate," in *Ecosystems of Florida*, ed. Ronald L. Myers and John J. Ewel (Orlando: University of Central Florida Press, 1990), 11–34; and William H. Martin and Stephen G. Boyce, "Introduction: The Southeastern Setting," in *Biodiversity of the Southeastern United States: Lowland Terrestrial Communities*, ed. William H. Martin, Stephen G. Boyce, and Arthur C. Echternacht (New York: John Wiley and Sons, 1993), 2: 1–46.

Precipitation patterns. Alabama Maps, "Average Annual Rainfall [map]," University of Alabama, accessed July 2010, http://alabamamaps.ua.edu/contemporarymaps/alabama/climate/index.html; and PRISM Climate Group and Oregon Climate Service, Oregon State University, "*Average Annual Precipitation, 1971–2000, Alabama* [map]," accessed July 2010, http://www.prism.oregonstate.edu/state_products/index.phtml?id=AL.

El Niño and La Niña. NOAA, "What Is an El Niño?," accessed July 2010, http://www.pmel.noaa.gov/tao/elnino/el-nino-story.html; and NOAA, "What Is La Niña?," accessed July 2010, http://www.pmel.noaa.gov/tao/elnino/la-nina-story.html#publ.

Growing season in Alabama. Jere R. Gallup, *Climatic Features and Length of Growing Season in Alabama* (Auburn, AL: AAES, 1980).

Chapter 4

Fundamental geology. Edward J. Tarbuck and Frederick K. Lutgens, *Earth Science*, 12th edition (Upper Saddle River, NJ: Pearson Prentice-Hall, 2009). In addition, Jim Lacefield and David Ufnar have taught me much about basic geology and southeastern geology.

Cave Salamander. Roger Conant and Joseph T. Collins, *A Field Guide to the Reptiles and Amphibians of Eastern and Central North America* (Boston: Houghton Mifflin, 1991); and Robert H. Mount, *The Reptiles and Amphibians of Alabama* (Tuscaloosa: University of Alabama Press, 1975).

Chickamauga Limestone. Jim Lacefield, *Lost Worlds in Alabama Rocks: A Guide to the State's Ancient Life and Landscapes* (Tuscaloosa: Alabama Geological Society, 2000); and USGS, "Chickamauga Limestone," accessed July 2010, http://tin.er.usgs.gov/geology/state/sgmc-unit.php?unit=ALOc%3B3.

Plate tectonics and North America's drift. Lacefield, *Lost Worlds in Alabama Rocks*; National Museum of Natural History (NMNH), "Geologic Time: A Story of a Changing Earth," Smithsonian Institution, accessed July 2010, http://paleobiology.si.edu/geotime/main/index.html; Christopher R.

Scotese, "Plate Tectonic Maps and Continental Drift Animations," PA-LEOMAP Project, accessed July 2010, www.scotese.com; and Tarbuck and Lutgens, *Earth Science*.

Soil affinities and plant requirements. Michael G. Barbour, Jack H. Burk, and Wanna D. Pitts, *Terrestrial Plant Ecology*, 3rd edition (New York: Benjamin Cummings, 1999); L. Katherine Kirkman, Claud L. Brown, and Donald J. Leopold, *Native Trees of the Southeast* (Portland, OR: Timber Press, 2007); and Natural Resources Conservation Service (NRCS), "PLANTS Database," National Plant Data Team, accessed July 2010, http://plants.usda.gov/index.html.

Geologic map of Alabama. W. Edward Osborne, Michael W. Szabo, Charles W. Copeland Jr., and Thornton L. Neathery, *Geologic Map of Alabama: Special Map 221* (Tuscaloosa: Geological Survey of Alabama [GSA], 1989).

Physiographic provinces. Mike Neilson, "Physiographic Sections of Alabama," Encyclopedia of Alabama, accessed July 2010, http://www.encyclopediaofalabama.org/face/Article.jsp?id=h-1362; and USGS, "A Tapestry of Time and Terrain: The Union of Two Maps—Geology and Topography," accessed July 2010, http://tapestry.usgs.gov/Default.html.

Chapter 5

Geologic time. Dates and names of geologic time units were adopted from USGS, "Divisions of Geologic Time—Major Chronostratigraphic and Geochronologic Units: USGS Fact Sheet 2010-3059," accessed July 2012, http://pubs.usgs.gov/fs/2010/3059/pdf/FS10-3059.pdf.

Geologic history. Ron Blakey, "Paleogeography and Geologic Evolution of North America," Northern Arizona University, accessed July 2010, http://jan.ucc.nau.edu/~rcb7/nam.html; Paul A. Delcourt and Hazel Delcourt, "Paleoclimates, Paleovegetation, and Paleofloras of North America North of Mexico during the Late Quaternary," in *Flora of North America*, ed. Flora of North America Editorial Committee, (New York: Oxford University Press, 1993) 1: 71–94; Paul A. Delcourt, Hazel R. Delcourt, Dan F. Morse, and Phyllis A. Morse, "History, Evolution, and Organization of Vegetation and Human Culture," in *Biodiversity of the Southeastern United States: Lowland Terrestrial Communities*, ed. William H. Martin, Stephen G. Boyce, and Arthur C. Echternacht (New York: John Wiley and Sons, 1993), 2: 47–80; Jim Lacefield, *Lost Worlds in Alabama Rocks: A Guide to the State's Ancient Life and Landscapes* (Tuscaloosa: Alabama Geological Society, 2000); NMNH, "Geologic Time: A Story of a Changing Earth,"

Smithsonian Institution, accessed July 2010, http://paleobiology.si.edu/geotime/main/index.html; Christopher R. Scotese, "Plate Tectonic Maps and Continental Drift Animations," PALEOMAP Project, accessed July 2010, www.scotese.com; and Edward J. Tarbuck and Frederick K. Lutgens, *Earth Science*, 12th edition (Upper Saddle River, NJ: Pearson Prentice-Hall, 2009).

Post-Paleozoic uplift of Southern Appalachians. Jim Lacefield, *Lost Worlds in Alabama Rocks: A Guide to the State's Ancient Life and Landscapes*, 2nd edition (manuscript in preparation); D. C. Prowell and R. A. Christopher, "Evidence for Cenozoic Uplift of the Appalachian Mountains in the Southeastern United States," *Geological Society of America, Southeastern Section Meeting, Abstracts with Programs* 25 (1993): 4; D. C. Prowell and R. A. Christopher, "Evidence for Late Cenozoic Uplift in the Southern Appalachian Mountains from Isolated Sediment Traps," *Geological Society of America, Southeastern Section Meeting, Abstracts with Programs* 38 (2006): 27; and USGS, "Geologic Provinces of the United States: Appalachian Highlands Province," accessed July 2010, http://geomaps.wr.usgs.gov/parks/province/appalach.html.

Fall Line origins and biogeography. Herbert T. Boschung and Richard L. Mayden, *Fishes of Alabama* (Washington, DC: Smithsonian Books, 2004); Lacefield, *Lost Worlds in Alabama Rocks*; Maurice F. Mettee, Patrick E. O'Neil, and J. Malcolm Pierson, *Fishes of Alabama and the Mobile Basin* (Birmingham, AL: Oxmoor House, 1996); Robert H. Mount, *The Reptiles and Amphibians of Alabama* (Tuscaloosa: University of Alabama Press, 1975); Michael J. Neilson, "Fall Line," Encyclopedia of Alabama, accessed July 2010, http://eoa.auburn.edu/face/Article.jsp?id=h-1618; Prowell and Christopher, "Evidence for Late Cenozoic Uplift"; and James D. Williams, Arthur E. Bogan, and Jeffrey T. Garner, *Freshwater Mussels of Alabama and the Mobile Basin in Georgia, Mississippi, and Tennessee* (Tuscaloosa: University of Alabama Press, 2008).

K-Pg extinction. See above sources for "Geologic history," and also Richard Cowen, *History of Life* (Cambridge, MA: Blackwell Scientific Publications, 2000).

Central American land bridge. Delcourt and Delcourt, "Paleoclimates, Paleovegetation, and Paleofloras"; Earth Observatory, "Panama: Isthmus That Changed the World," National Aeronautics and Space Administration, accessed July 2010, http://earthobservatory.nasa.gov/IOTD/view.php?id=4073; and NMNH, "Geologic Time."

Pleistocene megafauna. Miles Barton, Nigel Bean, Stephen Dunleavy, Ian

Gray, and Adam White, *Prehistoric America: A Journey through the Ice Age and Beyond* (New Haven, CT: Yale University Press, 2002); and Charles C. Mann, *1491: New Revelations of the Americas before Columbus* (New York: Vintage Books, 2006). See above sources for "Geologic history."

Pre-Columbian peoples. Delcourt and Delcourt, "Paleoclimates, Paleovegetation, and Paleofloras," provide a thorough summary, and I adopted their delineations of pre-Columbian cultural periods. Other sources included Ian W. Brown, *Bottle Creek: A Pensacola Culture Site in South Alabama* (Tuscaloosa: University of Alabama Press, 2003); Charles Hudson, *The Southeastern Indians* (Knoxville: University of Tennessee Press, 1976); Mann, *1491*; Southeast Archeological Center, "Outline of Prehistory and History: Southeastern North America and the Caribbean," United States National Park Service (USNPS), accessed July 2010, http://www.nps.gov/seac/; and John A. Walthall, *Prehistoric Indians of the Southeast: Archaeology of Alabama and the Middle South* (Tuscaloosa: University of Alabama Press, 1980).

Chapter 6

Evolution and speciation. Basic information on the processes of evolution and speciation can be found in a variety of introductory biology texts such as Carl T. Bergstrom and Lee A. Dugatkin, *Evolution* (New York: W. W. Norton, 2012).

Fossil record. A careful summary is provided by Paul A. Delcourt and Hazel Delcourt, "Paleoclimates, Paleovegetation, and Paleofloras of North America North of Mexico during the Late Quaternary," in *Flora of North America*, ed. Flora of North America Editorial Committee (New York: Oxford University Press, 1993), 1: 71–94.

Salamander diversity and origins. Basic details about salamander diversity and evolution from Robert H. Mount, *The Reptiles and Amphibians of Alabama* (Tuscaloosa: University of Alabama Press, 1975); and Bruce A. Stein, Lynn S. Kutner, and Jonathan S. Adams, *Precious Heritage: The Status of Biodiversity in the United States* (New York: Oxford University Press, 2000). A recent model of plethodontid speciation is David R. Vieites, Mi-Sook Min, and David B. Wake, "Rapid Diversification and Dispersal during Periods of Global Warming by Plethodontid Salamanders," *Proceedings of the National Academy of Sciences* 102 (2007): 19903–19907.

Relationships between North American and Asian plants. Alan Graham, "History of North American Vegetation: Cretaceous (Maastrichtian)—

Tertiary," in *Flora of North America*, ed. Flora of North America Editorial Committee (New York: Oxford University Press, 1993), 1: 57–70; Stein, Kutner, and Adams, *Precious Heritage*; and Jun Wen, "Evolution of Eastern Asian and Eastern North American Disjunct Distributions in Flowering Plants," *Annual Review of Ecology and Systematics* 30 (1999): 421–455.

Cave biodiversity and origins. David C. Culver, Lawrence L. Master, Mary C. Christman, and Horton H. Hobbs III, "Obligate Cave Fauna of the 48 Contiguous United States," *Conservation Biology* 14 (2000): 386–401; David C. Culver, Louis Deharveng, Anne Bedos et al., "The Mid-latitude Biodiversity Ridge in Terrestrial Cave Fauna," *Ecography* 29 (2006): 120–128; and Stein, Kutner, and Adams, *Precious Heritage*.

Freshwater fish diversity and origins. Summaries about the extent and origins of freshwater fish diversity in the Southeast and Alabama are provided by Herbert T. Boschung and Richard L. Mayden, *Fishes of Alabama* (Washington, DC: Smithsonian Books, 2004); Maurice F. Mettee, Patrick E. O'Neil, and J. Malcolm Pierson, *Fishes of Alabama and the Mobile Basin* (Birmingham, AL: Oxmoor House, 1996); and Stein, Kutner, and Adams, *Precious Heritage*. Details about the Mountain Shiner are from Boschung and Mayden, *Fishes of Alabama*. A seminal overview of stream piracy in the Southeast is Wayne C. Starnes and David A. Etnier, "Drainage Evolution and Fish Biogeography of the Tennessee and Cumberland Rivers Drainage Realm," in *The Zoogeography of North American Freshwater Fishes*, ed. C. H. Hocutt and E. O. Wiley III (New York: John Wiley and Sons, 1986), 325–361. There has been much scientific debate about the circumstances and timing of southeastern freshwater fish diversification. Early influential publications on this were Richard L. Mayden, "Vicariance Biogeography, Parsimony, and Evolution in North American Freshwater Fishes," *Systematic Zoology* 37 (1988): 329–355; C. C. Swift, C. R. Gilbert, S. A. Bortone, G. H. Burgess, and R. W. Yerger, "Zoogeography of the Freshwater Fishes of the Southeastern United States: Savannah River to Lake Pontchartrain," in *The Zoogeography of North American Freshwater Fishes*, ed. C. H. Hocutt and E. O. Wiley III (New York: John Wiley and Sons, 1986), 325–361; and E. O. Wiley and Richard L. Mayden, "Species and Speciation in Phylogenetic Systematics, with Examples from the North American Fish Fauna," *Annals of the Missouri Botanical Garden* 72 (1985): 596–635. The molecular clock studies mentioned were Thomas J. Near and Michael F. Benard, "Rapid Allopatric Speciation in Logperch Darters (Percidae: *Percina*)," *Evolution* 58 (2004): 2798–2808; and Thomas J. Near, Todd W. Kassler,

Jeffrey B. Koppelman, Casy B. Dillman, and David P. Philipp, "Speciation in North American Black Basses, *Micropterus* (Actinopterygii: Centrarchidae)," *Evolution* 57 (2003): 1610–1621.

Chapter 7

Hot spot for freshwater biodiversity. Charles Lydeard and Richard L. Mayden, "A Diverse and Endangered Aquatic Ecosystem of the Southeast United States," *Conservation Biology* 9 (1995): 800–805; Maurice F. Mettee, Patrick E. O'Neil, and J. Malcolm Pierson, *Fishes of Alabama and the Mobile Basin* (Birmingham, AL: Oxmoor House, 1996); and Bruce A. Stein, "States of the Union: Ranking America's Biodiversity" (Arlington, VA: NatureServe, 2002).

77,000 miles of rivers and streams. Mettee, O'Neil, and Pierson, *Fishes of Alabama and Mobile Basin.*

The Aquatic State. The suggestion by Edward O. Wilson that Alabama be known as the aquatic state appears in James D. Williams, Arthur E. Bogan, and Jeffrey T. Garner, *Freshwater Mussels of Alabama and the Mobile Basin in Georgia, Mississippi, and Tennessee* (Tuscaloosa: University of Alabama Press, 2008).

Origins of aquatic biodiversity. The factors causing high freshwater fish biodiversity in Alabama (see above sources for "Freshwater fish diversity and origins," chapter 6) also caused the proliferation of other freshwater taxa in the state. See also Williams, Bogan, and Garner, *Freshwater Mussels of Alabama.*

River continuum concept. This model is nicely described in Colbert E. Cushing and J. David Allan, *Streams: Their Ecology and Life* (New York: Academic Press, 2001).

Effects of dams. General effects are described by J. David Allan, *Stream Ecology: Structure and Function of Running Waters* (Boston: Kluwer Academic, 1995); and Cushing and Allan, *Streams.* Role of dams in driving extinctions in Alabama is summarized by USFWS, "The Mobile River Basin," accessed August 2009, http://www.fws.gov/southeast/pubs/esmobile.pdf.

Biology of darters and shiners. Herbert T. Boschung and Richard L. Mayden, *Fishes of Alabama* (Washington, DC: Smithsonian Books, 2004); and Mettee, O'Neil, and Pierson, *Fishes of Alabama and Mobile Basin.* Counts of Alabama's endangered darters and shiners taken from USFWS, "Species Reports: Listings and Occurrences for Alabama," accessed August 2009, http://ecos.fws.gov/tess_public/pub/stateListingAndOccurrenceIndividual.jsp?state=AL.

Mussel biology. Williams, Bogan, and Garner's tome *Freshwater Mussels of Alabama* has an excellent introduction to mussel biology. Filtering rates from ADCNR, "AABC Fact Sheet," accessed August 2009, http://www.outdooralabama.com/research-mgmt/aquatic/factsheet.cfm. Mussel valves as hoe blades from Robert E. Warren, "Native American Tools," Illinois State Museum, accessed August 2009, http://www.museum.state.il.us/OHIA/htmls/land/index.html.

Large fishes. Basic biology of the fishes described in the "River Monsters, *Really?*" section comes from Boschung and Mayden, *Fishes of Alabama*. Additional sources are listed here by species. Blue Catfish: Jan H. Brunvand, ed. *American Folklore: An Encyclopedia* (New York: Garland, 1996). Paddlefish: ADCNR, "Paddlefish," accessed August 2009, http://www.outdooralabama.com/fishing/freshwater/fish/paddlefish/; and USGS, "Paddlefish Study Project: Frequently Asked Questions," accessed August 2009, http://www.umesc.usgs.gov/aquatic/fish/paddlefish/faq.html. Gulf Sturgeon: Associated Press, "Jumping sturgeon breaks girl's leg," Tampa Bay Times, accessed August 2009, http://www.sptimes.com/2007/06/26/State/Jumping_sturgeon_brea.shtml; and Robert W. Hastings and Frank M. Parauka, "Gulf Sturgeon," in *Alabama Wildlife, Volume 2: Imperiled Aquatic Mollusks and Fishes*, ed. Ralph E. Mirarchi, Jeffrey T. Garner, and Maruice F. (Scott) Mettee (Tuscaloosa: University of Alabama Press, 2004), 204–205.

American Alligator. Basic biology provided by Robert H. Mount, *The Reptiles and Amphibians of Alabama* (Tuscaloosa: University of Alabama Press, 1975). History of alligator conservation in Alabama from ADCNR, "Alligator Hunting Season in Alabama," accessed August 2009, http://outdooralabama.com/hunting/game/alligatorhunthome/. Analysis of alligator attacks on humans from Ricky L. Langley, "Alligator Attacks on Humans in the United States," *Wilderness and Environmental Medicine* 16 (2005): 119–124.

Freshwater biodiversity and endemism in Mobile River Basin. The importance of southeastern aquatic biodiversity was brought to the world's attention by Lydeard and Mayden's famous paper "A Diverse and Endangered Aquatic Ecosystem of the Southeast United States." I first heard the comparison of the Mobile River Basin to the Colorado River Basin from Paul Johnson, project supervisor of the Alabama Aquatic Biodiversity Center (AABC). The number of fishes in the latter basin is from USGS, "Southwest," in *Status and Trends of the Nation's Biological Resources*, project director Michael J. Mac (Reston, VA: USGS, 1998), 2: 543–592.

Table of watershed statistics. Sources for these data include Boschung and Mayden, *Fishes of Alabama;* Mettee, O'Neil, and Pierson, *Fishes of Alabama and Mobile Basin*; Ralph E. Mirarchi, ed., *Alabama Wildlife, Volume 1: A Checklist of Vertebrates and Selected Invertebrates: Aquatic Mollusks, Fishes, Amphibians, Reptiles, Birds, and Mammals* (Tuscaloosa: University of Alabama Press, 2004); Guenter A. Schuster, Christopher A. Taylor, and John Johansen, "An Annotated Checklist and Preliminary Designation of Drainage Distributions of the Crayfishes of Alabama," *Southeastern Naturalist* 7 (2008): 493–504; and Williams, Bogan, and Garner, *Freshwater Mussels of Alabama.* Area estimates come from a variety of sources, whose methods in calculating watershed area may have varied. No clear size estimate of the Mobile River watershed was found; area data presented are only for the delta and exclude uplands draining to the bay and delta. Chattahoochee River data include data for the Chipola River. Fish tallies include species migrating between rivers and estuaries. No up-to-date and comprehensive listing of snail species by basin or watershed exists. Snail data were taken from Mirarchi et al., eds., *Alabama Wildlife, Volume 2,* and include all gastropods; snails distributed "throughout Alabama" were assigned to all watersheds. Crayfish data were only available collectively for the Conecuh, Blackwater, and Yellow rivers.

Watershed descriptions. Three books were particularly helpful for providing information about Alabama's river basins and watersheds: Boschung and Mayden, *Fishes of Alabama*; Mettee, O'Neil, and Pierson, *Fishes of Alabama and Mobile Basin*; and Williams, Bogan, and Garner, *Freshwater Mussels of Alabama.* Additional sources are listed by watershed. Cahaba: USFWS, "Final Environmental Assessment: Habitat Management Plan Cahaba River National Wildlife Refuge" (Anniston, AL: USFWS, 2007). Coosa: ADCNR, "Boating, Fishing and Fish in the Coosa River and Its Impoundments," accessed August 2009, http://www.outdooralabama.com/fishing/freshwater/where/rivers/coosa/; Alabama Power, "Jordan Dam," accessed August 2009, http://www.alabamapower.com/lakes/jordan.asp; Alabama Power, "Lay Dam," accessed August 2009, http://www.alabamapower.com/lakes/lay.asp; and Paul D. Johnson and Robert S. Butler, "Freshwater Biodiversity of the Upper Coosa Watershed [poster]" (Montgomery, AL: ADCNR, n.d.); Tallapoosa: ADCNR, "Fish and Fishing on Lake Martin," accessed August 2009, http://www.outdooralabama.com/fishing/freshwater/where/reservoirs/martin/; and Alabama Power, "R. L. Harris Dam," accessed August 2009, http://www.alabamapower.com/lakes/harris.asp.

Chapter 8

General description of Alabama coast. W. Everett Smith, "Geomorphology of Coastal Baldwin County, Alabama, Bulletin 124" (Tuscaloosa: GSA, 1986).

Omernik ecoregion classification system. Commission for Environmental Cooperation (CEC), "Ecological Regions of North America: Toward a Common Perspective" (Montréal, Canada: CEC, 1997); and Nationalatlas.gov, "Map Layer Info: Omernik's Level III Ecoregions of the Continental United States," accessed February 2009, http://www.nationalatlas.gov/mld/ecoomrp.html.

NatureServe's ecological systems. NatureServe, "Ecological Systems of the United States: Executive Summary," accessed February 2009, http://www.natureserve.org/publications/usEcologicalsystems.jsp. I used NatureServe's database (NatureServe Explorer, accessed June 2010, http://www.natureserve.org/index.jsp) on ecological systems extensively for details about most ecosystems described in this book.

Snowy Plover and other barrier island breeding birds. Much of what I describe is from personal observations of these breeding birds. Other details from Roger Clay, "Snowy Plover," in *Alabama Wildlife, Volume 3: Imperiled Amphibians, Reptiles, Birds, and Mammals*, ed. Ralph E. Mirarchi, Mark A. Bailey, Thomas M. Haggerty, and Troy L. Best (Tuscaloosa: University of Alabama Press, 2004), 107–108; and Howard Horne, email to author, July 2009.

Barrier islands and bird migration. Much of what is described here is from personal experience and the teachings of veteran birders of the Gulf Coast, especially my parents, Robert A. and Lucy R. Duncan. Details about bird migration and weather along the Gulf Coast from Robert Duncan.

Origins of strandplains and interdunal wetlands. Numerous scientific papers have been published that offer evidence as to the timing of the formation of the strandplains along the northern Gulf Coast. Two articles in support of the Holocene highstand hypothesis are Michael D. Blum, Tamara J. Misner, Eric S. Collins, David B. Scott, Robert A. Morton, and Andres Aslan, "Middle Holocene Sea-Level Rise and Highstand at +2 M, Central Texas Coast," *Journal of Sedimentary Research* 71 (2001): 581–588; and Robert A. Morton, Jeffrey G. Paine, and Michael D. Blum, "Responses of Stable Bay-Margin and Barrier-Island Systems to Holocene Sea-Level Highstands, Western Gulf of Mexico," *Journal of Sedimentary Research* 70 (2000): 478–490. Two articles in support of the Pleistocene origin hypothesis are Ervin G. Otvos, "Holocene Gulf Levels: Recognition Issues and an Updated Sea-level Curve," *Journal of Coastal Research* 20 (2004): 680–699; and Ervin G. Otvos, "Multiple Pliocene-Quaternary Marine Highstands, Northeast Gulf Coastal Plain—Fallacies and Facts," *Journal of Coastal Research* 11 (1995): 984–1002.

History and future of Dauphin Island and nearby islands. Carl R. Froede Jr., "The Return of 1717 Isle Dauphine (USA)," *Journal of Coastal Research* 25 (2009): 793–795; Robert A. Morton, "Historical Changes in the Mississippi-Alabama Barrier-Island Chain and the Roles of Extreme Storms, Sea Level, and Human Activities," *Journal of Coastal Research* 24 (2008): 1587–1600; and Ervin G. Otvos and Gregory A. Carter, "Hurricane Degradation—Barrier Development Cycles, Northeastern Gulf of Mexico: Landform Evolution and Island Chain History," *Journal of Coastal Research* 24 (2008): 463–478. Numerous conversations with David Ufnar also shaped my understanding of coastal geology.

Salt marshes, seagrass meadows, and oyster reefs. Robert H. Chabreck, *Coastal Marshes: Ecology and Wildlife Management* (Minneapolis: University of Minnesota Press, 1988); H. Dickson Hoese and Richard H. Moore, *Fishes of the Gulf of Mexico: Texas, Louisiana, and Adjacent Waters* (College Station: Texas A&M University Press, 1977); Robert J. Livingston, "Inshore Marine Habitats," in *Ecosystems of Florida*, ed. Ronald L. Myers and John J. Ewel (Orlando: University of Central Florida Press, 1990), 549–573; Clay L. Montague and Richard G. Wiegert, "Salt Marshes," in *Ecosystems of Florida,* ed. Myers and Ewel, 481–516; NatureServe Explorer; and Marvin F. Smith Jr., "Ecological Characterization Atlas of Coastal Alabama: Map Narrative" (Washington, DC: USFWS and United States Minerals Management Service [USMMS], 1984).

Marine invertebrates of marsh. Richard W. Heard, *Guide to Common Tidal Marsh Invertebrates of the Northeastern Gulf of Mexico* (Ocean Springs, MS: Mississippi-Alabama Sea Grant Consortium [MASGC], 1982).

American Oystercatchers. David A. Sibley, *The Sibley Guide to Bird Life and Behavior* (New York: Alfred A. Knopf, 2001).

Mississippi Diamondback Terrapin. Kurt Buhlmann, Tracey Tuberville, and Whit Gibbons, *Turtles of the Southeast* (Athens: University of Georgia Press, 2008); and Ken R. Marion and David H. Nelson, "Mississippi Diamondback Terrapin," in *Alabama Wildlife, Volume 3,* ed. Mirarchi et al., 52–53.

Grand Bay geology. Morton, Paine, and Blum, "Responses of Stable Bay-Margin and Barrier-Island Systems"; Keil Schmid, "Shoreline Erosion Analysis of Grand Bay Marsh" (Jackson: Mississippi Department of Environmental Quality, 2000); W. Everett Smith, "Geomorphology of Coastal Mobile County, Alabama" (Tuscaloosa: GSA, Bulletin 162, 1997); and W. Everett Smith, "Regimes Contributory to Progressive Loss of Alabama Coastal Shoreline and Wetlands" (Tuscaloosa: GSA, Reprint Series 85, 1991).

Coastal shell middens. Lawrence E. Aten, "Shell Mounds and Shellfish: Staff of Prehistoric Life?" University of Texas at Austin, accessed July 2009, http://www.texasbeyondhistory.net/coast/prehistory/images/shellfish.html; Center for Archaeological Studies, "Dauphin Island Shell Mounds," University of South Alabama, accessed July 2009, http://www.southalabama.edu/archaeology/dauphin-island-shell-mounds.html; R. Barry Lewis, "Sea-Level Rise and Subsidence Effects on Gulf Coast Archaeological Site Distributions," *American Antiquity* 65 (2000): 525–541; and Steve B. Wimberly, "Indian Pottery from Clarke County and Mobile County, Southern Alabama" (Tuscaloosa: Alabama Museum of Natural History, Museum Paper 36, 1960).

Near-coast pine flatwoods. Warren G. Abrahamson and David C. Hartnett, "Pine Flatwoods and Dry Prairies," in *Ecosystems of Florida,* ed. Myers and Ewel, 103–149; Bill Finch (The Nature Conservancy [TNC]), discussion with author, April 2009; and NatureServe Explorer.

Mississippi Sandhill Crane. Thomas A. Imhof, *Alabama Birds*, 2nd edition (Tuscaloosa: University of Alabama Press, 1976); and USFWS, "Recovery Plan: Mississippi Sandhill Crane" (Atlanta, GA: USFWS, 1991).

Statistics about Mobile Bay. NOAA, "Final Evaluation Findings Alabama Coastal Area Management Program, December 2003 through November 2007" (Washington, DC: NOAA, 2008).

Tidal wooded swamp. NatureServe Explorer.

West Indian Manatee. Ruth H. Carmichael (Dauphin Island Sea Lab [DISL]), discussion with author, August 2009; D. Fertl, A. J. Schiro, G. T. Regan et al., "Manatee Occurrence in the Northern Gulf of Mexico, West of Florida," *Gulf and Caribbean Research* 17 (2005): 69–94; Jo A. Lewis, "West Indian Manatee," in *Alabama Wildlife, Volume 3,* ed. Mirarchi et al., 189–190; MASGC, "West Indian Manatees: Protection and Conservation," MASGC, accessed August 2009, http://www.masgc.org/pdf/masgp/08-015.pdf; Claire M. Pabody, Ruth H. Carmichael, Lauren Rice, and Monica Ross, "A New Sighting Network Adds to 20 Years of Historical Data on Fringe West Indian (*Trichechus manatus*) Manatee Populations in Alabama Waters," *Gulf of Mexico Science* 27 (2009): 52–61; John E. Reynolds III and Daniel K. Odell, *Manatees and Dugongs* (New York:

Facts on File, 1991); and USFWS, "West Indian Manatee (*Trichechus manatus*) 5-Year Review: Summary and Evaluation" (Jacksonville, FL: USFWS, 2007).

Jubilees. Harold Loesch, "Sporadic Mass Shoreward Migrations of Demersal Fish and Crustaceans in Mobile Bay, Alabama," *Ecology* 41 (1960): 292–298; Edwin B. May, "Extensive Oxygen Depletion in Mobile Bay," *Limnology and Oceanography* 18 (1976): 353–366; Ben Raines, "Frequency May Signal Early Dead Zone in Bay," *Press-Register*, July 5, 2009; and David Ranier, "Jubilees Spice Life on Eastern Shore," ADCNR, accessed August 2009, http://www.outdooralabama.com/oaonline/jubilee08.cfm.

Gaillard Island. Mike Odum, "Bird Rookery Continues to Flourish," *Press-Register*, June 9, 2008; and Ben Raines, "Return to Gaillard Island," *Press-Register*, accessed August 2009, http://cosee-central-gom.org/online_presentations/2005/06reading2.pdf; and Judy P. Stout, "Preliminary Characterization of Habitat Loss: Mobile Bay National Estuary Program" (Dauphin Island, AL: Mobile Bay National Estuary Program [MBNEP], 1998).

Brown Pelicans. USFWS, "Brown Pelican (*Pelecanus occidentalis*)," accessed August 2009, http://library.fws.gov/Pubs/brown_pelican1109.pdf; and USFWS, "Species Program: Brown Pelican (*Pelecanus occidentalis*)," accessed August 2009, http://ecos.fws.gov/speciesProfile/profile/speciesProfile.action?spcode=B02L.

Mobile River Delta. The 5 Rivers—Alabama's Delta Resource Center in Spanish Fort, Alabama, provides a wealth of resources for learning about and exploring the delta. Some of the facts about the delta are from the center's displays (visited April 2009). Another valuable source is the *Press-Register*'s special report "A Wilderness Despite Us: The Magnificent Mobile-Tensaw Delta, How We Have Diminished It, and How We Can Restore It," accessed August 2009, http://www.al.com/specialreport/mobileregister/?delta2.html. Other sources included NatureServe Explorer; NOAA, "Final Evaluation Findings"; and Everett Smith, "Mobile Tensaw Delta," Encyclopedia of Alabama, accessed August 2009, http://www.encyclopediaofalabama.org/face/Article.jsp?id=h-1201.

Black Bears in delta. Bill Finch, "Delta's Furtive Bears," *Press-Register*, January 12, 1999; Kent R. Hersey, Andrew S. Edwards, and Joseph D. Clark, "Assessing American Black Bear Habitat in the Mobile-Tensaw Delta of Southwestern Alabama," *Ursus* 16 (2005): 245–254; Chris Jaworowski, "The Truth about Black Bears in Alabama," ADCNR, accessed July 2009, http://outdooralabama.com/hunting/hunterresources/articles/truth-bears.cfm; Michale S. Mitchell, "Black Bear," in *Alabama Wildlife, Volume 3,* ed. Mariarchi et al., 187–189; Tennessee Wildlife Resources Agency, "Black Bears in Tennessee," accessed July 2009, http://www.state.tn.us/twra/bearmain.html; and Daniel Toole, "Black Bear," ADCNR, accessed July 2009, http://www.outdooralabama.com/watchable-wildlife/what/mammals/Carnivores/bb.cfm.

Reptiles of delta. Buhlmann, Tuberville, and Gibbons, *Turtles of the Southeast*; and Smith, "Mobile-Tensaw Delta."

Red-bellied Turtle. Alabama Department of Archives and History, "Official Symbols and Emblems of Alabama: Official Alabama Reptile," accessed July 2009, http://www.archives.state.al.us/emblems/st_rept.html; Michael C. Bolton, "Fence to Save Red-bellied Turtle?" *Press-Register*, October 23, 2008; David H. Nelson and William M. Turner Jr., "Alabama Red-bellied Turtle," in *Alabama Wildlife, Volume 3,* ed. Mirarchi et al., 54–55; USFWS, "Alabama Red-bellied Turtle Recovery Plan" (Atlanta, GA: USFWS, 1990); and USFWS, "Species Profile: Alabama Red-Belly Turtle (*Pseudemys alabamensis*)," accessed July 2009, http://ecos.fws.gov/speciesProfile/profile/speciesProfile.action?spcode=C01W. Note that earlier sources list the species as endemic to Alabama, but it is now known from southeastern Mississippi.

Waterfowl in Mobile Bay and delta. Sam Hodges, "When Ducks Blocked the Sun," *Press-Register*, January 12, 1999; and Marvin F. Smith Jr., "Ecological Characterization Atlas of Coastal Alabama."

Swallow-tailed Kites. Imhof, *Alabama Birds*; and Eric C. Soehren, "Swallow-tailed Kite," in *Alabama Wildlife, Volume 3,* ed. Mirarchi et al., 128–129.

Conservation concerns in delta. DISL Coastal Policy Center, "The Mobile Bay Causeway: Existing Conditions and Enhancement Opportunities" (Dauphin Island, AL: DISL, 2002); Bill Finch, "Causeway Constricts the Delta Estuary," *Press-Register*, January 12, 1999; Bill Finch, "Cut-rate Golden Gate," *Press-Register*, January 8, 1999; Ben Raines, "Scientists Say Causeway Hurting the Bay; Propose Experiment," *Press-Register*, July 29, 2009.

Perdido River and Perdido Bay. ADCNR, "Perdido River Complex," accessed July 2009, http://www.outdooralabama.com/public-lands/stateLands/foreverWild/FWTracts/PerdidoRvrComplex/; Colbert E. Cushing and J. David Allan, *Streams: Their Ecology and Life* (New York: Academic Press, 2001); Gulfbase, "Perdido Bay," accessed July 2009, http://www.gulfbase.org/bay/view.php?bid=perdido; NatureServe Explorer; W. Everett Smith, "Geomorphology of Coastal Baldwin County"; and TNC, "Betty and

Crawford Rainwater Perdido River Preserve," accessed July 2009, http://
www.nature.org/ourinitiatives/regions/northamerica/unitedstates/ala-
bama/placesweprotect/perdido-river-nature-preserve.xml.

Chapter 9

Southeastern Plains' high species diversity. Taylor H. Ricketts, Eric Di-
nerstein, David M. Olson et al., *Terrestrial Ecoregions of North America:
A Conservation Assessment* (Washington, DC: Island Press, 1999); and
Bruce A. Stein, Lynn S. Kutner, and Jonathan S. Adams, *Precious Heritage:
The Status of Biodiversity in the United States* (New York: Oxford Univer-
sity Press, 2000).

Southeastern Plains as Pleistocene refugium. Miles Barton, Nigel Bean,
Stephen Dunleavy, Ian Gray, and Adam White, *Prehistoric America: A
Journey through the Ice Age and Beyond* (New Haven, CT: Yale University
Press, 2002); Paul A. Delcourt and Hazel Delcourt, "Paleoclimates, Paleovegetation, and Paleofloras of North America North of Mexico during
the Late Quaternary," in *Flora of North America*, ed. Flora of North America Editorial Committee (New York: Oxford University Press, 1993) 1:
71–94; and Bruce A. Sorrie and Alan S. Weakley, "Coastal Plain Vascular
Plant Endemics: Phytogeographic Patterns" *Castanea* 66 (2001): 50–82.

Geologic history of the Southeastern Plains. Glenn E. Griffith et al., *Ecoregions of Alabama and Georgia* [color poster with map, descriptive text,
summary tables, and photographs] (Reston, VA: United States Geological Survey [USGS], 2001); Jim Lacefield, *Lost Worlds in Alabama Rocks:
A Guide to the State's Ancient Life and Landscapes* (Tuscaloosa: Alabama
Geological Society, 2000); and multiple discussions with Jim Lacefield.

Longleaf Pine woodlands. Information comes from dozens of research papers, seminars, courses, conferences, conversations with scientists and foresters, and my own observations and research, and would be impossible
to summarize here. However, for anyone wanting to learn more about
these forests, I highly recommend Lawrence S. Earley, *Looking for Longleaf* (Chapel Hill: University of North Carolina Press, 2004). This book
should be read by all serious students of southeastern natural history.

Conecuh National Forest. Alice S. Christenson, "National Forests of Alabama," Encyclopedia of Alabama, accessed June 2009, http://www.ency-
clopediaofalabama.org/face/Article.jsp?id=h-1354; and USFWS, "National Forests in Alabama," accessed June 2009, http://www.fs.fed.us/r8/
alabama/.

Upland communities in the Conecuh National Forest. Robert E. Carter,

Mark D. MacKenzie, and Dean H. Gjerstad, "Ecological Land Classification in the Southern Loam Hills of South Alabama," *Forest Ecology and Management* 114 (1999): 395–404; and Robert E. Carter, Mark D. MacKenzie, Dean H. Gjerstad, and David Jones, "Species Composition of Fire Disturbed Ecological Land Units in the Southern Loam Hills of South Alabama," *Southeastern Naturalist* 3, no. 1 (2004): 297–308. In addition, information from James and Sierra Stiles and Mark A. Bailey (Conservation Southeast, Inc.) helped me plan my visit to the Conecuh (April 2009).

Red-cockaded Woodpecker. James W. Tucker Jr. and W. Douglas Robinson, "Red-cockaded Woodpecker," in *Alabama Wildlife, Volume 3: Imperiled Amphibians, Reptiles, Birds, and Mammals*, ed. Ralph E. Mirarchi, Mark A. Bailey, Thomas M. Haggerty, and Troy L. Best (Tuscaloosa: University of Alabama Press, 2004), 112–113; USFWS, "Red-cockaded Woodpecker," accessed June 2009, http://library.fws.gov/Pubs4/redcockadedwp02.pdf; and USFWS, "Red-cockaded Woodpecker Recovery," accessed June 2009, http://www.fws.gov/rcwrecovery/.

Southeastern Pocket Gopher. J. Ralph Jordan Jr., "Southeastern Pocket Gopher," in *Alabama Wildlife, Volume 3*, ed. Mirarchi et al., 204–205.

Gopher Tortoise. Matthew J. Aresco and Craig Guyer, "Gopher Tortoise," in *Alabama Wildlife, Volume 3*, ed. Mirarchi et al., 82–83; and Wendell A. Neal, "Gopher Tortoise Recovery Plan" (Jackson, MS: USFWS, 1990).

Eastern Diamondback Rattlesnake. D. Bruce Means, "Eastern Diamondback Rattlesnake," in *Alabama Wildlife, Volume 3*, ed. Mirarchi et al., 73–74 ; Robert H. Mount, *The Reptiles and Amphibians of Alabama* (Tuscaloosa: University of Alabama Press, 1975); Henry R. Mushinsky and Alan H. Savitzky, "Position of the American Society of Ichthyologists and Herpetologists Concerning Rattlesnake Conservation and Roundups," American Society of Ichthyologists and Herpetologists, accessed June 2009, http://www.asih.org/; Jay Reeves, "Opp Rattlesnake Rodeo and Other Southern Snake Hunts Pressured by Environmentalists to End," Al.com, accessed June 2010, http://blog.al.com/spotnews/2010/04/opp_rattlesnake_rodeo_and_othe.html; and Thomas Spencer, "Eastern Diamondback Rattler Might Get Protection," *The Birmingham News*, May 10, 2012.

Eastern Indigo Snakes. James Altiere, "Eastern Indigo Snake," ADCNR, accessed June 2009, http://alabamaforeverwild.com/watchable-wildlife/what/Reptiles/Snakes/eis.cfm; James C. Godwin, "Eastern Indigo Snake," in *Alabama Wildlife, Volume 3*, ed. Mirarchi et al., 40–41; Mount, *Reptiles and Amphibians of Alabama*; and USFWS, "Species Profile: Eastern Indigo Snake (*Drymarchon corais couperi*)," accessed June 2009, http://ecos.

fws.gov/speciesProfile/profile/speciesProfile.action?spcode=C026#re-covery.

Wetlands of the Dougherty Plain and Southern Pine Plains and Hills ecoregions. Information about Blue Pond at the Solon Dixon Forestry Education Center was provided by Mark Hainds (Longleaf Alliance) during a visit there (April 2009). Other useful references included Diane De Steven and Maureen M. Toner, "Vegetation of Upper Coastal Plain Depression Wetlands: Environmental Templates and Wetland Dynamics Within a Landscape Framework," *Wetlands* 24 (2004): 23–42; Florida Natural Areas Inventory (FNAI) and Florida Department of Natural Resources (FDNR), "Guide to the Natural Communities of Florida" (Tallahassee, FL: FNAI and FDNR, 1990); L. Katherine Kirkman, P. Charles Goebel, Larry West, Mark B. Drew, Brian J. Palik, "Depressional Wetland Vegetation Types: A Question of Plant Community Development," *Wetlands* 20 (2000): 373–385; and NatureServe Explorer, accessed June 2010, http://www.natureserve.org/index.jsp. Discussions with Alfred Schotz (Alabama Natural Heritage Program [ANHP], July 2009) and Mark A. Bailey (Conservation Southeast, Inc., July 2009) were also very helpful. The existence of Citronelle ponds was brought to the attention of ecologists by George W. Folkerts, "Citronelle Ponds: Little-known Wetlands of the Central Gulf Coastal Plain, USA," *Natural Areas Journal* 17 (1997): 6–16.

Gopher Frogs. Mark A. Bailey and D. Bruce Means, "Gopher Frog and Mississippi Gopher Frog," in *Alabama Wildlife, Volume 3,* ed. Mirarchi et al., 15–16; and W. Boyd Blihovde, "Terrestrial Movements and Upland Habitat Use of Gopher Frogs in Central Florida," *Southeastern Naturalist* 5 (2006): 265–276.

Shrub bogs and carnivorous plants. Ecology of the bogs largely from NatureServe Explorer. The tally of carnivorous plants in the Southeastern Plains ecoregion obtained from NRCS, "PLANTS Database," National Plant Data Team, accessed November 2010, http://plants.usda.gov/java/. A discussion and visit to TNC's Splinter Hill Bog Preserve with Bill Finch (TNC) was very helpful (April 2009). Life history information about carnivorous plants from Botanical Society of America, "Carnivorous Plants/ Insectivorous Plants," accessed June 2009, http://www.botany.org/Carnivorous_Plants/; and Robert K. Godfrey and Jean W. Wooten, *Aquatic and Wetland Plants of Southeastern United States* (Athens, GA: University of Georgia Press, 1981).

Pine Barrens Treefrog. D. Bruce Means, "Pine Barrens Treefrog," in *Alabama Wildlife, Volume 3,* ed. Mirarchi et al., 25–26.

Henslow's Sparrow. James W. Tucker Jr., "Henslow's Sparrow," ADCNR, accessed June 2009, http://www.outdooralabama.com/watchable-wildlife/what/birds/passerines/hs.cfm; and James W. Tucker Jr. and W. Douglas Robinson, "Influence of Season and Frequency of Fire on Henslow's Sparrows (*Ammodramus henslowii*) Wintering on Gulf Coast Pitcher Plant Bogs," *The Auk* 120 (2003): 96–106.

Fish distributions in the hilly regions of Southeastern Plains. Herbert T. Boschung and Richard L. Mayden, *Fishes of Alabama* (Washington, DC: Smithsonian Books, 2004); and Maurice F. Mettee, Patrick E. O'Neil, and J. Malcolm Pierson, *Fishes of Alabama and the Mobile Basin* (Birmingham, AL: Oxmoor House, 1996).

Red Hills geology and ecology. Glenn Griffith (US Environmental Protection Agency [USEPA]), email to author, June 2009; Griffith et al., *Ecoregions of Alabama and Georgia*; Roland M. Harper, *Forests of Alabama* (Tuscaloosa: GSA, Monograph 10, 1943); Lacefield, *Lost Worlds in Alabama Rocks*; Charles T. Mohr, *Plant Life of Alabama* (Montgomery, AL: Brown Printing, 1901); NatureServe Explorer; and W. Edward Osborne, Michael W. Szabo, Charles W. Copeland Jr., and Thornton L. Neathery, *Geologic Map of Alabama: Special Map 221* (Tuscaloosa: Geological Survey of Alabama [GSA], 1989).

Red Hills Salamander. ADCNR, "Red Hills Tract," accessed June 2009, http://www.nature.org/wherewework/northamerica/states/alabama/press/press4481.html; and D. Bruce Means and Mark A. Bailey, "Red Hills Salamander," in *Alabama Wildlife, Volume 3*, ed. Mirarchi et al., 34–35; and Mount, *Reptiles and Amphibians of Alabama*.

Southern Hilly Gulf Coastal Plain geology and ecology, including cuestas. Griffith et al., *Ecoregions of Alabama and Georgia*; Harper, *Forests of Alabama*; Lacefield, *Lost Worlds in Alabama Rocks*; Mohr, *Plant Life of Alabama*; NatureServe Explorer; Mike Neilson, "East Gulf Coastal Plain Physiographic Section," Encyclopedia of Alabama, accessed June 2009, http://encyclopediaofalabama.org/face/Article.jsp?id=h-1256; and Osborne et al., *Geologic Map of Alabama*.

Pike County Pocosin. David Borland, discussion with author, July 2009; Alvin R. Diamond Jr., Michael Woods, James A. Hall, and Brian H. Martin, "The Vascular Flora of the Pike County Pocosin Nature Preserve, Alabama," *Southeastern Naturalist* 1, no. 1 (2002): 45–54; Roland M. Harper, *Economic Botany of Alabama, Part 1: Geographical Report on Forests* (Tuscaloosa: GSA, Monograph 8, 1913); and Harper, *Forests of Alabama*.

Prairies (all) of Southeastern Plains ecoregion. These ecosystems were among the most challenging to research since little is known about their original condition. John A. Barone, "Historical Presence and Distribution of Prairies in the Black Belt of Mississippi and Alabama," *Castanea* 70 (2005): 170–183; John A. Barone and Jovonn G. Hill, "Herbaceous Flora of the Blackland Prairie Remnants in Mississippi and Western Alabama," *Castanea* 72 (2007): 226–234; Bill Finch (TNC), discussion with author, June 2009; Harper, *Forests of Alabama*; Roland M. Harper, "The Limestone Prairies of Wilcox County, Alabama," *Ecology* 1 (1920): 198–203; Lacefield, *Lost Worlds in Alabama Rocks*; Mohr, *Plant Life of Alabama*; L. P. Moran, David E. Pettry, Richard E. Switzer, Sidney T. McDaniel, and Ronald G. Wieland, "Soils of Native Prairie Remnants in the Jackson Prairie Region of Mississippi" (Starkville: Mississippi Agriculture and Forestry Experiment Station, Bulletin 1067, 1997); NatureServe Explorer; Osborne et al., *Geologic Map of Alabama*; Evan Peacock and Timothy Schauwecker, *Blackland Prairies of the Gulf Coastal Plain: Nature, Culture, and Sustainability* (Tuscaloosa: University of Alabama Press, 2003); Alfred Schotz (ANHP), discussion with author, July 2009; Eugene A. Smith, "Report for the Years 1881 and 1882, Embracing an Account of the Agricultural Features of the State" (Montgomery: GSA, 1883); and Eugene A. Smith, "Report on the Cotton Production of the State of Alabama, with a Discussion of the General Agricultural Features of the State" (Washington, DC: United States Department of the Interior [USDOI], 1884).

American Bison. Troy L. Best, "Bison," ADCNR, accessed July 2009, http://www.outdooralabama.com/watchable-wildlife/what/mammals/ungulates/b.cfm; and Erhard Rostlund, "The Geographic Range of the Historic Bison in the Southeast," *Annals of the Association of American Geographers* 50 (1960): 395–407.

Fall Line Hills and Transition Hills geology and ecology. Relatively little has been written about these regions or their ecosystems. Some of the ecology presented is from my personal observations. Griffith et al., *Ecoregions of Alabama and Georgia*; James Lacefield, discussion with author, May 2009; and NatureServe Explorer.

Mitchell's Satyr. Sara Bright and Paulette H. Ogard, *Butterflies of Alabama* (Tuscaloosa: University of Alabama Press, 2010); and USFWS, "Mitchell's Satyr: Fact Sheet," accessed February 2009, http://www.fws.gov/midwest/endangered/insects/misa_fctsht.html.

Alabama Canebrake Pitcher Plant. TNC, "Roberta Case Pine Hills Preserve," accessed February 2009, http://www.nature.org/ourinitiatives/regions/northamerica/unitedstates/alabama/placesweprotect/roberta-case-pine-hills-preserve.xml; and USFWS, "Alabama Canebrake Pitcher Plant (*Sarracenia rubra* ssp. *alabamensis*) Recovery Plan" (Atlanta, GA: USFWS, 1992).

Bald Eagle and Warbling Vireo. Thomas A. Imhof, *Alabama Birds*, 2nd edition (Tuscaloosa: University of Alabama Press, 1976); and Dennis Sherer, "Shoals Prime Spot for Viewing Bald Eagles," *TimesDaily*, January 25, 2010.

Chapter 10

Ridge and Valley geology and geography. Herbert T. Boschung and Richard L. Mayden, *Fishes of Alabama* (Washington, DC: Smithsonian Books, 2004); Glenn E. Griffith et al., *Ecoregions of Alabama and Georgia* [color poster with map, descriptive text, summary tables, and photographs] (Reston, VA: United States Geological Survey [USGS], 2001); Jim Lacefield, *Lost Worlds in Alabama Rocks: A Guide to the State's Ancient Life and Landscapes* (Tuscaloosa: Alabama Geological Society, 2000); Mike Neilson, "Tennessee Valley and Ridge Physiographic Section," Encyclopedia of Alabama, accessed June 2009, http://www.encyclopediaofalabama. org/face/Article.jsp?id=h-1308; and W. Edward Osborne, Michael W. Szabo, Charles W. Copeland Jr., and Thornton L. Neathery, *Geologic Map of Alabama: Special Map 221* (Tuscaloosa: Geological Survey of Alabama [GSA], 1989). See above sources for "Post-Paleozoic uplift of Southern Appalachians," chapter 5.

Southern Sandstone Ridges and Southern Dissected Ridges and Knobs geology and geography. Glenn Griffith, James Omernik, and Sandra Azevedo, *Ecoregions of Tennessee* (Nashville: Tennessee Department of Environment and Conservation; Reston, VA: USGS; Washington, DC: USEPA; and Washington, DC: NRCS, 1999); Griffith et al., *Ecoregions of Alabama and Georgia*; and Jim Lacefield, *Lost Worlds in Alabama Rocks: A Guide to the State's Ancient Life and Landscapes*, 2nd edition (manuscript in preparation).

Oak Mountain State Park. Most of what I have written is based on my experiences studying and exploring the park's forests. G. Daniel Irwin, W. Edward Osborne, and Willard E. Ward, *Geologic Map of the Chelsea 7.5-Minute Quadrangle, Shelby County, Alabama* (Tuscaloosa: GSA, 2002); NatureServe Explorer, accessed June 2010, http://www.natureserve.org/index. jsp; and W. Edward Osborne and Willard E. Ward, *Geologic Map of the Helena 7.5-Minute Quadrangle, Jefferson and Shelby Counties, Alabama* (Tuscaloosa: GSA, 1996).

Montane Longleaf Pine woodlands. Adam M. Bale, "Fire Effects and Litter Accumulation Dynamics in a Montane Longleaf Pine Ecosystem," Master's thesis, University of Missouri-Columbia, 2009; Arvind Bhuta, discussions with author, December 2009; Jeff Gardner (United States Forest Service [USFS]), discussion with author, May 2009; and J. Morgan Varner III, John S. Kush, and Ralph S. Meldahl, "Structural Characteristics of Frequently-Burned Old-Growth Longleaf Pine Stands in the Mountains of Alabama," *Castanea* 68 (2003): 211–221.

Boynton's Oak and Georgia Oak. Flora of North America Editorial Committee, ed., *Flora of North America,* vol. 3 (New York: Oxford University Press, 1997).

Sandstone glades of the Ridge and Valley. The sandstone glades of the ecoregion haven't been carefully studied and I found no sources on its ecology. An excellent introduction to outcrop ecosystems and other ecosystems featured in this book is Roger C. Anderson, James S. Fralish, and Jerry M. Baskin, *Savannas, Barrens, and Rock Outcrop Plant Communities of North America* (New York: Cambridge University Press, 1999). Discussion (January 2011) with Alfred Schotz (ANHP) was also helpful.

Red-cockaded Woodpeckers in Oak Mountain State Park. Harriet Wright, discussion with author, March 2005.

Shale Valleys and Limestone Valleys geology. Griffith et al., *Ecoregions of Alabama and Georgia;* and H. H. Hobbs, "Caves and Springs," in *Biodiversity of the Southeastern United States: Aquatic Communities,* ed. Courtney T. Hackney, S. Marshall Adams, and William H. Martin (New York: Wiley, 1992), 59–132.

Watercress Darter. Boschung and Mayden, *Fishes of Alabama*; R. Scot Duncan, Chad P. Elliott, Brook L. Fluker, and Bernard R. Kuhajda, "Habitat Use of the Watercress Darter (*Etheostoma nuchale*): An Endangered Fish in an Urban Landscape," *American Midland Naturalist* 164, no. 1 (2010): 9–21; USFWS, "Watercress Darter (*Etheostoma nuchale*) Recovery Plan" (Jackson, MS: USFWS, 1992); Christopher J. Walsh, Allison H. Roy, Jack W. Feminella, Peter D. Cottingham, Peter M. Groffan, and Raymond P. Morgan II, "The Urban Stream Syndrome: Current Knowledge and the Search for a Cure," *Journal of the North American Benthological Society* 24, no. 3 (2005): 706–723.

Pygmy Sculpin. Walton Beacham, Frank V. Castronova, and Suzanne Sessine, *Beacham's Guide to the Endangered Species of North America,* vol. 2 (Detroit, MI: Gale Group, 2001); Boschung and Mayden, *Fishes of Alabama*; Fred G. Thompson and John E. McCaleb, "A New Freshwater Snail from a Spring in Eastern Alabama," *American Midland Naturalist* 100 (1978): 350–358; and USFWS, "Pygmy Sculpin (*Cottus pygmaeus*) Recovery Plan" (Jackson, MS: USFWS, 1991).

Coosa flatwoods and Alabama Leather Flower. It was difficult finding information about the pre-Columbian condition of the Coosa flatwoods because they were cleared long before the first biologists studied them. Linda G. Chafin, *Field Guide to the Rare Plants of Georgia* (Athens: University of Georgia Press, 2007); Roland M. Harper, *Economic Botany of Alabama, Part 1: Geographical Report on Forests* (Tuscaloosa: GSA, Monograph 8, 1913); Roland M. Harper, *Forests of Alabama* (Tuscaloosa: GSA, Monograph 10, 1943); NatureServe Explorer; USFWS, "Alabama Leather Flower (*Clematis socialis*) Recovery Plan" (Jackson, MS: USFWS, 1989); and USFWS, "Alabama Leather Flower, *Clematis socialis*, 5-Year Review: Summary and Evaluation" (Jackson, MS: USFWS, 2010).

Coosa Valley prairies and Giant Whorled Sunflower. Little is known about the original ecology of the Coosa Valley prairies. The stated role of fire and distribution of the prairies is based on my own deductions and was informed by Wayne Barger (ADCNR), discussion with author, January 2011; Chafin, *Field Guide to the Rare Plants of Georgia*; NatureServe Explorer; David F. Salisbury, "For More Than 100 Years the Whorled Sunflower Was Considered Extinct. Now, a Genetic Study Has Confirmed That It Is a Distinct Native Species," Vanderbilt University, accessed July 2011, http://www.vanderbilt.edu/exploration/text/index.php?action=view_section&id=1327&story_id=318&images=; and Alfred Schotz (ANHP), discussion with author, January 2011.

Bibb County Glades. James R. Allison and Timothy E. Stevens, "Vascular Flora of Ketona Dolomite Outcrops in Bibb County, Alabama," *Castanea* 66 (2001): 154–205; R. Scot Duncan, Corinna B. Anderson, Heather N. Sellers, and Erin E. Robbins, "The Effect of Fire Reintroduction on Endemic and Rare Plants of a Southeastern Glade Ecosystem," *Restoration Ecology* 16, no. 1 (2008): 39–49; and NatureServe Explorer.

Talladega National Forest and Dugger Mountain Wilderness Area. Jeff Gardner (USFS) kindly spent several hours showing me around the Shoal Creek Ranger District of the Talladega National Forest (May 2009) and providing much of the information presented here. Francine Hutchinson and Bruce Hutchinson, who along with Peter Conroy were the driving force in establishing the wilderness area, generously showed me portions of the national forest (May 2009) and wilderness area. Francine N. Hutchinson, "A Vascular Flora of Dugger Mountain, Alabama," Master's thesis,

Jacksonville State University, 1998; and USFS, "Shoal Creek Ranger District," USFS, accessed January 2011, http://www.fs.usda.gov/detail/alabama/about-forest/districts/?cid=fsbdev3_002557.

Mountain Longleaf National Wildlife Refuge. Steve Miller (USFWS) provided information during a field trip and discussion at the refuge (May 2009). USFWS, "Mountain Longleaf National Wildlife Refuge," accessed January 2011, http://www.fws.gov/refuges/profiles/index.cfm?id=43666).

Forever Wild Land Trust. Information about the tracts described, and tracts throughout the state, is found at the trust's website: ADCNR, "Forever Wild Land Trust," accessed January 2011, http://www.alabamaforever-wild.com/.

Cahaba River National Wildlife Refuge. USFWS, "Final Environmental Assessment: Habitat Management Plan Cahaba River National Wildlife Refuge" (Anniston, AL: USFWS, 2007); and James D. Williams, Arthur E. Bogan, and Jeffrey T. Garner, *Freshwater Mussels of Alabama and the Mobile Basin in Georgia, Mississippi, and Tennessee* (Tuscaloosa: University of Alabama Press, 2008).

Chapter 11

Geologic origins of Southwestern Appalachians. Glenn E. Griffith et al., *Ecoregions of Alabama and Georgia* [color poster with map, descriptive text, summary tables, and photographs] (Reston, VA: United States Geological Survey [USGS], 2001); Jim Lacefield, *Lost Worlds in Alabama Rocks: A Guide to the State's Ancient Life and Landscapes* (Tuscaloosa: Alabama Geological Society, 2000); James Lacefield, discussion with author, May 2009; Mike Neilson, "Cumberland Plateau Physiographic Section," Encyclopedia of Alabama, accessed March 2009, http://encyclopediaofalabama.org/face/Article.jsp?id=h-1301; and USNPS, "Geology and History of the Cumberland Plateau," accessed March 2009, http://www.nps.gov/biso/planyourvisit/upload/webgeo.pdf.

Little River Canyon and the Southern Table Plateaus. Griffith et al., *Ecoregions of Alabama and Georgia*; W. Edward Osborne, Michael W. Szabo, Charles W. Copeland Jr., and Thornton L. Neathery, *Geologic Map of Alabama: Special Map 221* (Tuscaloosa: Geological Survey of Alabama [GSA], 1989); Thomas V. Ress, "Little River Canyon National Preserve," Encyclopedia of Alabama, accessed August 2009, http://encyclopediaofalabama.org/face/Article.jsp?id=h-2354; and USNPS, "Little River Canyon National Preserve," accessed March 2009, www.nps.gov/liri/.

Tableland forests. Roland M. Harper, *Forests of Alabama* (Tuscaloosa: GSA, Monograph 10, 1943); and NatureServe Explorer, accessed June 2010, http://www.natureserve.org/index.jsp. Ideas presented about the role and frequency of fire are my deductions based on tableland forest composition and structure and what is known about the prehistoric role of fire in Alabama's mountains.

Green Pitcher Plants. NatureServe Explorer; and USFWS, "Green Pitcher Plant Recovery Plan" (Jackson, MS: USFWS, 1994). Brittney Hughes (Alabama State Parks [ASP]) and Jack Johnston have shown me several bogs hosting this species.

Sandstone glades of the Southwestern Appalachians. In contrast to those of the Ridge and Valley ecoregion, these glades have been studied carefully. Elise Quarterman, Madeline P. Burbanck, and Donald J. Shure, "Rock Outcrop Communities: Limestone, Sandstone, and Granite," in *Biodiversity of the Southeastern United States: Lowland Terrestrial Communities*, ed. William H. Martin, Stephen G. Boyce, and Arthur C. Echternacht (New York: John Wiley and Sons, 1993), 2: 35–86; and Alfred Schotz, "An Account of Limestone and Sandstone Glades in William Bankhead National Forest" (Montgomery, AL: ANHP, 2006).

DeSoto State Park. Thomas V. Ress, "DeSoto State Park," Encyclopedia of Alabama, accessed May 2010, http://encyclopediaofalabama.org/face/Article.jsp?id=h-2573.

Green Salamander. George R. Cline, "Green Salamander," in *Alabama Wildlife, Volume 3: Imperiled Amphibians, Reptiles, Birds, and Mammals*, ed. Ralph E. Mirarchi, Mark A. Bailey, Thomas M. Haggerty, and Troy L. Best (Tuscaloosa: University of Alabama Press, 2004), 29–30; and Robert H. Mount, *The Reptiles and Amphibians of Alabama* (Tuscaloosa: University of Alabama Press, 1975).

Rockhouses and cliffs. NatureServe Explorer.

Imperiled species of Little River Canyon. USFWS, "Kral's Water-plantain Recovery Plan" (Jackson, MS: USFWS, 1991); and USNPS, "Little River Canyon."

Jackson County Mountains and mesophytic forests. Griffith et al., *Ecoregions of Alabama and Georgia*; C. Ross Hinkle, Paul A. Schmalzer, William C. McComb, and John Marcus Safley, "Mixed Mesophytic Forests," in *Biodiversity of the Southeastern United States*, ed. Martin, Boyce, and Echternacht, 2: 203–253; NatureServe Explorer; and Osborne at al., *Geologic Map of Alabama*.

Walls of Jericho. Gregory M. Lein, "Walls of Jericho: A Long Awaited Expectation Made Forever Wild," *Outdoor Alabama*, October 2005. I learned much about the Walls while participating in a Bioblitz sponsored by ADCNR, June 2006.

Cerulean Warbler. Paul B. Hamel, "Cerulean Warbler (*Setophaga cerulea*)," The Birds of North America Online, accessed January 2009, http://bna. birds.cornell.edu/bna/species/511/articles/introduction; and Thomas A. Imhof, *Alabama Birds*, 2nd edition (Tuscaloosa: University of Alabama Press, 1976).

Ruffed Grouse. Imhof, *Alabama Birds*; Donald H. Rusch, Stephen Destefano, Michael C. Reynolds, and David Lauten, "Ruffed Grouse (*Bonasa umbellus*)," The Birds of North America Online, accessed January 2009, http://bna.birds.cornell.edu/bna/species/515/articles/introduction; and Jim Schrenkel, "Ruffed Grouse," ADCNR, accessed January 2009, http://www.outdooralabama.com/watchable-wildlife/what/Birds/grouse/rg.cfm.

Allegheny Woodrat. Travis H. Henry, "Allegheny Woodrat," in *Alabama Wildlife, Volume 3,* ed. Mirarchi et al., 206–207; and Petra B. Wood, "Characteristics of Allegheny Woodrat (*Neotoma magister*) Habitat in the New River Gorge National River, West Virginia" (Glen Jean, WV: Final Project Report submitted to USNPS, 2001).

Sinkhole walls. NatureServe Explorer. Bill Finch (TNC) and Doug Fears (TNC) took me to see sink habitats (April 2010) in TNC's Sharp-Bingham Preserve in the Jackson County Mountains.

Caves. See above sources for "Cave biodiversity and origins," chapter 6. Details about Cathedral Caverns State Park from my visit in March 2009; and Thomas V. Ress, "Cathedral Caverns State Park," Encyclopedia of Alabama, accessed August 2010, http://encyclopediaofalabama.org/face/Article.jsp?id=h-2582).

Gray Bats and white-nose syndrome. Troy L. Best, "Gray Myotis," in *Alabama Wildlife, Volume 3,* ed. Mirarchi et al., 179–180; USFWS, "Gray Bat (*Myotis grisescens*), 5-Year Review: Summary and Evaluation" (Columbia, MO: USFWS, 2009). USFWS, "White Nose Syndrome," accessed June 2011, http://www.fws.gov/whitenosesyndrome/; and USGS, "White-Nose Syndrome Threatens the Survival of Hibernating Bats in North America," accessed June 2011, http://www.fort.usgs.gov/wns/. Details about emergences from Sauta Cave based on my visit there in May 2009.

Guntersville Lake and Sequatchie Valley. The lake is well-known among birders for the quantity and diversity of overwintering birds. Griffith et al., *Ecoregions of Alabama and Georgia*; Osborne et al., *Geologic Map of Ala-*

bama; and Angela C. Otts, "Guntersville Dam and Lake," Encyclopedia of Alabama, accessed August 2010, http://encyclopediaofalabama.org/face/Article.jsp?id=h-2621).

Shale Hills geology and ecology. Prescott Atkinson, "Tracks from the Past," *Outdoor Alabama*, February 2006; James S. Day, "Coal Mining," Encyclopedia of Alabama, accessed May 2010, http://encyclopediaofalabama.org/face/Article.jsp?id=h-1473; Griffith et al., *Ecoregions of Alabama and Georgia*; Harper, *Forests of Alabama*; and Lacefield, *Lost Worlds in Alabama Rocks*.

Dissected Plateau geology and ecology. Griffith et al., *Ecoregions of Alabama and Georgia*; Lacefield, *Lost Worlds in Alabama Rocks*; and Neilson, "Cumberland Plateau Physiographic Section."

Giant Tulip-poplar in Sipsey Wilderness. Statistics from Alabama Forestry Commission, "Alabama's Champion Tree Program," accessed May 2010, http://www.forestry.alabama.gov/ChampionTreeProgram.aspx?bv=5&s=1.

Dismalites. Thomas V. Ress, "Dismals Canyon," Encyclopedia of Alabama, accessed January 2012, http://encyclopediaofalabama.org/face/Article.jsp?id=h-3172; and John Sivinski, "Prey Attraction by Luminous Larvae of the Fungus Gnat *Orfelia fultoni*," *Ecological Entomology* 7 (1982): 443–446.

Eastern Hemlock and Black Birch. L. Katherine Kirkman, Claud L. Brown, and Donald J. Leopold, *Native Trees of the Southeast* (Portland, OR: Timber Press, 2007).

Bankhead National Forest and Mary Burks. Alice S. Christenson, "National Forests of Alabama," Encyclopedia of Alabama, accessed June 2009, http://www.encyclopediaofalabama.org/face/Article.jsp?id=h-1354; John N. Randolf, "Mary Ivy Burks," Encyclopedia of Alabama, accessed March 2009, http://www.encyclopediaofalabama.org/face/Article.jsp?id=h-1129; and USFWS, "National Forests in Alabama," accessed March 2009, see above).

Flattened Musk Turtle. Kurt Buhlmann, Tracey Tuberville, and Whit Gibbons, *Turtles of the Southeast* (Athens: University of Georgia Press, 2008); Christopher J. Fonnesbeck and C. Kenneth Dodd Jr., "Estimation of Flattened Musk Turtle (*Sternotherus depressus*) Survival, Recapture, and Recovery Rate during and after a Disease Outbreak," *Journal of Herpetology* 37 (2003): 602–607; Ken R. Marion and Mark A. Bailey, "Flattened Musk Turtle," in *Alabama Wildlife, Volume 3,* ed. Mirarchi et al., 80–81; and Mount, *Reptiles and Amphibians of Alabama*.

Darters of the Dissected Plateau. Herbert T. Boschung and Richard L. Mayden, *Fishes of Alabama* (Washington, DC: Smithsonian Books, 2004); and Maurice F. Mettee, Patrick E. O'Neil, and J. Malcolm Pierson, *Fishes of Alabama and the Mobile Basin* (Birmingham, AL: Oxmoor House, 1996).

Black Warrior Waterdog. Mark A. Bailey and Craig Guyer, "Black Warrior Waterdog," in *Alabama Wildlife, Volume 3,* ed. Mirarchi et al., 36–37; and Michelle C. D. Moreno, Craig Guyer, and Mark A. Bailey, "Distribution and Population Biology of the Black Warrior Waterdog, *Necturus alabamensis*," *Southeastern Naturalist* 5 (2006): 69–84.

Chapter 12

Piedmont geology and geography. *Alabama Atlas and Gazetteer*, 1st edition (Yarmouth, ME: DeLorme, 1998); Glenn E. Griffith et al., *Ecoregions of Alabama and Georgia* [color poster with map, descriptive text, summary tables, and photographs] (Reston, VA: United States Geological Survey [USGS], 2001); Jim Lacefield, *Lost Worlds in Alabama Rocks: A Guide to the State's Ancient Life and Landscapes* (Tuscaloosa: Alabama Geological Society, 2000); Mike Neilson, "Piedmont Upland Physiographic Section," Encyclopedia of Alabama, accessed December 2009, http://www.encyclopediaofalabama.org/face/Article.jsp?id=h-1309; W. Edward Osborne, Michael W. Szabo, Charles W. Copeland Jr., and Thornton L. Neathery, *Geologic Map of Alabama: Special Map 221* (Tuscaloosa: Geological Survey of Alabama [GSA], 1989); and USGS, "A Tapestry of Time and Terrain: The Union of Two Maps—Geology and Topography," accessed July 2010, http://tapestry.usgs.gov/Default.html.

Talladega Upland forests. NatureServe Explorer, accessed June 2010, http://www.natureserve.org/index.jsp; and James N. Skeen, Phillip D. Doerr, and David H. Van Lear, "Oak-hickory-pine Forests," in *Biodiversity of the Southeastern United States: Lowland Terrestrial Communities*, ed. William H. Martin, Stephen G. Boyce, and Arthur C. Echternacht (New York: John Wiley and Sons, 1993), 2: 1–33. See above sources for "Montane Longleaf Pine woodlands," chapter 10. This section was also informed by my own explorations of this ecoregion.

Talladega National Forest and Pinhoti Trail. Jeff Gardner (USFS), discussion with author, May 2009; Thomas Spencer, "Appalachian Trail Expansion Gets Closer to Linking with Alabama," *The Birmingham News*, February 13, 2010; Thomas Spencer, "Conservation Fund Buys 762 Acres to Connect Alabama's Pinhoti Trail with Flagg Mountain," *The Birmingham News*, February 3, 2011; USFS, "Shoal Creek Ranger District," accessed

December 2009, http://www.fs.usda.gov/detail/alabama/about-forest/districts/?cid=fsbdev3_002557; USFS, *Talladega National Forest: Talladega and Shoal Creek Ranger District* [map] (Washington, DC: USFS, 2008); and USFS, "Talladega Ranger District," accessed December 2009, http://www.fs.usda.gov/detail/alabama/about-forest/districts/?cid=fsbdev3_002555.

Eastern Turkeybeard. Center for Plant Conservation, "*Xerophyllum asphodeloides*," accessed December 2009, http://www.centerforplantconservation.org/collection/cpc_viewprofile.asp?CPCNum=6634; and Jeff Gardner of the USFS, who took me to see the plant.

Imperiled species of Talladega National Forest. Herbert T. Boschung and Richard L. Mayden, *Fishes of Alabama* (Washington, DC: Smithsonian Books, 2004); Jeff Gardner (USFS), discussion with author, May 2009; and James D. Williams, Arthur E. Bogan, and Jeffrey T. Garner, *Freshwater Mussels of Alabama and the Mobile Basin in Georgia, Mississippi, and Tennessee* (Tuscaloosa: University of Alabama Press, 2008).

Red Crossbills. Curtis S. Adkisson, "Red Crossbill (*Loxia curvirostra*)," The Birds of North America Online, accessed December 2009, http://bna.birds.cornell.edu/bna/species/256/articles/introduction; and Thomas A. Imhof, *Alabama Birds*, 2nd edition (Tuscaloosa: University of Alabama Press, 1976).

Black Bears in Piedmont. Jeff Gardner (USFS), discussion with author, May 2009; Georgia Wildlife Resources Division, "Black Bear Fact Sheet," accessed December 2009, http://www.georgiawildlife.com/BlackBearFacts; and Bruce and Francine Hutchinson, conversations with the author. See above sources for "Black Bears in delta," chapter 8.

Florida Panthers. David Rainer, "Panthers in Alabama: Fact or Folklore," ADCNR, accessed December 2009, http://www.outdoorsalabama.org/oaonline/panthers08.cfm; USFWS, "Florida Panther Recovery Plan (*Puma concolor coryi*), Third Revision" (Atlanta, GA: USFWS, 2008); and USFWS, "Troup County Panther Was a Florida Panther—Wildlife CSI: High-Tech Genetic Testing Used to Determine Cat's Parentage," accessed December 2009, http://www.fws.gov/news/newsreleases/showNews.cfm?newsId=F48F71C9-FD69-5C24-650FDEB42F9E60DF.

Lower Piedmont geology and geography. Griffith et al., *Ecoregions of Alabama and Georgia*; Lacefield, *Lost Worlds in Alabama Rocks*; Mike Neilson, "Piedmont Upland Physiographic Section," Encyclopedia of Alabama, accessed December 2009, http://www.encyclopediaofalabama.org/face/Article.jsp?id=h-1309; and Osborne et al., *Geologic Map of Alabama*.

Lower Piedmont's original forests. C. Mark Cowell, "Historical Change in Vegetation and Disturbance on the Georgia Piedmont," *American Midland Naturalist* 140 (1998): 78–89; Michael S. Golden, "Forest Vegetation of the Lower Alabama Piedmont," *Ecology* 60 (1979): 770–782; and NatureServe Explorer.

Agricultural history and soils. Michael A. Godfrey, *Field Guide to the Piedmont: The Natural Habitats of America's Most Lived-in Region, From New York City to Montgomery, Alabama* (Chapel Hill: University of North Carolina Press, 1997); Griffith et al., *Ecoregions of Alabama and Georgia*; Michael W. Higgins, *The Structure, Stratigraphy, Tectonostratigraphy, and Evolution of the Southernmost Part of the Appalachian Orogen* (Reston, VA: USGS, 1988); and Skeen, Doerr, and Van Lear, "Oak-hickory-pine Forests," 1–33.

Succession and plantation forestry. Henry J. Oosting, "An Ecological Analysis of the Plant Communities of Piedmont, North Carolina," *American Midland Naturalist* 28 (1942): 1–126; and Skeen, Doerr, and Van Lear, "Oak-hickory-pine Forests," 1–33.

Wind Creek State Park. Forrest Bailey (ASP), discussion with author, May 2011; and Thomas V. Ress, "Wind Creek State Park," Encyclopedia of Alabama, accessed October 2010, http://www.encyclopediaofalabama.org/face/Article.jsp?id=h-2918.

Horseshoe Bend National Military Park. USDOI and United States Department of Agriculture, "National Park Service Implements Prescribed Fire in Rare Mountain Long Leaf Pine Ecosystem Horseshoe Bend National Military Park, Alabama," accessed December 2009, http://www.forestsandrangelands.gov/success/stories/2007/nfp_2007_q1_al_nps_hobe_fuelsreduction.shtml; and USNPS, "Horseshoe Bend National Military Park," accessed December 2009, http://www.nps.gov/hobe/index.htm).

Forever Wild tracts in Lower Piedmont. ADCNR, "Forever Wild Program Land Tracts," accessed December 2009, http://www.outdooralabama.com/public-lands/stateLands/foreverWild/FWTracts/.

Chewacla State Park. Thomas V. Ress, "Chewacla State Park," Encyclopedia of Alabama, accessed October 2010, http://www.encyclopediaofalabama.org/face/Article.jsp?id=h-2574. Details provided also from my own explorations of this park (April 2009).

Granite outcrops. Lawrence J. Davenport, discussion with author, March 2013. Godfrey, *Field Guide to the Piedmont*; NatureServe Explorer; Donald J. Shure and Harvey L. Ragsdale, "Patterns of Primary Succession on Granite Outcrop Surfaces," *Ecology* 58 (1977): 993–1006; and Eugene A. Smith and Henry McCalley, "Index to the Mineral Resources of Alabama," Bulletin 9, GSA (Montgomery, AL: Brown Printing, 1904).

Piedmont's rivers and their biodiversity. Boschung and Mayden, *Fishes of Alabama*; Paul D. Johnson and Robert S. Butler, "Freshwater Biodiversity of the Upper Coosa Watershed [poster]" (Montgomery, AL: ADCNR, n.d.); Carol E. Johnston, "Shoal Bass," in *Alabama Wildlife, Volume 2: Imperiled Aquatic Mollusks and Fishes*, ed. Ralph E. Mirarchi, Jeffrey T. Garner, and Maruice F. (Scott) Mettee (Tuscaloosa: University of Alabama Press, 2004), 224 ; Carol E. Johnston and Bernard R. Kuhajda, "Halloween Darter," in *Alabama Wildlife, Volume 2,* ed. Mirarchi et al., 201–202; Maurice F. Mettee, Patrick E. O'Neil, and J. Malcolm Pierson, *Fishes of Alabama and the Mobile Basin* (Birmingham, AL: Oxmoor House, 1996); J. Malcolm Pierson, "Holiday Darter," in *Alabama Wildlife, Volume 2,* ed. Mirarchi et al., 187; J. Malcolm Pierson, "Lipstick Darter," in *Alabama Wildlife, Volume 2,* ed. Mirarchi et al., 228–229; and Williams, Bogan, and Garner, *Freshwater Mussels of Alabama.*

Chapter 13

Alabama Cavefish. Herbert T. Boschung and Richard L. Mayden, *Fishes of Alabama* (Washington, DC: Smithsonian Books, 2004); Bernard R. Kuhajda, "Alabama Cavefish," in *Alabama Wildlife, Volume 2: Imperiled Aquatic Mollusks and Fishes*, ed. Ralph E. Mirarchi, Jeffrey T. Garner, and Maruice F. (Scott) Mettee (Tuscaloosa: University of Alabama Press, 2004), 181–182; and Bernard R. Kuhajda and Richard L. Mayden, "Status of the Federally Endangered Alabama Cavefish, *Speoplatyrhinus poulsoni* (Amblyopsidae), in Key Cave and Surrounding Caves, Alabama," *Environmental Biology of Fishes* 62 (2001): 215–222.

Interior Plateau geology and geography. William S. Bryant, William C. McComb, and James S. Fralish, "Oak-hickory Forests (Western Mesophytic/Oak-Hickory Forests)," in *Biodiversity of the Southeastern United States: Lowland Terrestrial Communities*, ed. William H. Martin, Stephen G. Boyce, and Arthur C. Echternacht (New York: John Wiley and Sons, 1993), 2:143–201; Glenn E. Griffith et al., *Ecoregions of Alabama and Georgia* [color poster with map, descriptive text, summary tables, and photographs] (Reston, VA: United States Geological Survey [USGS], 2001); Jim Lacefield, *Lost Worlds in Alabama Rocks: A Guide to the State's Ancient Life and Landscapes* (Tuscaloosa: Alabama Geological Society, 2000); Maurice F. Mettee, Patrick E. O'Neil, and J. Malcolm Pierson, *Fishes of Alabama and the Mobile Basin* (Birmingham, AL: Oxmoor House, 1996); Mike Neilson, "Highland Rim Physiographic Section," Encyclopedia of

Alabama, accessed January 2011, http://encyclopediaofalabama.org/face/ Article.jsp?id=h-1311; W. Edward Osborne, Michael W. Szabo, Charles W. Copeland Jr., and Thornton L. Neathery, *Geologic Map of Alabama: Special Map 221* (Tuscaloosa: Geological Survey of Alabama [GSA], 1989); and USGS, "A Tapestry of Time and Terrain: The Union of Two Maps—Geology and Topography," accessed July 2010, http://tapestry. usgs.gov/Default.html.

Tennessee River. Boschung and Mayden, *Fishes of Alabama*; Mettee, O'Neil, and Pierson, *Fishes of Alabama and Mobile Basin*; US Army Corps of Engineers (USACE), "Nashville District History—The Construction of Wilson Lock and Dam," accessed January 2011, http://www.lrn.usace.army. mil/history/wilson_lock_and_dam.htm; and James D. Williams, Arthur E. Bogan, and Jeffrey T. Garner, *Freshwater Mussels of Alabama and the Mobile Basin in Georgia, Mississippi, and Tennessee* (Tuscaloosa: University of Alabama Press, 2008).

Tennessee River fishes. Boschung and Mayden, *Fishes of Alabama*; Steve Doyle, "Researchers Seeking Protection for Sunfish," *The Huntsville Times*, November 30, 2009; Carol E. Johnston, "Slackwater Darter," in *Alabama Wildlife, Volume 2*, ed. Mirarchi et al., 186; Mettee, O'Neil, and Pierson, *Fishes of Alabama and Mobile Basin*; Peggy Shute, "Boulder Darter," in *Alabama Wildlife, Volume 2*, ed. Mirarchi et al., 194–195; USFWS, "Boulder Darter (*Etheostoma wapiti*), 5-Year Review: Summary and Evaluation" (Cookeville, TN: USFWS, 2009); Melvin L. Warren Jr., "Spring Pygmy Sunfish," in *Alabama Wildlife, Volume 2*, ed. Mirarchi et al., 184–185; and James D. Williams and Ronald M. Nowak, "Vanishing Species in Our Own Backyard: Extinct Fish and Wildlife of the United States and Canada," in *The Last Extinction*, edited by Les Kaufman, 115–148 (Cambridge: Massachusetts Institute of Technology Press, 1993).

Wheeler National Wildlife Refuge. USFWS, "Wheeler National Wildlife Refuge," accessed January 2011, http://www.fws.gov/wheeler/. Additionally, Dwight Cooley, refuge manager, was kind enough to be interviewed and show me portions of the refuge (May 2009).

North American Beavers. Steve Boyle and Stephanie Owens, "North American Beaver (*Castor canadensis*): A Technical Conservation Assessment" (Golden, CO: USFS, 2007).

Grassland birds of Interior Plateau. My own observations and Dwight Cooley (USFWS), discussion with author, May 2009; and Thomas A. Imhof, *Alabama Birds*, 2nd edition (Tuscaloosa: University of Alabama Press, 1976).

Forests of Interior Plateau. Bryant, McComb, and Fralish, "Oak-hickory Forests," 143–201; Roland M. Harper, *Economic Botany of Alabama, Part 1: Geographical Report on Forests* (Tuscaloosa: GSA, Monograph 8, 1913); Roland M. Harper, *Forests of Alabama* (Tuscaloosa: GSA, Monograph 10, 1943); and NatureServe Explorer, accessed June 2010, http://www.natureserve.org/index.jsp.

Eastern and Western Highland Rim geology. Griffith et al., *Ecoregions of Alabama and Georgia*; Lacefield, *Lost Worlds in Alabama Rocks*; Mettee, O'Neil, and Pierson, *Fishes of Alabama and Mobile Basin*; Neilson, "Highland Rim Physiographic Section"; and Osborne et al., *Geologic Map of Alabama*.

Limesinks. Thomas L. Chrisman, "Natural Lakes of the Southeastern United States: Origin, Structure, and Function," in *Biodiversity of the Southeastern United States: Aquatic Communities*, ed. Courtney T. Hackney, S. Marshall Adams, and William H. Martin (New York: Wiley, 1992), 475–538; and NatureServe Explorer.

Limestone glades. Roger C. Anderson, James S. Fralish, and Jerry M. Baskin, *Savannas, Barrens, and Rock Outcrop Plant Communities of North America* (New York: Cambridge University Press, 1999); Jerry M. Baskin and Carol C. Baskin, "The Vascular Flora of Cedar Glades of the Southeastern United States and Its Phytogeographical Relationships," *Journal of the Torrey Botanical Society* 130 (2003): 101–118; James Lacefield, discussion with author, May 2009; NatureServe Explorer; and TNC, "Prairie Grove Glades," accessed January 2011, http://www.nature.org/ourinitiatives/regions/northamerica/unitedstates/alabama/placesweprotect/prairie-grove-glades.xml.

Outer Nashville Basin. Griffith et al., *Ecoregions of Alabama and Georgia*; and Osborne et al., *Geologic Map of Alabama*.

Paint Rock River. Bill Finch (TNC) and Doug Fears (TNC) kindly gave me a tour of the Paint Rock River Valley and TNC's restoration projects. Boschung and Mayden, *Fishes of Alabama*; TNC, "Roy B. Whitaker Preserve," accessed January 2011, http://www.nature.org/ourinitiatives/regions/northamerica/unitedstates/alabama/placesweprotect/whitaker-preserve.xml; and Williams, Bogan, and Garner, *Freshwater Mussels of Alabama*.

French's Shootingstar and Cane Creek Canyon. Steven R. Hill, "Conservation Assessment for French's Shootingstar (*Dodecatheon frenchii*) (Vasey) Rydb" (Milwaukee, WI: USFS, 2002); and Thomas V. Ress, "Cane Creek Canyon Nature Preserve," Al.com, accessed June 2011, http://blog.al.com/keeping-alabama-forever-wild/2011/04/cane_creek_canyon_nature_prese.html.

Little Mountain geology. Griffith et al., *Ecoregions of Alabama and Georgia*; Lacefield, *Lost Worlds in Alabama Rocks*; Mettee, O'Neil, and Pierson, *Fishes of Alabama and Mobile Basin*; Neilson, "Highland Rim Physiographic Section"; W. Edward Osborne, Michael W. Szabo, Charles W. Copeland Jr., and Thornton L. Neathery, *Geologic Map of Alabama: Special Map 221* (Tuscaloosa: Geological Survey of Alabama [GSA], 1989); and William A. Thomas and Greg H. Mack, "Paleogeographic Relationship of a Mississippian Barrier-island and Shelf-bar System (Hartselle Sandstone) in Alabama to the Appalachian-Ouachita Erogenic Belt," *Geological Society of America Bulletin* 93 (1982): 6–19.

Chapter 14

Gulf of Mexico statistics. USEPA, "General Facts about the Gulf of Mexico," accessed June 2010, http://www.epa.gov/gmpo/about/facts.html.

Headlines. Jody Bourton, "Giant Deep Sea Jellyfish Filmed in Gulf of Mexico," BBC Earth News, accessed June 2010, http://news.bbc.co.uk/earth/hi/earth_news/newsid_8638000/8638527.stm; Mike Brantley, "Whale Sharks Approach Alabama Coast in Unprecedented Numbers," *Press-Register*, August 17, 2009; John A. MacDonald, "Biologist Discusses Big Fish Story about Killer Whales off Alabama's Gulf Coast," *The Birmingham News*, March 19, 2009; and Jasmin Melvin, "US Scientists Net Giant Squid in Gulf of Mexico," Reuters, accessed June 2010, http://www.reuters.com/article/2009/09/22/us-usa-giantsquid-idUSTRE58K4OF20090922.

Gulf geography, topography, and currents. Steve Gittings, "Liquid Wind," NOAA, accessed June 2010, http://oceanexplorer.noaa.gov/explorations/islands01/background/wind/wind.html; Google Earth 5.1.3535.3218 (Mountain View, CA: Google, Inc.); Robert H. Gore, *The Gulf of Mexico: A Treasury of Resources in the American Mediterranean* (Sarasota, FL: Pineapple Press, 1992); NOAA, *Gulf of Mexico Coastal and Ocean Zones: Strategic Assessment Data Atlas* (Washington, DC: NOAA, 1985); and L.-Y. Oey, T. Ezer, and H.-C. Lee, "Loop Current, Rings and Related Circulation in the Gulf of Mexico: A Review of Numerical Models and Future Challenges," in *Circulation in the Gulf of Mexico: Observations and Models*, Geophysical Monograph Series, ed. W. Sturges and A. Lugo-Fernandez (Washington, DC: American Geophysical Union, 2005), 161: 31–56.

Whale Sharks and plankton. Mike Brantley, "Whale Sharks: Previously Rare in Local Waters, World's Largest Fish Recently Spotted with Frequency off Alabama Coast," *Press-Register*, August 16, 2009; Mike Brantley, "Whale Sharks: Swimming Alongside One of the Giants of the Oceanic World,"

Press-Register, August 17, 2009; FishBase, "Whale Shark," accessed February 2010, http://www.fishbase.org/summary/Rhincodon-typus.html; and Gore, *The Gulf of Mexico*.

Mackerel. FishWatch, "King Mackerel," NOAA, accessed February 2010, http://www.fishwatch.gov/seafood_profiles/species/mackerel/species_pages/king_mackerel.htm; FishWatch, "Spanish Mackerel," NOAA, accessed February 2010, http://www.fishwatch.gov/seafood_profiles/species/mackerel/species_pages/spanish_mackerel.htm; and NOAA, *Gulf of Mexico Coastal and Ocean Zones*.

Tunas. Barbara A. Block, Steven L. H. Teo, Andreas Walli et al. "Electronic Tagging and Population Structure of Atlantic Bluefin Tuna," *Nature* 434 (2005): 1121–1127; Eugene H. Buck, "Atlantic Bluefin Tuna: International Management of a Shared Resource" (Washington, DC: Congressional Research Service, 1995); FishWatch, "Atlantic Yellowfin Tuna," NOAA, accessed February 2010, http://www.fishwatch.gov/seafood_profiles/species/tuna/species_pages/atl_yellowfin_tuna.htm; FishWatch, "Western Atlantic Bluefin Tuna," NOAA, accessed February 2010, http://www.fishwatch.gov/seafood_profiles/species/tuna/species_pages/atl_bluefin_tuna.htm; Jean-Marc Fromentin and Joseph E. Powers, "Atlantic Bluefin Tuna: Population Dynamics, Ecology, Fisheries and Management," *Fish and Fisheries* 6 (2005): 281–306; International Commission for the Conservation of Atlantic Tunas (ICCAT), "Atlantic Bluefin Tuna, Executive Summary," accessed June 2010, http://www.iccat.int/en/assess.htm; NOAA, "Atlantic Bluefin Tuna Proposal Not Adopted after Intense Debate," accessed June 2010, http://www.noaanews.noaa.gov/stories2010/20100318_tuna.html; and David Randall, Warren Burggren, and Kathleen French, *Eckert Animal Physiology: Mechanisms and Adaptations* (New York: W. H. Freeman, 2002).

Marlins. Barbara A. Block, David Booth, and Francis G. Carey, "Direct Measurement of Swimming Speeds and Depth of Blue Marlin," *Journal of Experimental Biology* 166 (1992): 267–284; Susie Gardieff, "Atlantic Blue Marlin," Florida Museum of Natural History (FMNH), accessed February 2010, http://www.flmnh.ufl.edu/fish/gallery/descript/bluemarlin/bluemarlin.html; Susie Gardieff, "White Marlin," FMNH, accessed February 2010, http://www.flmnh.ufl.edu/fish/gallery/descript/whitemarlin/whitemarlin.html; ICCAT, "Report of the 2006 ICCAT Billfish Stock Assessment," *Collective Volume of Scientific Papers ICCAT*, 60 (2007): 1431–1546; Eric D. Prince, Robert K. Cowen, Stacy A. Lucy et al. "Movements and Spawning of White Marlin and Blue Marlin off Punta Cana,

Dominican Republic," *US Fishery Bulletin* 103, no. 4 (2005): 659–669; and White Marlin Biological Review Team, "Atlantic White Marlin Status Review: Report to National Marine Fisheries Service (NMFS), Southeast Regional Office" (Silver Spring, MD: NMFS, 2002).

Swordfish. FishWatch, "North Atlantic Swordfish," NOAA, accessed February 2010, http://www.fishwatch.gov/seafood_profiles/species/swordfish/species_pages/north_atlantic_swordfish.htm; Susie Gardieff, "Swordfish," FMNH, accessed February 2010, http://www.flmnh.ufl.edu/fish/gallery/descript/swordfish/swordfish.html; ICCAT, "Atlantic Swordfish, Executive Summary," accessed February 2010, http://www.iccat.int/en/assess.htm; and Brian E. Luckhurst, Eric D. Prince, Joel K. Llopiz, Derke Snodgrass, and Edward B. Brothers, "Evidence of Blue Marlin (*Makaira nigricans*) Spawning in Bermuda Waters and Elevated Mercury Levels in Large Specimens," *Bulletin of Marine Science* 79, no. 3 (2006): 691–704.

Sharks. Information on shark species from the databases provided by FishBase, http://www.fishbase.org/home.htm; FMNH Ichthyology Department, http://www.flmnh.ufl.edu/fish/Education/bioprofile.htm; and NOAA, http://www.fishwatch.gov/, all accessed May 2010. Mark Grace and Terry Henwood, "Assessment of the Distribution and Abundance of Coastal Sharks in the US Gulf of Mexico and Eastern Seaboard, 1995 and 1996," *Marine Fisheries Review* 59 (1997): 23–32; International Union for the Conservation of Nature (IUCN), "More Oceanic Sharks Added to the IUCN Red List," accessed May 2010, http://www.iucn.org/knowledge/news/?103/More-oceanic-sharks-added-to-the-IUCN-Red-List; IUCN, "Release of the 2006 IUCN Red List of Threatened Species Reveals Ongoing Decline of the Status of Plants and Animals," accessed May 2010, http://www.flmnh.ufl.edu/fish/organizations/ssg/2006mayredlist.pdf; IUCN, "Third of Open Ocean Sharks Threatened with Extinction," accessed May 2010, http://www.iucn.org/?3362/Third-of-open-ocean-sharks-threatened-with-extinction; Ransom A. Myers, Julia K. Baum, Travis D. Shepherd, Sean P. Powers, and Charles H. Peterson, "Cascading Effects of the Loss of Apex Predatory Sharks from a Coastal Ocean," *Science* 315, no. 5280 (2007): 1846–1850; NOAA, "Fish Stocks in the Gulf of Mexico: Fact Sheet," accessed May 2010, http://sero.nmfs.noaa.gov/sf/deepwater_horizon/Fish_economics_FACT_SHEET.pdf/; and J. D. Stevens, R. Bonfil, N. K. Dulvy, and P. A. Walker, "The Effects of Fishing on Sharks, Rays, and Chimaeras (Chondrichthyans), and the Implications for Marine Ecosystems," *ICES Journal of Marine Science* 57 (2000): 476–494.

Shark attack information from International Shark Attack File, FMNH, accessed May 2010, http://www.flmnh.ufl.edu/fish/sharks/ISAF/ISAF.htm.

Cetaceans. Mark F. Baumgartner, Keith D. Mullin, L. Nelson May, and Thomas D. Leming, "Cetacean Habitats in the Northern Gulf of Mexico," *US Fisheries Bulletin* 99 (2001): 219–239; Randall W. Davis, Joel G. Ortega-Ortiza, Christine A. Ribic et al., "Cetacean Habitat in the Northern Oceanic Gulf of Mexico," *Deep-Sea Research I* 49 (2002): 121–142; William B. Davis and David J. Schmidly, "Mammals of Texas Online Edition," Texas Parks and Wildlife Department, accessed May 2010, http://www.nsrl.ttu.edu/tmot1/; William Lang, "Gulf of Mexico—Home to Whales and Dolphins," USMMS, accessed May 2010, http://www.gomr.mms.gov/homepg/regulate/environ/marmam/Cetacean.pdf; NMFS, "Bryde's Whale (*Balaenoptera edeni*)," accessed May 2010, http://www.nmfs.noaa.gov/pr/species/mammals/cetaceans/brydeswhale.htm; NMFS, "Draft Recovery Plan for the Sperm Whale (*Physeter macrocephalus*)" (Silver Spring, MD: NMFS, 2006); NMFS, "Killer Whale (*Orcinus orca*)," accessed May 2010, http://www.nmfs.noaa.gov/pr/species/mammals/cetaceans/killerwhale.htm; NMFS, "Killer Whale (*Orcinus orca*): Northern Gulf of Mexico Stock," accessed May 2010, http://www.nmfs.noaa.gov/pr/pdfs/sars/ao2009whki-gmxn.pdf; NMFS, "Sperm Whales (*Physeter macrocephalus*)," accessed May 2010, http://www.nmfs.noaa.gov/pr/species/mammals/cetaceans/spermwhale.htm; NMFS, "Sperm Whale (*Physeter macrocephalus*): Northern Gulf of Mexico Stock," accessed May 2010, http://www.nmfs.noaa.gov/pr/pdfs/sars/ao2009whsp-gmxn.pdf; Gordon T. Waring, Elizabeth Josephson, Carol P. Fairfield, and Katherine Maze-Foley, eds, "US Atlantic and Gulf of Mexico Marine Mammal Stock Assessments—2006" (Silver Spring, MD: NMFS, 2007); and David W. Weller, Andrew J. Schiro, Victor G. Cockcroft, Wang Ding, "First Account of a Humpback Whale (*Megaptera novaeangliae*) in Texas Waters, with a Re-evaluation of Historical Records from the Gulf of Mexico," *Marine Mammal Science* 12, no. 1 (1996): 1133–1137.

Sargassum. Tara L. Casazza and Steve W. Ross, "Fishes Associated with Pelagic *Sargassum* and Open Water Lacking *Sargassum* in the Gulf Stream off North Carolina," *Fishery Bulletin* 106 (2008): 348–363; Dave Ferrell, "Sargassum: Weed of Life," *Marlin: The International Sportfishing Magazine*, July 31, 2001; Jim Gower and Stephanie King, "Satellite Images Show the Movement of Floating Sargassum in the Gulf of Mexico and Atlantic Ocean," Nature Precedings, accessed February 2010, http://hdl.handle.

net/10101/npre.2008.1894.1; and W. Randolph Taylor, "Sketch of the Character of the Marine Algal Vegetation of the Shores of the Gulf of Mexico," in *Gulf of Mexico: Its Origin, Waters, and Marine Life*, Fishery Bulletin 89, ed. Paul S. Galtsoff (Washington, DC: USFWS, 1954), 177–192.

Continental Shelf habitats. James M. Brooks and Charles P. Giammona, eds., "Mississippi-Alabama Continental Shelf Ecosystem Study Data Summary and Synthesis, Volume I: Executive Summary" (New Orleans, LA: USMMS, 1991); and Triniti A. Dufrene, "Geological Variability and Holocene Sedimentary Record on the Northern Gulf of Mexico Inner to Mid-Continental Shelf," Master's thesis, Louisiana State University, 2005.

The Pinnacles. Gregory S. Boland, James Sinclair, and Susan A. Childs, "Gulf of Mexico Offshore Oases: A Teacher's Companion" (New Orleans, LA: USMMS, 2006); Brooks and Giammona, eds., "Mississippi-Alabama Continental Shelf"; Continental Shelf Associates, Inc., "Mississippi-Alabama Shelf Pinnacle Trend Habitat Mapping Study" (New Orleans, LA: USMMS, 1992); and William W. Sager, William W. Schroeder, J. Scott Laswell, Kenneth S. Davis, Richard Rezak, and Stephen R. Gittings, "Mississippi-Alabama Outer Continental Shelf Topographic Features Formed during the Late Pleistocene-Holocene Transgression," *Geo-Marine Letters* 12, no. 1 (1992): 41–48.

Red Snapper. Cathleen Bester, "Northern Red Snapper," FMNH, accessed May 2010, http://www.flmnh.ufl.edu/fish/gallery/descript/redsnapper/redsnapper.html; Brooks and Giammona, eds., "Mississippi-Alabama Continental Shelf"; FishWatch, "Red Snapper," NOAA, accessed May 2010, http://www.fishwatch.gov/seafood_profiles/species/snapper/species_pages/red_snapper.htm; NMFS, "Gulf of Mexico Red Snapper Frequently Asked Questions," accessed May 2010, http://sero.nmfs.noaa.gov/sf/pdfs/red%20snapper%20general%20FAQs%20032410.pdf; David Rainer, "Snapper Stocks Updates; Increased Quota Delayed," *Outdoor Alabama Newsletter*, February 2010; Southeast Data, Assessment, and Review (SEDAR), "Stock Assessment Report of SEDAR 7: Gulf of Mexico Red Snapper" (Charleston, SC: SEDAR, 2005); Shannon Tompkins, "Reel in Those Expectations: Population of Snapper Has Improved, but Strict Regulations Remain," *Houston Chronicle*, February 10, 2010; and Ian K. Workman and Daniel G. Foster, "Occurrence and Behavior of Juvenile Red Snapper, *Lutjanus campechanus*, on Commercial Shrimp Fishing Grounds in the Northeastern Gulf of Mexico," *Marine Fisheries Review* 56 (1994): 9–11.

Shrimp. FishWatch, "Brown Shrimp," NOAA, accessed May 2010, http://

www.fishwatch.gov/seafood_profiles/species/shrimp/species_pages/
brown_shrimp.htm; FishWatch, "Pink Shrimp," NOAA, accessed May
2010, http://www.fishwatch.gov/seafood_profiles/species/shrimp/spe-
cies_pages/pink_shrimp.htm; FishWatch, "White Shrimp," NOAA, accessed May 2010, http://www.fishwatch.gov/seafood_profiles/species/
shrimp/species_pages/white_shrimp.htm; Leslie D. Hartman, "Alabama's
Shrimp from Plankton to Plate," in "Where Rivers Meet the Sea: A Report
from the Mobile Bay National Estuary Program," MBNEP, accessed May
2010, http://www.mobilebaynep.com/images/uploads/library/Where_
Rivers_Meet_the_Sea.pdf; NMFS, *Fisheries Economics of the United
States, 2009* (Silver Spring, MD: NMFS, 2010); Thomas P. O'Connor
and Gary C. Matlock, "Shrimp Landing Trends as Indicators of Estuarine
Habitat Quality," *Gulf of Mexico Science* 23 (2005): 192–196; Susan-Marie
Stedman and Jeanne Hanson, "Habitat Connections: Wetlands, Fisheries and Economics; Part Four: Wetlands, Fisheries, and Economics in the
Gulf of Mexico," NMFS, accessed May 2010, http://www.nmfs.noaa.gov/
habitat/habitatconservation/publications/habitatconections/num4.htm;
and Richard Wallace, "Shrimp in Alabama" (Auburn: Alabama Cooperative Extension System [ACES], 1997).

Trawling. Kristen M. Fletcher, "Bycatch Reduction Device Rule in Gulf,"
Mississippi-Alabama Sea Grant Legal Program, accessed May 2010, http://
masglp.olemiss.edu/Water%20Log/WL18/brds.htm; Food and Agriculture Organization of the United Nations (FAO), "Shrimp Otter Trawling," accessed May 2010, http://www.fao.org/fishery/fishtech/1021/en;
FAO, "Shrimp Outrigger Trawling," accessed May 2010, http://www.
fao.org/fishery/fishtech/1022/en; Louisiana Sea Grant, "Definition and
History: Bycatch Reduction Device (BRD)," accessed May 2010, http://
www.seagrantfish.lsu.edu/management/TEDs&BRDs/brds.htm; and Peter Sheridan and Phillip Caldwell, "Compilation of Data Sets Relevant to
the Identification of Essential Fish Habitat on the Gulf of Mexico Continental Shelf and for the Estimation of the Effects of Shrimp Trawling Gear
on Habitat" (Galveston, TX: NOAA, 2002).

Artificial reefs. ADCNR, "Alabama's Artificial Reefs: A Fishing Information
Guide," accessed May 2010, http://www.outdooralabama.com/fishing/
saltwater/where/artificial-reefs/reef_brochure.pdf; Gary D. Grossman,
Geoff P. Jones, and William J. Seaman Jr., "Do Artificial Reefs Increase Regional Fish Production? A Review of Existing Data," *Fisheries* 22 (1997):
17–23; William F. Patterson and James H. Cowen, "Site Fidelity and Dis-
persion of Red Snapper Associated with Artificial Reefs in the Northern

Gulf of Mexico," *American Fisheries Society Symposium* 36 (2003): 181–193; Megan B. Peabody and Charles A. Wilson, "Fidelity of Red Snapper (*Lutjanus campechanus*) to Petroleum Platforms and Artificial Reefs in the Northern Gulf of Mexico" (New Orleans, LA: USMMS, 2006); Daniel J. Sheehy, "The Use of Designed and Prefabricated Artificial Reefs in the United States," *Marine Fisheries Review* 44 (1982): 4–15; and Stephen T. Szedlmayer, "Artificial Reefs: Design, Placement, and Permitting" (Auburn, AL: ACES, 2000).

Continental slope habitats. Boland, Sinclair, and Childs, "Gulf of Mexico Offshore Oases"; and Gilbert T. Rowe and Mahlon C. Kennicutt II, "Northern Gulf of Mexico Continental Slope Habitats and Benthic Ecology Study: Final Report" (New Orleans, LA: USMMS, 2009).

Chemosynthetic communities. Boland, Sinclair, and Childs, "Gulf of Mexico Offshore Oases"; Jim Lacefield, *Lost Worlds in Alabama Rocks: A Guide to the State's Ancient Life and Landscapes* (Tuscaloosa: Alabama Geological Society, 2000); and Ian R. MacDonald, William W. Schroeder, and James M. Brooks, "Chemosynthetic Ecosystems Studies Final Report, Volume 1: Executive Summary" (New Orleans, LA: USMMS, 1995).

Production platforms reefs. Ann Bull et al., "Islands of Life: A Teacher's Companion" (New Orleans, LA: USMMS, 2005); Peabody and Wilson, "Fidelity of Red Snapper"; and USMMS, "Islands of Life" [poster] (New Orleans, LA: USMMS, 2006).

Chapter 15

Extinctions. ANHP, "Alabama Inventory List: The Rare, Threatened, and Endangered Plants and Animals of Alabama" (Montgomery, AL: ANHP, 2010); Ralph E. Mirarchi, ed., *Alabama Wildlife, Volume 1: A Checklist of Vertebrates and Selected Invertebrates: Aquatic Mollusks, Fishes, Amphibians, Reptiles, Birds, and Mammals* (Tuscaloosa: University of Alabama Press, 2004); Bruce A. Stein, "States of the Union: Ranking America's Biodiversity" (Arlington, VA: NatureServe, 2002).

Imperiled species. ADCNR, "Nongame Vertebrates Protected by Alabama Regulations," accessed July 2011, http://www.outdooralabama.com/watchable-wildlife/regulations/nongame.cfm; ANHP, "Alabama Inventory List"; Center for Biological Diversity, "The Southeast Freshwater Extinction Crisis," accessed July 2011, http://www.biologicaldiversity.org/programs/biodiversity/1000_species/the_southeast_freshwater_extinction_crisis/index.html; Stein, "States of the Union"; USFWS, "Candidate Species in Alabama Based on Published Population Data," accessed

July 2011, http://ecos.fws.gov/tess_public/pub/stateListingIndividual. jsp?state=AL&status=candidate; USFWS, "Candidate Species: Section 4 of the Endangered Species Act," accessed July 2011, http://www.fws. gov/endangered/esa-library/pdf/candidate_species.pdf; and USFWS, "Species Listed in Alabama Based on Published Historic Range and Population," accessed July 2011, http://ecos.fws.gov/tess_public/pub/stateListingIndividual.jsp?state=AL&status=listed.

Damaged and lost ecosystems. ADCNR, "ACWCS" (Montgomery, AL: ADCNR, 2005); and Reed F. Noss, Edward T. LaRoe III, and J. Michael Scott, "Endangered Ecosystems of the United States: A Preliminary Assessment of Loss and Degradation" (Washington, DC: United States National Biological Service, 1995).

Non-native invasive species. Many of the species mentioned are well-known pests in Alabama. ADCNR, "ACWCS"; Alabama Cogongrass Control Center, "Alabama Cogongrass Control Center—Home," accessed June 2012, http://www.alabamacogongrass.com/cogongrass/index.php; Nico Dauphiné and Robert J. Cooper, "Impacts of Free-Ranging Domestic Cats (*Felis catus*) on Birds in the United States: A Review of Recent Research with Conservation and Management Recommendations," *Proceedings of the Fourth International Partners in Flight Conference: Tundra to Tropics*, 2009: 205–219; David K. Nelson and M. Keith Causey, "Feral Hogs in Alabama," *Alabama's Treasured Forests*, Spring 2001; Noss, LaRoe, and Scott, "Endangered Ecosystems of the United States"; Ben Raines, "Insufferable Snails: Invasive Amazonian Apple Snail Difficult to Eradicate from Delta," *Press-Register*, December 3, 2011; Ben Raines, "Lionfish Sighted off Alabama Coastline; Could Pose Threat to Some Native Species," *Press-Register*, September 16, 2010; The Wildlife Society, "Ecological Impacts of Feral Cats," accessed June 2012, http://joomla.wildlife.org/documents/cats_ecological_impacts.pdf.

Overharvested species. Several of the species mentioned were described in earlier chapters; see above for sources. ADCNR, "ACWCS"; Troy L. Best, "Red Wolf," in *Alabama Wildlife, Volume 3: Imperiled Amphibians, Reptiles, Birds, and Mammals*, ed. Ralph E. Mirarchi, Mark A. Bailey, Thomas M. Haggerty, and Troy L. Best (Tuscaloosa: University of Alabama Press, 2004), 171–172; Mirarchi, ed., *Alabama Wildlife, Volume 1*; and Noss, LaRoe, and Scott, "Endangered Ecosystems of the United States."

Pollution. ADCNR, "ACWCS"; Alabama Department of Environmental Management (ADEM), "Draft Total Maximum Daily Load (TMDL) for the Cahaba River Watershed: Siltation, Pathogens and Other Habitat

Alteration" (Montgomery, AL: ADEM, 2003); Alabama Department of Public Health, "Alabama Department of Public Health Issues 2011 Fish Consumption Advisories," accessed July 2011, http://www.adph.org/tox/ assets/FishAdvNR2011.pdf; J. Bryan Atkins, "Water Quality in the Mobile River Basin: Alabama, Georgia, Mississippi, and Tennessee 1999–2001" (Reston, VA: USGS, 2004); Gregory C. Johnson, "Environmental Setting and Water-Quality Issues of the Mobile River Basin, Alabama, Georgia, Mississippi, and Tennessee" (Reston, VA: USGS, 2002); Rob Kerth and Shelley Vinyard, *Wasting Our Waterways 2012: Toxic Industrial Pollution and the Unfulfilled Promise of the Clean Water Act* (Boston: Environment America Research & Policy Center, 2012); David K. Mueller and Dennis R. Helsel, "Nutrients in the Nation's Waters—Too Much of a Good Thing?" (Reston, VA: USGS, 1996); Noss, LaRoe, and Scott, "Endangered Ecosystems of the United States"; Ben Raines, "ADEM Could Lose Authority to Issue Water Permits Due to Funding Cuts," *Press-Register*, June 28, 2012; USEPA, "Mercury," accessed July 2011, http://www.epa. gov/hg/about.htm; USEPA, "Polychlorinated Biphenyl (PCB)," accessed July 2011, http://www.epa.gov/osw/hazard/tsd/pcbs/pubs/about.htm; and USGS, "Nutrient Delivery to the Gulf of Mexico among Highest Measured," accessed July 2011, http://www.usgs.gov/newsroom/article. asp?ID=2240.

Deepwater Horizon. Associated Press, "Scientists Lower Gulf Health Grade Following Spill," Al.com, accessed July 2011, http://blog.al.com/ wire/2010/10/scientists_lower_gulf_health_g.html; Cara Bales, "Low Seafood Catches Remain a Mystery," *Daily Comet*, April 15, 2012; M. McNutt, George Guthrie, George Guthrie et al., "Assessment of Flow Rate Estimates for the Deepwater Horizon/Macondo Well Oil Spill" (Washington, DC: USDOI, 2011); Ben Raines, "Nearly 4 Times as Much Sea Life in Alabama, Mississippi Coastal Waters since 2010 Gulf Oil Spill," *Press-Register*, May 26, 2012; and Brian Walsh, "The BP Oil Spill, One Year Later: How Healthy Is the Gulf Now?" *Time*, accessed June 2012, http://www.time.com/time/health/article/0,8599,2066031,00.html.

Climate Change. L. J. Davenport, "Climate Change and Its Potential Effects on Alabama's Plant Life" (Birmingham, AL: Vulcan Materials Center for Environmental Stewardship and Education, Samford University, 2007); Richard B. Primack, *Essentials of Conservation Biology*, 5th edition (Sunderland, MA: Sinauer Associates, 2010); and Robert R. Twilley, "Confronting Climate Change in the Gulf Coast Region: Prospects for Sustaining Our Ecological Heritage" (Cambridge, MA: Union of Concerned

Scientists; and Washington, DC: Ecological Society of America, 2001).

Ecological services. Bradley J. Cardinale, "The Functional Role of Producer Diversity in Ecosystems," *American Journal of Botany* 98 (2011): 572–592; and Millennium Ecosystem Assessment, "Ecosystems and Human Well-being: Biodiversity Synthesis" (Washington, DC: World Resources Institute, 2005).

Moral, ethical, spiritual, and economic arguments for saving biodiversity. Summaries of these arguments for protecting biological diversity can be found in M. J. Groom, G. K. Meffe, and C. Ronald Carroll, *Principles of Conservation Biology*, 3rd edition (Sunderland, MA: Sinauer Associates, 2006); G. Tyler Miller Jr., *Living in the Environment*, 15th edition (Belmont, CA: Thompson Learning, 2007); and Primack, *Essentials of Conservation Biology*.

Ecotourism. Forever Wild, "Top 20 Forever Wild Facts," accessed June 2012, http://www.alabamiansforforeverwild.org/uploadedFiles/File/Top_20_Forever_Wild_Facts.pdf; Gregory M. Lein, "Wild Recreation," *Outdoor Alabama*, April 2012; and David Rouse and LaDon Swann, "From Coastal Birding to Primeval Forest—Alabama Has It All!" *Action*, Winter 2006.

Boulder Darter. Herbert T. Boschung and Richard L. Mayden, *Fishes of Alabama* (Washington, DC: Smithsonian Books, 2004).

Recovered species. Chris Cook, "White-tailed Deer," ADCNR, accessed July 2011, http://www.outdooralabama.com/watchable-wildlife/what/mammals/Ungulates/wtd.cfm; USFWS, "Delisting Report," accessed July 2011, http://www.fws.gov/ecos/ajax/tess_public/DelistingReport.do; and USFWS, "Tulotoma Snail on the Road to Recovery," accessed July 2011, http://www.fws.gov/southeast/news/2010/r10-043.html.

Recovery programs and recovering species. ADCNR, "Alabama Aquatic Biodiversity Center," accessed July 2011, http://www.outdooralabama.com/research-mgmt/aquatic/; ADCNR, "Nongame Vertebrates"; Alabama Plant Conservation Alliance, "About the Alliance," accessed July 2011, http://gump.auburn.edu/boyd/apca/About_Us.html; Roger Clay, "Up on the Roof," ADCNR, accessed July 2011, http://www.outdooralabama.com/watchable-wildlife/watchablearticles/roof.cfm; Cornell Lab of Ornithology, "Cooper's Hawk," accessed July 2011, http://www.allaboutbirds.org/guide/coopers_hawk/lifehistory#; Howard Horne, email to author, May 2011; "Interrupted Rocksnail Reintroduced to the Coosa River," *Outdoor Alabama*, February 2004; Paul Johnson (AABC), discussion with author, May 2011; Claire M. Pabody, Ruth H. Carmichael, Lauren Rice, and Monica Ross, "A New Sighting Network Adds to 20 Years of Historical Data on Fringe West Indian (*Trichechus manatus*) Manatee Populations in Alabama Waters," *Gulf of Mexico Science* 27 (2009): 52–61; David Ranier, "Eastern Indigo Snakes Get Another Chance," ADCNR, accessed July 2011, http://www.outdooralabama.com/oaonline/indigo10.cfm; and USFWS, "Recovery Plan for the Red-cockaded Woodpecker (*Picoides borealis*): Second Revision" (Atlanta, GA: USFWS, 2003). See above sources for "Black Bears in Piedmont," chapter 12.

River restoration. ADCNR, "Boating, Fishing and Fish in the Coosa River and Its Impoundments," accessed July 2011, http://www.outdooralabama.com/fishing/freshwater/where/rivers/coosa/; ADCNR, "Restoring Mussels a Priority at Aquatic Biodiversity Center," accessed July 2011, http://www.outdooralabama.com/oaonline/biodiversity10.cfm; ADCNR, "Walls of Jericho Complex," accessed July 2011, http://www.outdooralabama.com/public-lands/stateLands/foreverWild/FWTracts/WallsJerichoComplex/; Katherine Bouma, "Cahaba River Dam Removal Restores River Life," *The Birmingham News*, June 24, 2008; and Maurice F. Mettee, Patrick E. O'Neil, Thomas E. Shepard, and Stuart W. McGregor, "A Study of Fish Movements and Fish Passage at Claiborne and Millers Ferry Locks and Dams on the Alabama River, Alabama" (Tuscaloosa: GSA, 2005). See above sources for "Paint Rock River," chapter 13.

Non-riverine ecosystem restoration and creation. See above sources for "Gaillard Island" and "Artificial reefs," chapters 8 and 14, respectively. Jeff Dute, "Conservation Groups Unveil Plan to Restore Gulf Oyster Reefs, Marshes," *Press-Register*, September 15, 2010; Groom, Meffe, and Carroll, *Principles of Conservation Biology*; Debbie M. Lord, "Controlled Burn Set for Wednesday on 48 Acres of Gulf State Park," *Press-Register*, February 12, 2012; and Primack, *Essentials of Conservation Biology*.

Land preservation. ADCNR, "Forever Wild Land Trust," accessed June 2012, http://alabamaforeverwild.com/; ADCNR, "The Forever Wild Land Trust: An Interim Report to the Citizens of Alabama—1992 through 2009" (Montgomery, AL: ADCNR, 2009); ASP, "Alabama State Parks," accessed July 2011, http://www.alapark.com/; Chad Love, "Alabama Lawmakers Split over Extending Land Preservation Program," Field and Stream, accessed July 2011, http://www.fieldandstream.com/blogs/field-notes/2011/04/alabama-lawmakers-split-over-extending-land-preservation-program; Mary Orndorf, "US Fish and Wildlife Service Shelves Plans to Expand Cahaba River National Wildlife Refuge," *The Birmingham News*, May 6, 2011; Thomas V. Ress, "State Parks of Alabama," Encyclopedia of Alabama, accessed July 2011, http://www.encyclopediaofalabama.

org/face/Article.jsp?id=h-1642; Southeast Regional Partnership for Planning and Sustainability, "About Us," accessed July 2011, http://serppas.org/About.aspx; TNC, "The Nature Conservancy: Alabama," accessed July 2011, http://www.nature.org/ourinitiatives/regions/northamerica/unitedstates/alabama/index.htm; USACE, "US Army Corps of Engineers: Mobile District," accessed July 2011, http://www.sam.usace.army.mil/; USFS, "National Forests in Alabama," accessed July 2011, http://www.fs.usda.gov/alabama; USFWS, "National Wildlife Refuges by State," accessed July 2011, http://www.fws.gov/southeast/refuges/refuges-by-state.html; and USNPS, "National Park Service: Alabama," accessed July 2011, http://www.nps.gov/state/al/index.htm?program=all.

Species and ecological discoveries. The sources for the new discoveries are too numerous to list. Most were found through a search of Biological Abstracts, a scientific publication database, with Alabama as a topic and searching only articles related to taxonomy and systematics. Some recent discoveries are from newspapers, science magazines, and discussions with taxonomic experts. Discoveries discussed at more length from Alvin R. Diamond Jr., Hanan El Mayas, and Robert S. Boyd, "*Rudbeckia auriculata* Infected with a Pollen-mimic Fungus in Alabama," *Southeastern Naturalist* 5 (2006): 103–112; Becky J. Nichols and Keith R. Langdon, "The Smokies All Taxa Biodiversity Inventory: History and Progress," *Southeastern Naturalist*, Special Issue I (2001): 27–34; F. P. Parauka, M. S. Duncan, and P. A. Lang, "Winter Coastal Movement of Gulf of Mexico Sturgeon throughout Northwest Florida and Southeast Alabama," *Journal of Applied Ichthyology* 27 (2011): 343–350; Ben Raines, "'Pink Meanie,' New Species of Giant Jellyfish, Identified in Gulf of Mexico," *Press-Register*, January 12, 2011; Eric Soehren (ADCNR), March 16, 2011, posting on Albirds (http://groups.yahoo.com/group/albirds/); Thomas Spencer, "Snail Listed as Extinct Found in Cahaba River," *The Birmingham News*, August 9, 2012; and Karen M. Tenaglia, Jeffrey L. Van Zant, and Michael C. Wooten, "Genetic Relatedness and Spatial Associations of Jointly Captured Alabama Beach Mice (*Peromyscus polionotus ammobates*)," *Journal of Mammalogy* 88 (2007): 580–588.

Index

physiographic district, 348n2

Birmingham-Southern College, 1, 240

bison, 191. *See also* American Bison; Long-horned Bison

black bear. *See* American Black Bear

Black Belt Prairie. *See* Blackland Prairie ecoregion; Southern Coastal Plain Blackland Prairie and Woodland

Black Birch, 65, 239, 241

Black Cherry, 193

Black Needlerush, 133, 134, 135, 137, 141

Black Oak, 192

Black Rail, 339

Black Walnut, 72, 227, 271

Black Warrior River, 72, 89, 90, 111–13, 114, 236, 240, 241

Black Warrior River Valley, 37

Black Warrior Waterdog, 241

Blackfoot Quillwort, 257

Blackgum, 193, 197

Blackjack Oak, 175, 195, 215

blackland prairie. *See* Southern Coastal Plain Blackland Prairie and Woodland

Blackland Prairie ecoregion, 181, 183, 184

Blacknose Shark, 292

Blacktip Shark, 292, *293*

Blackwater River, *112–13*, 118

blackwater rivers, *119*–20, 151–52, 171. *See also* Blackwater River; Perdido River

Blackwell Creek, 268–69

Blackwell Swamp, 268–69

bladderwort, 172–73, 347n6 (chap. 9). *See also* Horned Bladderwort

Blakeley River, 145

blazing star. *See* Smallhead Blazing Star

Blount Mountain, 213–15

Blue Catfish, 106

Blue Crab, 135, 324

Blue Grosbeak, 159, 230

Blue Marlin, 290

Blue Pond, 166

Blue Ridge ecoregion, 189, 239, 244

Blue Ridge physiographic province, 115, 117

Blue Shark, 292

Blue Shiner, 225

Blue Spring, 164–66, *165*

Blue Spring Wildlife Management Area, 158

Blue Stem Palmetto, 147, 177

blue water (oceanic), 283, 290, 292

Blue Whale, 297

Blueface Darter, 103

bluestem, 181. *See also* Gulf Bluestem; Little Bluestem

Bluestripe Shiner, 102

Blue-winged Warbler, 230

Bobcat, 135

bog. *See* seepage bog

Boll Weevil, 252–53

Bon Secour National Wildlife Refuge, 10, 27, 335

Bosch's Filmy Fern, 278

Bottle Creek Native American settlement, 73, 147

bottleneck effect. *See* genetic bottleneck

Bottlenose Dolphin, 296, 301, 314

Boulder Darter, 264, 327, 328

box canyons, 6, 237–39

Boynton's Oak, *196*, 197, 348n4

brachiopods, 44, 57
Brasher, Larry, *238*
Brindley Mountain, 213–15
Brown Anole, 319
Brown Pelican, 10, 144, 328
Brown Shrimp, 306–7
brownwater rivers, 151
Bryde's Whale, 297
bryozoans, 44, 309
Buck's Pocket State Park, 226, 350n2
Buckwheat Tree, 171
buffalo. *See* bison
Buhrstone Hills Subdistrict, 347n7 (chap. 9)
Buhrstone/Lime Hills ecoregion, 174–76, 348n8 (chap. 9)
Bulblet Bladderferns, 231
Bull Shark, xvii, 108, 293, 350n4
bunting, 127. *See also* Indigo Bunting; Painted Bunting
Burgess, George, 293–94
Burks, Mary Ivy, 240
Butterfly Milkweed, *182*
butterwort, 172, 347n6 (chap. 9)
bycatch, 291, 294, 308
bycatch reduction devices, 309

cactus, 18, 28, 220. *See also* Eastern Pricklypear
caddisflies, 96
Cahaba Daisy Fleabane, 208, 348n8 (chap. 10)
Cahaba Lily, xvii, 98, 210
Cahaba Paintbrush, 208, 348n8 (chap. 10)
Cahaba Pebblesnail, 339
Cahaba Prairie Clover, 348n8 (chap. 10)

Cahaba Ridges physiographic district, 348n2
Cahaba River, xvii, 89, 90, 98, *112–13*, 114–15, 210–11, *323*, 332, 338
Cahaba River National Wildlife Refuge, 98, 210–11, 335, 336
Cahaba River Valley, 194
Cahaba River Wildlife Management Area, 210
Cahaba Torch, 348n8 (chap. 10)
Cahaba Valley physiographic district, 348n2
calcium, 175, 181, 184, 205, 226, 324, 346
calcium carbonate, 44, 48–49, 50, 103, 120, 344n2 (chap. 4)
California, 2, 39
Cambrian Period, 54–55, 207
Canada, 64, 105, 235
Canada Goose, 267
candidate species, 317, 348n7 (chap. 10)
Cane Creek Canyon Nature Preserve, 279
canebrake pitcher plant. *See* Alabama Canebrake Pitcher Plant
caprock, 221, 222, 223, 225, 227, 237–39, 241, 278, 279
Carboniferous Period, 57, 344n2. *See also* Mississippian Subperiod; Pennsylvanian Subperiod
Caribbean Sea, 30, 34, 64, 283, 304
Caribbean continental plate, 64
carnivorous plants, 6, 140, 154, 171–73, 174, 216. *See also* bladderwort; butterworts; pitcher plant; sundew
Carolina Parakeet, 317, 321
Carolinas, 174, 244, 255

carrying capacity, 25, 71

cat. *See* Domestic Cat

catalpa, 86

Catawba Rhododendron, 221, *222*, 246

catfish, 106, 142, 236. *See also* Blue Catfish

Cathedral Caverns State Park, 233

cattle, 183, 271, 276. *See also* ranching

cave fauna. *See* biodiversity, in caves

Cave Salamander, 40–*41*, 43, 47, 49

Cave Spring, 264

cavefish. *See* Alabama Cavefish

caverns, 7, 41, 43, 47, 48–49, 57, 65, 83, 87–89, 166, 175, 200, 211, 212, 221, 222, 228, 231–35, 261–62, 304, 335, 338

caves. *See* caverns

caviar, 107

Cedar Mountain, 244

cedars, 137, 183, 227, 228, 349n4 (chap. 13). *See also* Eastern Red Cedar

Centerville, 114, 189

Central Basin: of Gulf of Mexico, 283, 311; of Nashville Basin, 274

Central Interior Highlands and Appalachian Sinkhole and Depression Ponds, 271

Central Interior Highlands Calcareous Glade and Barrens, 272

Central Prairie Belt. *See* Blackland Prairie ecoregion

Centre, 348n6

Cerulean Warbler, 230

cetaceans. *See* whale

Chain Pipefish, 133

chalk, 63, 154, 181–82, 183

chalk bluff. *See* East Gulf Coastal Plain Dry Chalk Bluff

Chambers County, 255

Chapman Oak, 14

Chattahoochee River, 37, *112–13*, 117–18, 258–59, 338

Chattanooga, 90, 213

Cheaha Mountain, 242–43, 245, 246

Cheaha Quartzite, 244, 245, 246

Cheaha State Park, 245, 350n2

Cheaha Wilderness Area, 245, 335

chemoautotrophic bacteria, 312

Cherokee County, 210

Cherrybark Oak, 203

Cherryfin Shiner, 102

chert, 56, 270. *See also* Fort Payne Chert

chestnut, 70, 72, 86

Chestnut Oak, 50, 186, 187, 193, 246

Chewacla State Park, 254, 350n2

Chickamauga, 41

Chickamauga Limestone, 41–43, 44–*45*, 46–47, 55–56

Chickenclaws, 133

Chicxulub, 62–63, 345n5

Chilhowee Group, 54

Chilton County, 186

Chinese Privet, 320

Chinese Tallow, 319

Chinkapin Oak, 50, 226, 270

Chipola River, 118

Chitwood Barrens Preserve, 219

Choccolocco Mountain, *38*, 209

Choccolocco Wildlife Management Area, 245

Choctaw County, 185

Choctawhatchee River, *112–13*, 118

286–91, 304–6, 308–11; oil and gas extraction in, 314–15; physical features, 281–84; primary productivity and, 285–86, 322; sharks of, 291–95; shrimp of, 304–9; whales of, 295–98. *See also* climate, influence of Gulf of Mexico on

Gulf Shores, 27, 340

Gulf State Park, 27, 127, 350n2

Gulf Sturgeon, 106–7, 320, 339–40

gulfweed. *See Sargassum*

Guntersville Lake, 235–36

Guntersville State Park, 126, 350n2

gyres, 295–96, 299

habitat degradation. *See* habitat loss

habitat loss: general, 318–19

Hadley cell, 30, 344n5

Haines Island Park, 176, 178, 335

Halawaka, 49

Halloween Darter, 103, 259

halophytes, 133

hardwood, 347n2

Harelip Sucker, 263

Harlequin Darter, 103

Harperella, 225

Hartselle Sandstone, 187, 277–78, 279

Hatchet Creek, 254

Hatchetigbee, 49

Hatchetigbee Dome, 175

Hatchetigbee Dome Subdistrict, 347n7 (chap. 9)

Hawaii, 2, 4, 317

Hawk's View Overlook, 40, 44, 50

hawthorns, 67

head-cutting, 238–39, 279

headwaters river zone, 94–97

heat: defined, 14; role in climate, 28–30, 32, 34, 35, 37–38, 324, 343n1 (chap. 3), 344n4; role in ecosystems, 17–18, 29

heath bluff forests, 220–21, 224

Heflin, 246

hematite, 56

hemlock. *See* Eastern Hemlock

Henslow's Sparrow, 174

hickory, 70. *See also* Mockernut Hickory; Pignut Hickory; Sand Hickory; Shagbark Hickory

Highland Rim physiographic section, 262, 344n3 (chap. 4)

Highway 90 Causeway, 150

Hillabee, 49

Hillabee Greenstone, 245

hog, 74–75, 147, 149, 191, 251, 320, 347n8

hogback ridges, 191

Hogpen Creek, 192

Holiday Darter, 249, 259

Hollins Wildlife Management Area, 245

Holocene epoch, 69, 175, 181, 239, 345n9

Holocene highstand, 129

Honeylocust, 67

Horn Mountain, 244

Horned Bladderwort, 257

Horned Lark, 269

horse latitudes, 30, 32

Horseblock Mountain, 246

Horseshoe Bend National Military Park, 253–54

Humpback Whale, 297

Huntsville, 263

hurricane, 21, 23, 25, 24, 26, 27, 33, *34*, 35, 121, 125, 139, 252,

Little Mountain ecoregion, 278, 279

Little River, *97*, 213, 215, 220, 221, 226

Little River Canyon, 213, 226

Little River Canyon Center, 216

Little River Canyon National Preserve, 213, *214*, 215–16, 218, 335

Little River Canyon Onion, 220

Little River Falls, 215

Live Oak, 127

Liverleaf, 221

livestock, 74–75, 143, 275–76, 321, 322, 326

Loblolly Pine, 50, 183, 186, 226, 237, 249, 253

Locust Fork, 114

Locust Fork Darter, 241

locusts (trees), 67

Loggerhead Sea Turtle, 314

logging, 139, 147, 148, 151, 152, 157, 170, 178, 193, 197, 199, 208, 209, 222, 239, 254, 316, 318. *See also* deforestation; timbering

logperches, 91

Lollypop Darter, 103

Long-horned Bison, 65. *See also* bison

Longleaf Pine, 138, 155–57, 159, 160, 161, 186, 187, 195, 198, 246, 249, 252, 347n2

Longleaf Pine woodland, 319, 328, 332; of Piedmont ecoregion, 246–47, 248–49, 252, 253–54, 260; of Ridge and Valley ecoregion, 195–99, *196*, 205, 206, 207, 208, 209, 210; of Southeastern Plains ecoregion, 154, 155–58, *156*, 160, 161, 164, 168, 170, 171, 172, 174, 175, 176, 179, 185. *See also* East Gulf Coastal Plain Interior Upland Longleaf Pine Woodland; East Gulf Coastal Plain Near-Coast Pine Flatwoods; Southeastern Interior Longleaf Pine Woodland

longline fishing, 290–91, 294

Longtail Salamander, *85*

Lookout Mountain, 39, 213–15, 220, 226

Loop Current, 30, 64, 283–84, 286–87, 295, 299, 302, 304

Lost Worlds in Alabama Rocks, 279

Louann Salt, 312

Louisiana, 4, 39, 52, 144, 267, 283, 296

Lower Little Cahaba River, 206

Lower Piedmont, 252–54, 260

lower reaches river zone, 94, 98–100, 105, 116

Lower Tombigbee River, 111

Lower Tombigbee River watershed, 111

Lynn Overlook, 218–19

Lyreleaf Bladderpod, 272

mackerel, 288, 290, 314. *See also* King Mackerel; Spanish Mackerel

Madison County, 227, 231

magma, 244, 255

magnesium, 175, 206, 344n2 (chap. 4)

magnolia, 147, 166. *See also* Bigleaf Magnolia; Southern Magnolia

Mahimahi, 290

maize, 70, 71–74

mako. *See* Short-finned Mako

mammoth. *See* Columbian Mammoth

manatee. *See* West Indian Manatee

mantle, 45–46, 60

maple, 166. *See also* Red Maple; Silver Maple; Sugar Maple

Marion, 330

maritime effect, 30

maritime forest. *See* East Gulf Coastal Plain Maritime Forest

marlin, 290–91. *See also* Blue Marlin; White Marlin

Marsh Periwinkle, 134, 135, 136

Marshall County, 227, 231

Martin Dam, 116

Martin Lake, 253

mastodon, 6, 65

Mayapple, 86, *87*

Mediterranean Sea, 87

megafauna (Pleistocene), 65–69, 70, 191

Menawa, Chief, 253

mercury, 287–88, 291, 321

mesic slope forests. *See* Southern Coastal Plain Mesic Slope Forest

mesophytic forests, 228–30, 239, 279, 349n3 (chap. 13). *See also* South-Central Interior Mesophytic Forest

Mesozoic Era, 60, 62, 65

metamorphic rock, 46, 58, 115, 191, 244

methylmercury. *See* mercury

Mexican Plum, 179, 194

Mexico, 30

microbes, 58, 81, 95, 133, 134, 232, 322

midden, 137, 345n8

mid-reaches river zone, 94, 97–98, 103, 115

Midwest, 181, 183, 204, 235, 253, 267, 284

military bases and conservation, 209, 335

Millers Ferry Dam, 331

Mimosa, 175, 319

mining's impact on ecosystems, 59, 107, 210, 236–37, 241

Minnesota, 3

minnows, 10, 94, 97, 134, 149. *See also* Cyprinidae

Miocene Epoch, 60, 86, 90, 91, 170

Miocene Series, 170

Mississippi, 3, 24, 39, 52, 110, 131, 144, 170, 181, 185, 186

Mississippi Diamondback Terrapin, 135–36, 320

Mississippi River, 2, 60, 108, 110, 283, 296, 297, 302

Mississippi River Basin, 110, *112–13*, 114, 116–17, 178, 262

Mississippi Sandhill Crane, 140–41

Mississippi Sandhill Crane National Wildlife Refuge, 140

Mississippi Sound, 131, 141, 339, 346n2 (chap. 8)

Mississippi Sound Fresh and Oligohaline Tidal Marsh, 145

Mississippi Sound Salt and Brackish Tidal Marsh, 131

Mississippi–Alabama barrier island chain, 125, 129

Mississippian Cultural Period, 71–75, 322

Mississippian Subperiod, 56–57,

217, 219, 221, 227, 236, 241, 257, 274, 276, 277

Spanish explorers, 74–76, 220

Spanish Mackerel, 10, 286–87

Spanish Moss, 147

Spanish River, 145

Sparkleberry, 50

sparrows. *See* Henslow's Sparrow; Lark Sparrow

spear points, 67, *68*, 70

speciation, 59, 155, 258, 345n1; allopatric, 80, 83; cave fauna, 87–89; centers of, 83; deciduous forest plants, 86–87; freshwater fishes, 89–91; plethodontids, 84–86; steps of, 80–81; techniques for studying, 82–84. *See also* biological species concept; phylogenetic species concept

species, 81–82

species diversity. *See* biodiversity

species new to science, 87, 174, 231, 311, 337–39, 348n8 (chap. 10)

species problem, 81–82

Speckled Crab, 301

Speckled Trout, 135

Sperm Whale, 296–97

spiders, 13, 88, 231, 232, 338

Splinter Hill Bog Preserve, 174

sponges, 302, 303, 309

sportfishing, 106, 290, 294–95, 309, 311

Spotted Phacelia, 257

Spotted Rocksnail, 330

spray cliff. *See* Southern Appalachian Spray Cliff

Spring Pygmy Sunfish, 56, 264–*65*

spring, 48–49, 57, 65, 83, 165, 175, 199, 200–201, 228, 262, 263, 264, 265–66. *See also* Beaverdam Spring; Big Spring; Blue Spring; Cave Spring; Coldwater Spring; Pryor Spring

Spruce Pine, 176

squid, 287, 290, 296, 297

St. Clair County, 251

St. Petersburg, 74

Stanfield-Worley Bluff Shelter, 349n5 (chap. 13)

state amphibian, 3, 177

State Lands Division, 336

state parks. *See* Alabama State Parks

state rankings: amount of public lands, 337; biodiversity, 1–2, 3, 39, 40, 49, 52, 87, 93; carcinogen releases, 321; coal production, 236; endemic species, 2; enforcement of pollution laws, 321; imperiled species, 317; land area, 3; toxic releases, 321

"States of the Union: Ranking America's Biodiversity," 1–3

Stein, Bruce, 1, 2, 3

Steven C. Minkin Paleozoic Trackway Site, 237

Stevens, Timothy, 206

Sticky Rosinweed, *208*, 348n8 (chap. 10)

stingray, 142. *See also* Atlantic Stingray

storm surge, 21, 26, 80, 125, 126, 127, 137, 138

stormwater, 115, 201, 225, 321

strandplain dunes, 128–29, *130*

stream capture, 90–91

stream piracy. *See* stream capture

Striped Mullet, *9*

Striped Skunk, 14

Tennessee Cave Salamander, 84

Tennessee physiographic section, 190

Tennessee River, 90–91, 111–14, *112–13*, 116–17, 186, 187, 227, 234, 235, 236, 261, 262, 263, 264, 266, 268, 269, 275, 326

Tennessee River Valley, 37, 69, 263, 264, 266, 267, *270*, 346n4 (chap. 7)

Tennessee River watershed, 90

Tennessee Valley Authority, 117, 236, 264

Tennessee-Tombigbee Waterway, 111, 114

Tenn-Tom Waterway. *See* Tennessee-Tombigbee Waterway

Tensaw River, 145

tern, 10, 126, 235, 287. *See also* Least Tern; Royal Tern

Tertiary Period, 63–64, 65, 67, 82, 89, 90, 96, 137, 154–55, 175, 178, 184, 346n5 (chap. 7), 347n1

Texas, 2, 39, 144, 269, 306, 325, 348n4

Theodore Ship Canal, 144

Thompson, Patrick, *205*

threatened species. *See* endangered species

Three-lined Salamander, *85*

thrust faulting, 189–90, 194

thunderstorms. *See* convective rainstorm

tidal swamp. *See* East Gulf Coastal Plain Tidal Wooded Swamp

tides, 34, 104, 124, 141, 143, 145, 152, 299–301, 306; in Mobile River Delta, 100, 145, 145; in salt marsh, 132–33, 134–35; in swamp forests, 141. *See also* storm surge

Tiger Shark, 293, 350n4

timbering, 174, 175, 179, 199, 204, 215, 237, 239–40, 249, 326, 335, 336. *See also* deforestation; logging

titi flat. *See* Southern Coastal Plain Seepage Swamps and Baygalls

Tombigbee River, 37, 110–14, *112–13*, 116, 145

topminnow. *See* Whiteline Topminnow

topographic features, xviii, 4, 40, 88, 118, 129, 189, 206, 252, 269, *284*, *303*; diversity of, 4, 39, 51–52, 55, 87, 89, 110, 190, 211, 213, 254; formation of, 189–90, 212, 241, 273; influence on ecosystems, 47–48, 94, 175, 178, 199, 215, 241. *See also* karst

tornados, 35

Tracy's Sundew, 172

Transition Hills ecoregion, 185, 186–87

trawl. *See* otter trawl; outrigger trawlers

tree island, 273

treefrog. *See* Pine Barrens Treefrog

Triassic Period, 60–61

troglobites, 88, 231

troglophiles, 88

trogloxenes, 88

tropical depressions. *See* cyclones

tropical storms. *See* cyclones

tropics, 4, 44–45, 55, 56, 127, 147, 166, 230, 261, 277, 283, 297,